# Surface Chemistry and Interface Phenomena

# Surface Chemistry and Interface Phenomena

**Dr. Kamlesh Bilgaiyan**

ARCLER PRESS

www.arclerpress.com

**Surface Chemistry and Interface Phenomena**

*Dr. Kamlesh Bilgaiyan*

**Arcler Press**
4164 Lakeshore Road
Burlington, ON L7L 1A4
Canada
www.arclerpress.com
Email: orders@arclereducation.com



ISBN: 978-1-77956-957-8 (Hardcover)

# ABOUT THE AUTHOR

**Dr. Kamlesh Bilgaiyan (Ph D & NET)** a well known widely acclaimed assistant professor of Uttar Pradesh top rated Bundelkhand University, Jhansi (NAAC A+ Accredited University). Besides 10 U K design patents and two published patent, several publications. He has contributed pool of research by guiding two PhDs and more than 27 M Phil dissertations. He has published more than 20 papers in international and national reputed journals. Dr Bilgaiyan is having vast administrative experience of more than 20 years in various administrative positions. Presently he has working as Coordinator of B Sc, institute of Basic Science. He is also subject experts for Board of studies of various universities. Presently working on kinetics, material science and physicochemical properties of various plant extracts. Having more than 23 Years research experience.

# TABLE OF CONTENTS

# LIST OF FIGURES

# PREFACE

Surface and interface science is a part of modern science that gives fundamental information about how materials behave when they are exposed to the world around them. Regardless of whether one is considering paint adhesion, metal catalysis, bubble formation, or stabilization of emulsions, surface chemistry governs countless natural and technological processes. Surface phenomena control the behavior of solids, liquids, and gases in contact with each other, controlling process efficiency in industry, the effectiveness of drugs, and even the life of organisms in adverse environments. Prevention of corrosion, environmental cleanup, fabrication of electronic devices to nanomaterials, the knowledge of surfaces provides a tool for innovation and problem-solving across multiple disciplines.

Over the past century, researchers have made tremendous progress in comprehending the complexity of surface interactions. Progress is not only evident in theory but has extended to advanced experimental methodologies and applications found in everyday life. The creation of high-performance batteries, efficient catalysts for clean energy production, advanced coatings with unusual properties, and the control of colloids in medicine are all a direct debt to surface chemistry and interface phenomena. From the molecular scale to the macroscopic world, the subject integrates theoretical foundations with practicalities and is thus a scientific basis as well as a driving force behind technology.

The *Surface Chemistry and Interface Phenomena* book is designed to provide a comprehensive review of this active field. It follows a rational sequence beginning with the fundamentals followed by more technical issues such as adsorption, thermodynamics, colloidal systems, electrochemical processes, and nanotechnology. The goal is to not only provide a comprehensive summary of the subject but also indicate how these ideas are applied in industrial operations, high-performance materials, and environmental phenomena. Each chapter naturally succeeds the previous one, adding scientific explanations with applications that demonstrate the interaction between surface and interface science in academic and industrial applications.

Chapter 1 formulates the principles of surface chemistry. The chapter lays out basic concepts such as surface tension, free surface energy, and contact angles, which provide the background for the more sophisticated treatments of subsequent subsections. It formulates molecular interactions as governing surface behavior and shows why this is essential for the study of adhesion, wetting, and capillarity. Having established these basics, the chapter renders readers capable of following subsequent discussions of surface properties and phenomena.

Chapter 2 deals with solid surface structure and properties. The treatment in this chapter changes to atomic and molecular arrangement at solid surfaces, such as defects, crystallographic orientation, and the involvement of terraces, kinks, and steps. This structural image is connected to surface reactivity, adsorption sites, and physical and chemical behavior influenced by the arrangement of atoms. These ideas are used in the book to interpret technologies like semiconductors, thin films, and coatings.

Chapter 3 outlines adsorption phenomena, a process of outstanding importance in catalysis, pollution control, and material science. Physical and chemical adsorption are both treated, with emphasis on adsorption isotherms, surface coverage, and energetics. The text bridges theory and practice and illustrates how adsorption models are utilized to design effective catalysts, gas purification adsorbents, and drug delivery systems. In doing so, the chapter highlights adsorption as a pillar of surface chemistry.

Chapter 4 addresses surface thermodynamics and kinetics. Building on earlier descriptions, the chapter describes the thermodynamic concepts behind interfaces, including Gibbs adsorption equations and surface excess, and kinetic aspects of surface phenomena, including diffusion, nucleation, and growth. The concepts are applied to describe processes such as crystal growth, thin film deposition, and corrosion and how thermodynamics and kinetics provide predictive ability in design and control of surface processes.

Chapter 5 presents catalysis on surfaces, an area of utmost industrial and environmental significance. Herein, the role of the active sites, mechanisms of heterogeneous catalysis, and some examples such as hydrogenation, ammonia synthesis, and catalytic converters are presented. Catalyst design and factors controlling their activity and selectivity are described in great detail. Based on examples from industrial chemistry, the chapter shows how surface science is the basis for effectiveness and environmental appeal of catalytic systems.

Chapter 6 deals with colloids and dispersed systems, which are prevalent in natural and built milieus. The chapter explains classification, stability, and behavior of colloidal systems such as emulsions, foams, sols, and gels. Key concepts such as DLVO theory, zeta potential, and Brownian motion are explained, demonstrating their function in controlling stability and aggregation. Applications extend from food science, cosmetics, and pharmaceuticals to wastewater treatment, demonstrating the widespread impact of colloidal chemistry.

Chapter 7 addresses interfaces in liquids and solids, emphasizing how interfacial forces control wetting, spreading, and adhesion. The chapter addresses liquid-liquid, liquid-solid, and solid-solid interfaces, pointing out contact angle hysteresis, spreading coefficients, and lubrication. Such subjects bear a direct relation to technological applications like coatings, adhesion technology, and microfluidics. Through the bridge of theory and technology, this chapter encourages insight into how interfaces affect material performance.

Chapter 8 addresses electrochemical interfaces, which form the core of processes as varied as corrosion to energy storage. The chapter addresses electrical double layer structure, electrode potentials, and charge transfer kinetics. Batteries, fuel cells, and electroplating applications are highlighted as examples of how electrochemical interface understanding is converted into innovations in renewable energy and industrial processes. Corrosion control and protective coatings are also discussed, with emphasis on surface chemistry as a crucial element in durability and sustainability.

Chapter 9 illustrates surface characterization techniques. A wide range of contemporary techniques—electron microscopy, atomic force microscopy, X-ray photoelectron spectroscopy, and infrared spectroscopy—are explained as to their principles, benefits, and drawbacks. By outlining these methodologies, the chapter provides readers with the ability to understand how scientists and engineers achieve precise information about surfaces and interfaces. Case examples demonstrate how characterization techniques enable materials and device innovation.

Chapter 10 focuses on surface chemistry in materials science and nanotechnology. The chapter draws together much of what has been discussed previously, applying it to current challenges and possibilities in nanoscience. Surface effects in nanoparticles, quantum dots, carbon-based nanomaterials, and self-assembled systems are discussed. Particular focus is placed on the way surface chemistry produces unique nanoscale properties and enables applications in electronics, medicine, sensors, and energy. By connecting surface science to new technologies, the chapter emphasizes the continued pertinence of the field.

This book is intended for students beginning to learn physical chemistry and materials science through to scientists and engineers seeking an expert reference. It will be a valuable guide for graduate students in building advanced knowledge, and practitioners in chemical engineering, nanotechnology, and industrial chemistry will appreciate its focus on applications. Combining stringent explanation with relevant examples, the book is also suitable for educators seeking to introduce an organized and lucid guide to teaching. Finally, this book attempts to give readers both foundational knowledge and practical application ideas to be applied across disciplines and industries.

# CHAPTER 1

# FUNDAMENTALS OF SURFACE CHEMISTRY

## Contents



## INTRODUCTION

Surface chemistry is a vital sub-field that deals with the special properties and reactions happening in the interfaces of various materials. Surface chemistry needs to be learned in order to understand catalysis, adsorption, and surface reactions that are central to disciplines such as materials science, nanotechnology, and environmental chemistry. Specifically, surface interactions determine the physical and chemical behavior of solid and liquid materials by altering rates of reaction and product distribution. Classical bulk chemistry techniques tend to be inadequate in description of such interface phenomena, and specialized techniques and theories are added to explain surface processes. Through understanding of these intricate surface interactions, researchers are able to develop new materials and optimize existing ones for application in everything from industrial catalysis to sensor technologies, thereby confirming the underlying importance of surface chemistry in propelling modern science and technology. One of the most fundamental principles of surface chemistry, adsorption, is defined as the process by which atoms, ions, or molecules of a gas or liquid phase adhere to a solid surface. The process efficiency relies on a variety of factors that include temperature, pressure, and the type of adsorbate and adsorbent. Adsorption is crucial in applications such as catalysis, where reaction rate is critically reliant on the extent of binding of reactants to catalyst surfaces.

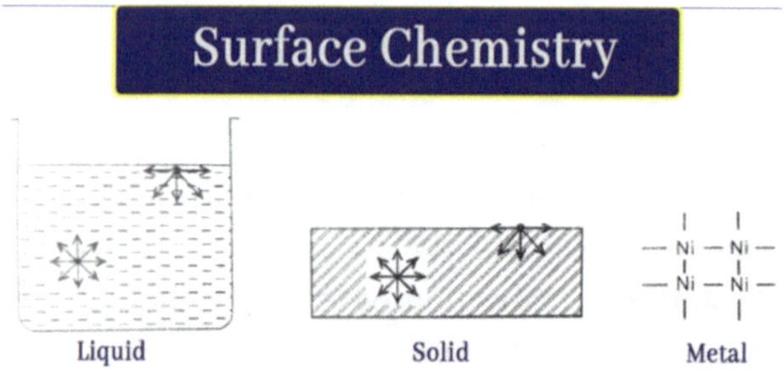


**Figure 1.1.** Surface chemistry.

*Source:* https://external-content.duckduckgo.com/iu/?u=https%3A%2F%2Fchemistrypage.in%2Fwp-content%2Fuploads%2F2021%2F03%2FSurface-Chemistry.jpg&f=1&nofb=1&ipt=ce82d445eec1556599190aaed4a7eefea0b661ad7a4060f53be4411e884cf319.

The Freundlich and Langmuir isotherms are two traditional models describing how adsorption occurs, which provide insights into community relationships and surface cover. Further, advances in spectroscopic techniques such as atomic force microscopy and scanning tunneling microscopy have enabled more precise studies of surface structures at the nanolevel, enhancing our understanding of adsorption processes and dynamics in various applications ranging from heterogeneous catalysis to environmental remediation. A further key area of surface chemistry is investigation of surface modification procedures for enhancing material performance. Techniques such as chemical vapor deposition, laser ablation, and self-assembled monolayers are among the versatile methods of surface property modification for specific application. These modifications can improve corrosion resistance, alter wetting characteristics, or introduce catalytic activity, thus expanding the range of material application in practical applications. For instance, the formation of hydrophobic surfaces is crucial in preventing fouling in aquatic systems, while anti-reflective coatings are essential in optimizing the efficiency of solar cells. As more industries require materials with distinctive surface properties, an in-depth understanding of surface chemistry is increasingly becoming indispensable. It enables the development of design and synthesis of advanced materials to suit new technologies according to worldwide goals of sustainability and efficiency (Figure 1.1).

## Definition and Importance of Surface Chemistry

Surface chemistry is the domain of physical and chemical behavior at interfaces between phases, particularly solid-liquid, solid-gas, and liquid-gas interfaces. Surface chemistry is deeply important since it governs material properties under varying applications, from catalysis to adhesion and corrosion. By governing the molecular interactions at surfaces, scientists and engineers can design materials for improved performance. For instance, the performance of catalysts can be enhanced by tuning surface properties to ensure maximum accessibility of reactants and active site concentration. Furthermore, the functionality of coatings relies heavily on the principles of surface chemistry because they must adhere well and resist degradation owing to the environment. A profound appreciation of these principles not only facilitates current technological advances but also releases potential for innovations that exploit the unique properties of materials at their surfaces. The significance of surface chemistry extends to the field of molecular electrochemistry, in which reactions are initiated by electron transfer at the electrode surface. In this case, the surface also plays an essential role in moderating chemical reaction kinetics. Not only are the electrodes electron conductors but also facilitators that affect the reactivity of the molecules involved in the electrochemical reactions. Reaction outcomes can be tailored through controlled modification of surface characteristics, pointing to the inherent interconnection between surface chemistry and electrochemical behavior. The progress in molecular electrochemistry drives many applications, such as the development of sensors and energy storage devices. The focus on the understanding of surface modification's effects on the interaction between molecules further shows the critical role that surface chemistry is to play in developing more efficient, sustainable technology in industrial as well as biological contexts, as it serves to show its versatility and necessity in research. Furthermore, the evaluation and control of the surface behavior are critical to industrial applications, such as papermaking. Water chemistry, according to related studies lock down, plays a significant role in papermaking, with pH controlling the interactions among fibers, additives, and the ambient water that ultimately determine paper quality and manufacturing efficiency (Pruszyński et al., 2022). An understanding of surface chemistry principles enables solutions which reverse normal problems, such as water contamination and variable water quality that are haunting the industry. Producers can optimize processes and improve product quality in general by focusing on the interplay between surface interactions and water chemistry. Such interdisciplinarity not only enhances industrial processes but also contributes to sustainability by allowing wise use of resources and thus making the significance of surface chemistry a focal point in scientific inquiry as well as practical applications in numerous fields.

## Historical Development of Surface Chemistry

Evolution of surface chemistry originated in the early 20th century with significant contributions arising from improvements in physical chemistry and material science. One very critical milestone came with the establishment of the conceptual difference between bulk and surface properties, which began to take shape with studies by scientists like Langmuir in the 1910s. His work on adsorption behavior led towards an understanding of molecular interactions with surfaces. Langmuir's isotherm, which defined how molecules that adsorb onto the surface, was made into a typical model in surface chemistry. Initial work demonstrated that surfaces possess unique qualities that can be very significant in affecting the chemical behavior of substances. Thus, the platform was set for fresh innovations, and researchers began to explore the role of surface interactions in catalysis, corrosion, and material development, ending up with a greater understanding of the intricate dynamics of surface reactions. At the end of the 20th century, new analytical techniques emerged, which revolutionized surface chemistry. Techniques such as scanning tunneling microscopy (STM) and atomic force microscopy (AFM) allowed scientists to observe surfaces down at the atomic scale, providing them with an unimaginable glimpse of surface structure and reactivity. The technology allowed complex phenomena, like the formation of surface films and the impact of surface defects on chemical reactions, to be studied. Further, surface research diversified into the synthesis of covalent organic frameworks (COFs), which are bonded together by covalent bonds to yield highly ordered materials with porosity and stability. The new approaches to the study of such frameworks highlighted the role of chemical bonds in the definition of surface properties, which in itself emphasized the need for a comprehensive understanding of surface chemistry both in functional materials and in the practice of sustainability in chemical engineering (Christian et al., 2017). The study of surface chemistry has also been influenced by broader discourse in terms of sustainability, especially in the scenario of global issues today. The idea of the planetary boundary (PB), which was introduced in 2009, seeks to delineate environmental sustainability's boundaries, underscoring the important role played by surface phenomena in driving processes like climate change and management of resources. Research has proven that chemical interactions at the surface contribute extensively to determining the fate of pollutants and their interaction with environmental processes. By integrating surface chemistry into sustainability, scientists can minimize and maximize the impacts of industrial operations and consumer products. This convergence of surface chemistry and sustainability highlights the field's ongoing evolution, toward a transition from applying surface research merely to enhance material properties to also sustaining environmental requirements and wise resource usage (Steffen et al., 2015) (Figure 1.2).

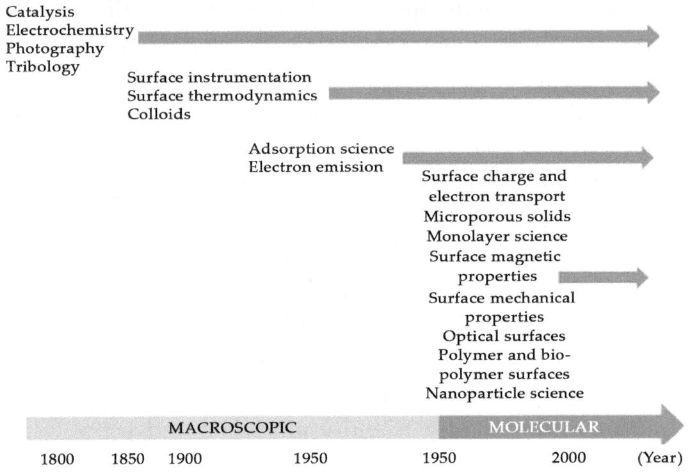


**Figure 1.2.** Timeline of the historical development of surface chemistry.

*Source:* https://www.researchgate.net/figure/Timeline-of-the-historical-development-of-surface-chemistry-1_fig1_281660822.

## Applications in Various Fields

Applications of surface chemistry are increasingly common in all fields, but most importantly, in nanotechnology, which is a subject of utmost concern over the last couple of years. Nanotechnology has enhanced many fields such as pharmaceuticals, agriculture, and food safety, among others. The unique properties of nanoparticles owing to their small dimensions and large surface area make them more reactive and functional. For instance, the synthesis of silver nanoparticles (AgNPs) through different routes, some of which are based on the green chemistry principles, reflects the importance of sustainable pathways in the production of material with better physiochemical properties. The nanoparticles are significant players in numerous applications, ranging from antimicrobial agents in medicines to catalysts of chemical reactions, thereby reflecting the great potential of surface chemistry developments (Husain et al., 2023). In medicine, surface chemistry-driven developments are profoundly revolutionary, particularly with the creation of nanotechnology products. Nanoparticles allow for intelligent sensing technology that enhances disease diagnosis and the effectiveness of treatment exponentially. Nanomaterials make molecular-level interactions possible, allowing them to confer unprecedented specificity and sensitivity in medical treatments. The application of nanotechnology in drug delivery systems is one instance of its potential for altering therapeutic techniques by bringing the medicine to the targeted tissues directly with less side effect. Moreover, the ability of nanomaterials to engage with biological systems also signifies their great possibilities in biomedical engineering and regenerative medicine and therefore having a wider range of available therapeutic choices today (Biswas et al., 2023). Besides medicine, surface chemistry has other applications in environmental management and sustainability. Surface chemistry is being used to develop new applications of nanomaterials as effective ways of addressing pollutants in water, air, and soil. The materials have high surface area and distinctive surface chemistry that facilitates efficient degradation and pollution removal, thus leaving the environments clean. Moreover, use of nanotechnology in industries like cosmetics and energy storage also indicates its applicability and universality. The global emphasis on the synthesis of sustainable and environmentally friendly materials and processes brings into prominence the pivotal role surface chemistry plays in avoiding environmental problems while propelling technology innovation across all disciplines. In as much as development continues, surface chemistry's multidisciplinary nature will indeed make even more revolutionary breakthroughs with significant effects on society (Husain et al., 2023).

## 1.1. BASIC CONCEPTS OF SURFACE CHEMISTRY

Surface chemistry is a basic foundation of knowledge for a variety of chemical interactions occurring at interfaces. Simple principles such as adsorption, desorption, and surface reactions form the basis of understanding how molecules interact with surfaces and dictate properties and behavior in various applications. The phenomenon of adsorption, for example, can be subdivided into physisorption and chemisorption, with different energy profiles and mechanisms. Physisorption is the result of weak van der Waals forces allowing reversible interaction between adsorbate and surface, whereas chemisorption entails the formation of stronger covalent or ionic bonds resulting in a more stable, often irreversible binding. This fundamental basis determines much surface-related process from catalysis to sensing technology, thereby highlighting the significance of surface chemistry knowledge to innovative applications and materials (Joshua et al., 2019). Surface energy and surface tension are two fundamental terms of surface chemistry that play a crucial role in determining the stability and behavior of systems. Surface energy is the excess energy at the material surface compared to its bulk because of the unsatisfied bonds present at the interface. This energy controls effects such as wetting, spreading, and adsorption and ultimately determines the functioning of coatings, emulsions, and nanostructured materials. Surface tension, directly related to surface energy, quantifies the forces of cohesion between molecules on the surface of a liquid and is of value in applications such as inkjet printing and surfactant design. It is important to understand and manage these properties in the optimization of material performance across most industries, justifying the importance of surface chemistry in technological advancement and daily applications (Crolla et al., 2015). Another central surface chemistry role is catalysis for acceleration of chemical reactions at interfaces. Catalysts typically work by enabling reaction routes that reduce the activation energy needed for surface reactions, thus being very much sought after in processes like industrial heterogeneous catalysis. In their case, reactants are on the catalyst surface where they undergo transformation before being released back to the surrounding environment. The creation of efficient catalysts is inseparable from surface chemistry; parameters such as surface area, density of active sites, and the type of surface interactions decisively influence catalytic efficacy. Current advances in nanoplasmonic sensor technologies bear witness to the potential of applying surface-enhanced spectroscopic approaches to explore such catalytic interactions at the nanoscale range, thereby further enriching the principles of basic surface chemistry with practical solutions for environmental science and biochemistry (Joshua et al., 2019).

### 1.1.1. Surface Energy and Surface Tension

The surface energy theory lies at the heart of surface chemistry since it captures the excess energy on the surface of a substance in comparison to the bulk. This phenomenon is due to the imbalance of intermolecular forces experienced by the molecules at the surface; unlike bulk molecules, which experience equal forces in all directions, surface molecules receive unbalanced forces from the inside, leading to an increase in energy. This high surface energy has considerable impact on many physical properties such as wettability, adhesion, and the formation of thin films. For coating applications, knowledge of this energy is essential, where the surface energy changes can enhance adhesion and wear resistance and thereby influence material performance from electronics to biomedicine. Further, microstructural engineering can also manage surface energy, designing material surfaces to meet a specific functional requirement and technological and design innovation. Surface tension, a term used for surface energy, describes the cohesive forces between liquid molecules at the air or second liquid surface. Surface tension is caused by the same unbalanced intermolecular forces that raise surface energy. It is the force per unit length on the surface that is acting, and it is commonly observed in the instances of droplet formation, walking of certain insects on water, and activity of detergents. Surface tension plays a crucial role in numerous applications, and one such instance is inkjet printing where surface tension plays a role in droplet formation and spreading over surfaces in order to achieve improved accuracy of printing. In addition, industries rely heavily on surfactants to regulate surface tension in a way that allows processes such as emulsification and foaming to occur, which are important in food technology, cosmetics, and pharmacy. The relation between surface tension and surface properties gives us an indication of how important this concept is in industrial processes and scientific research (Figure 1.3).

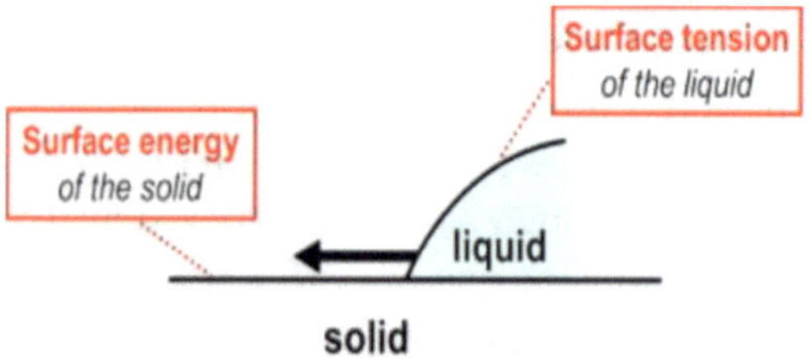


**Figure 1.3.** Surface energy and surface tension.

*Source:* https://external-content.duckduckgo.com/iu/?u=https%3A%2F%2Fqsstudy.com%2Fwp-content%2Fuploads%2F2017%2F08%2FSurface-Tension-and-Surface-Energy.jpg&f=1&nofb=1&ipt=68466dce08e9277540def8b1f147b377de69209fde35878235fbe532c5f96e07.

In surface chemistry in general, surface tension and surface energy are both interdependent on each other, influencing each other in various ways. As one illustration, the surface tension of a liquid is less when in contact with low surface energy substances. This interaction is used to affect how liquids wet surfaces, a matter that is relevant in coating processes and making functionalized surfaces. Material science advances are aimed at leveraging the science of surface energy and tension to develop the next generation of coatings and self-cleaning surfaces and enhance their utilization. And advances in nanotechnology have heightened the stakes for leveraging these surface effects down at the nanometer scale, where unique properties result due to the very large surface area-to-volume ratios. Consequently, a deep knowledge of surface energy and tension enables engineered surfaces to be tailored to meet some performance criteria in different fields, from environmental science to nanomedicine (Figure 1.4).

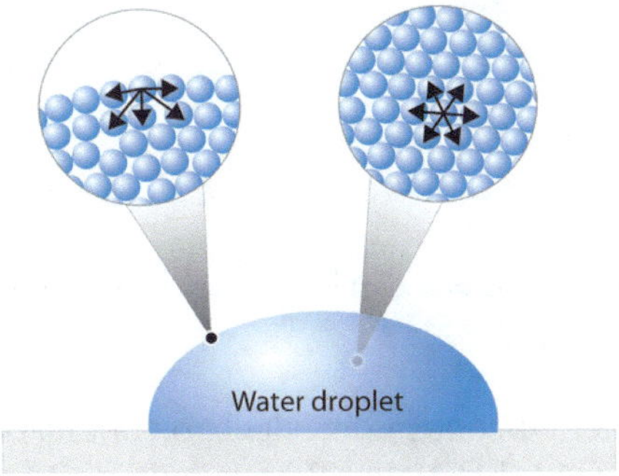


**Figure 1.4.** Molecules at the surface of water experience a net attraction to other molecules in the liquid, which holds the surface of the bulk sample together. In contrast, those in the interior experience uniform attractive forces.

*Source:* https://chem.libretexts.org/Bookshelves/Physical_and_Theoretical_Chemistry_Textbook_Maps/Supplemental_Modules_(Physical_and_Theoretical_Chemistry)/Physical_Properties_of_Matter/States_of_Matter/Properties_of_Liquids/Surface_Tension.

## 1.1.2. Adsorption Phenomena

The adsorption process is a fundamental surface chemistry process that describes the adhesion of molecules from a liquid or gas to the solid surface. This results due to various intermolecular forces like van der Waals forces, hydrogen bonds, and electrostatic forces, which act collectively to determine how and why molecules collect at the interface. Because surface area and surface function are significant factors in adsorption capacity, materials with higher surface area or chemically specifically tuned properties, e.g., functionalized materials treated with functional groups, can exhibit better adsorption characteristics. This becomes especially significant in uses as diverse as catalysis to sensor technology due to the reliance of such processes on the nanoproperties at the nanoscale surface. Identification of these interactions is very crucial for the logical design of material structures for particular applications in environmental observation and industrial catalysis. Adsorption procedures are of two main types: physisorption and chemisorption (Figure 1.5).

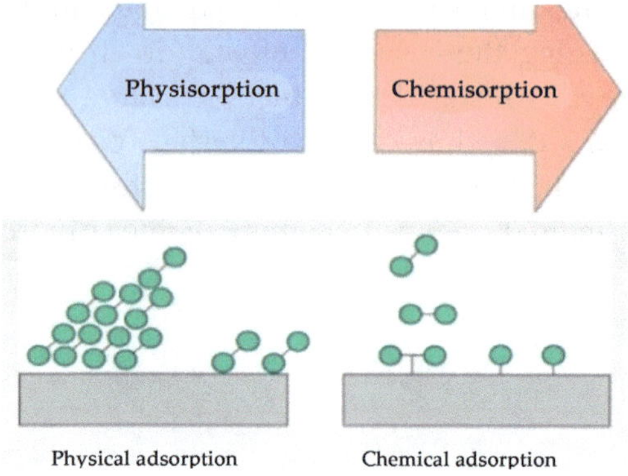


**Figure 1.5.** Adsorption phenomena can be categorized into two primary types: physisorption and chemisorption.

*Source:* https://external-content.duckduckgo.com/iu/?u=https%3A%2F%2Fi.ytimg.com%2Fvi%2F_G4AsbVQIng%2Fhqdefault.jpg&f=1&nofb=1&ipt=ab270ddfc60e4f026a4f9092ab0b1d04a3a847d34b52baadcac1147e97315385.

Physisorption is due to weak van der Waals interactions and usually reversible, with a relatively low energy barrier to desorb adsorbates. Chemisorption, on the other hand, is on the basis of the development of stronger chemical bonds that, as a general rule, provide a more permanent attachment of species onto the surface. The distinguishing features of these two types of adsorption show the importance of studying the interactions which take place on the molecular scale. For example, the role of oxygen species on the surface in carbon materials is a fertile area of investigation due to the fact that functional groups play a significant role in the redox stability and adsorption properties of carbon substrates. This is specifically relevant considering the observed differences in thermal stability and electronic properties of different oxygen-modified carbons, as underscored in recent studies (Dang et al., 2025). Additionally, the adsorption of biological molecules, such as amino acids (AAs), onto metal nanoparticles (NPs) has been of special interest in surface chemistry. The AA adsorption onto metallic substrates, particularly with silver and gold NPs, demonstrates the intricacy of molecular recognition processes that govern stability and structural fine-tuning in nanomaterials. Computational studies of the weak interactions underlying AA adsorption on Ag and Au surfaces reveal intricate features of the nature of the bonds, which can influence the overall performance of nanostructures in various applications, from drug delivery to biosensing. With advanced methodology, including quantum theory models and energy decomposition experiments, researchers have begun to unravel these weak binding effects and have opened windows of insight important to the design of better bioinorganic systems (Nicolau et al., 2025).

### 1.1.3. Types of Surfaces: Clean vs. Contaminated

The demarcation line between dirty and clean surfaces is inherent in surface chemistry, controlling much of the activity that follows at the molecular scale. Clean surfaces are characterized by the presence of no foreign matter, to provide maximum reactivity and interaction with adsorbates. Clean surfaces will facilitate chemical reactions like catalysis, where a clean substrate provides ideal conditions for reactants to interact and be converted. For instance, electrochemical reduction of nitrogen holds promise when conducted on clean surfaces, as seen in the emerging interest in ammonia as an renewable energy carrier and pointing toward the significance of surface cleanliness in achieving efficient chemical reactions (Choi et al., 2020). In contrast to this, dirty surfaces can hamper chemical processes and make a significant difference to physical properties. Contaminants, which might be organic molecules, dust, or surface oxides, preclude atomic or molecular structure at a surface, foreclosing critical interactions and affecting reactivity. In lithium-air batteries, parasitic reaction because of contamination at numerous interfaces—like the electrolyte-cathode interface—have been shown to severely limit performance and efficiency (Liu et al., 2020). Thus, the ability to understand clean and dirty surface dynamics is required in order to move technologies that depend on surface interaction forward from catalysis to battery efficiency. Not only is cleanliness of surfaces a necessity for achieving optimum chemical reactivity but also entails a variety of preparative and analytical techniques. Methods of cleaning, such as plasma treatment or solvent rinsing, become important in the preparation of the surface for a specific application, ensuring unwanted impurities are removed efficiently. Besides, surface cleanliness is generally quantified using techniques like XPS or SEM, which enable researchers to measure and characterize the amount of contaminants. With the analytical methods, increased understanding of surface chemistry concepts can be gained, ultimately leading to a better understanding of clean and dirty surfaces' behavior in practical applications, e.g., in efficient energy systems and advanced materials. By addressing the differences in surface types, researchers can develop better plans for optimizing the performance of various chemical applications.

## 1.2. ADSORPTION MECHANISMS

Adsorption is highly significant in many surface chemistry applications because it serves as a transition between a solid and a fluid phase where molecules or ions adsorb onto the surface of the solid. Mechanisms of adsorption play a very vital role if one needs to maximize catalytic processes like those utilized in energy conversion and storage systems. Physisorption and chemisorption represent the two dominant forms of adsorption. Physisorption results from weak van der Waals forces that allow reversible interaction, while chemisorption results from the formation of covalent bonds to establish a more stable bond. The distinct character of the two processes results in divergent energy profiles and interaction strengths with direct implications on the efficacy of surface interactions. The ratio of these two forms of adsorption is critical in catalysis and materials science, where desired outcomes depend on regulating surface interactions for the sake of achieving high efficiency and specificity. One critical aspect of adsorption processes is the heterogeneity of the surface in material, which influences the adsorption activity of reactants. Such influence typically originates from variations in surface charge, functional groups, or topographic properties that create different adsorption sites. For example, for graphdiyne (GDY), its structural complexity as well as the unbalanced charge distributions are accountable for its unique adsorption characteristics. These conditions allow GDY to accommodate a wide variety of transition metal catalysts, enhancing the latter's efficiency in catalytic reactions such as the oxygen evolution reaction. Scientists have explored how different atomic and electronic structures interact and influence adsorption behavior, which revealed that localized electronic states also influence adsorption energies and selectivities and provide new directions in catalyst development (He et al., 2022). Moreover, environmental forces along with the inherent properties of the adsorbate and substrate govern adsorption mechanism kinetics. Temperature, pressure, and the interactions between the solvent can significantly influence how rapidly adsorption occurs. For instance, an increase in temperature has the effect of increasing molecular mobility such that the process of adsorption becomes more convenient but, along with it, desorption occurs during physisorption. The conditions for such systems must be optimized, particularly in catalysis, since this is directly correlated with reaction rates as well as general effectiveness. And with the advancing surface chemistry field, new materials such as graphdiyne are also transforming ways of examining complex adsorption phenomena in the energy industry and elsewhere, further emphasizing the importance of being aware of adsorption mechanisms towards the enhancement of material sciences (Song et al., 2020).

### 1.2.1. Physisorption vs. Chemisorption

The adsorbate-substrate interactions lie at the heart of our understanding of surface chemistry, particularly in distinguishing between physisorption and chemisorption. Physisorption is characterized by weak van der Waals forces, resulting in relatively low-energy interactions in which the adsorbates can be readily desorbed without undergoing any significant change in their molecular structure. This sort of adsorption typically occurs at temperatures below chemisorption, where there is lower energy required because the adsorbate forms weaker bonds with the substrate—commonly covalent or ionic in nature. The difference in bond strength basically regulates the physisorbed compounds' stability and reactivity, which are observed to be more mobile on surfaces than chemisorbed counterparts, thereby affecting catalytic properties and surface reactions considerably. Investigative techniques play a significant role in elucidating physisorption and chemisorption characteristics, particularly in the observation of the influence exerted by the two processes over chemical bonding at surfaces. Spectroscopic techniques such as X-ray photoemission and X-ray absorption spectroscopy provide insights into the nature and intensity of adsorbate-substrate interaction. In the investigation of different regimes of interaction, researchers employ such spectroscopic methods for the characterization of molecular behavior in metal substrates. For instance, a study employing density functional theory revealed that weak physisorption yields minimal spectral change, but weak chemisorption can yield charge transfer and significant spectral changes in features, more or less digging up a more complex binding mechanism at play in diverse surface interactions (Samuel et al., 2022) (Figure 1.6).

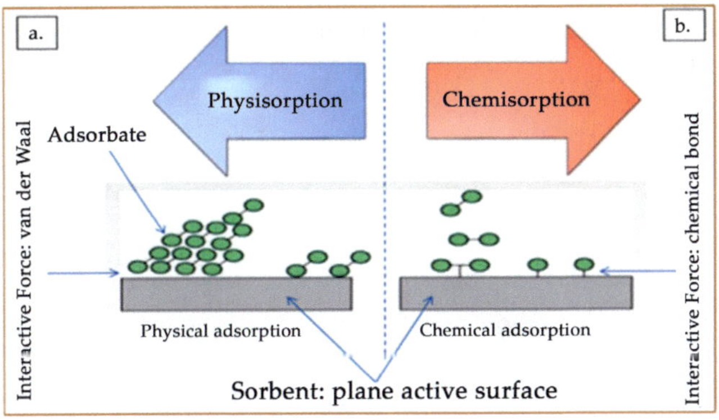


**Figure 1.6.** *Adsorption mechanisms of (a) physisorption; and (b) chemisorption.*

*Source: https://www.researchgate.net/figure/Adsorption-mechanisms-of-a-physisorption-and-b-chemisorption_fig3_354813605.*

The effects of physisorption and chemisorption extend beyond definition; they are key to technical application to catalysis, sensor technology, and materials science. The strength difference governs reactivity and stability, and hence may govern the operation of a catalytic process. Furthermore, an understanding of these interactions allows predictive models to be developed for surface reactions to optimize reaction conditions and improve material properties. Notably, advances in complex techniques within density functional theory have facilitated improved simulations of these processes, as seen in recent studies that consider a broad set of training data to maximize models for physisorption and chemisorption kinetics (Trepte et al., 2022). This multi-dimensioned understanding continues to be crucial to fostering innovations in surface chemistry and its applications.

## 1.2.2. Isotherms: Langmuir and Freundlich Models

In surface chemistry, isotherms give insight into the process of substances' adsorption onto solid surfaces. Two fundamental isotherm equations that are applied to describe the process are the Langmuir and Freundlich models. Langmuir's model, based on monolayer adsorption, is described as adsorption occurring in fixed points on a surface, with one molecule occupying each point. This model is particularly appropriate for systems where the adsorbate forms uniform layers and possesses a definite number of adsorption sites, allowing one to compute maximum adsorption capacity. The applicability of the model is cited in some research, for example, that for adsorption of phosphate onto magnetite, which claims its application in the prediction of adsorption mechanism and capacities under varied environmental conditions (Hou et al., 2020) (Figure 1.7).

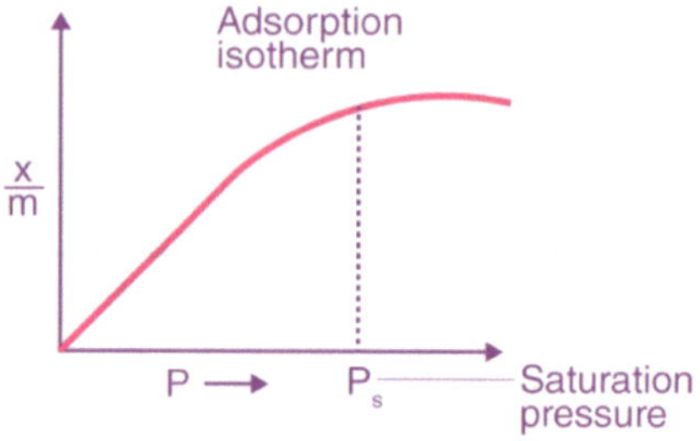


**Figure 1.7.** Adsorption isotherm.

*Source: https://byjus.com/chemistry/adsorption-isotherms-applications/.*

Conversely, Freundlich model provides a generalized model with accommodation of surface energies that are heterogeneous and various adsorbate interactions. The semi-empirical model provides for the intricacy of the adsorption process, especially where there is more than one layer or various affinities for adsorption sites. Freundlich equation typically represents a non-linear relation and is therefore particularly suitable for describing adsorption in systems whose surface comprises a mix of active sites with unlike energies. Its applicability is exhibited in various studies using various adsorbents, like using activated carbon in herbicide adsorption, in which Freundlich model can describe well the experimental data at equilibrium and adsorption behavior (Kuśmierek et al., 2020). The flexibility of the Freundlich model to examine complex systems of adsorption is exemplified by this applicability. In general, choice between Langmuir and Freundlich model depends primarily on the nature of adsorption process and physicochemical properties of the adsorbate and adsorbent. While Langmuir model is most appropriately applied to monolayer favoring systems, Freundlich model is most appropriately applied to heterogeneous surface systems. Understanding the limitations and strengths of these models enables surface chemistry scientists to frame their methodologies based on specific experimental conditions and outcomes. In addition, with continuous progress in research, enhancing the models through experiment validation makes them more suitable for application in actualistic processes, such as wastewater treatment and environmental remediation (Hou et al., 2020; Kuśmierek et al., 2020) (Figure 1.8).

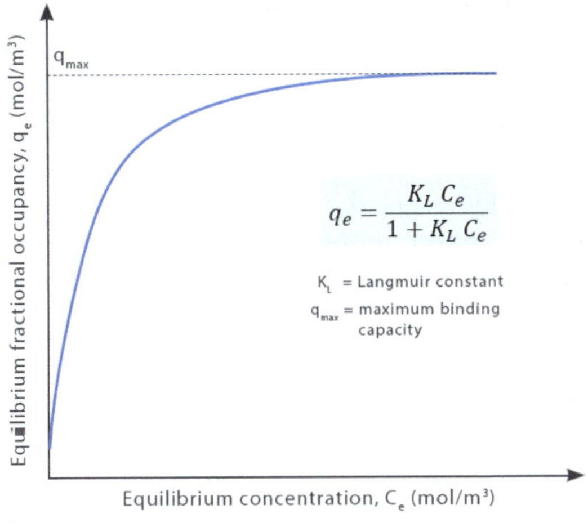


**Figure 1.8.** The Langmuir isotherm.

*Source:* https://www.chemistrylearner.com/adsorption/langmuir-isotherm.

## 1.2.3. Factors Affecting Adsorption

The process of adsorption, one of the main subjects of surface chemistry, is regulated by several principal factors that govern the interaction between adsorbents and adsorbates. Surface area is one of the main factors; substances with larger surface area typically have more sites for adsorption. For instance, open-textured adsorbents like zeolites and metal-organic frameworks can enhance adsorption capacity due to their high surface area and tailored porosities, which offer greater contact with liquid or gaseous reactants (Jonathan et al., 2022). There are also chemical properties involved from both the adsorbate and adsorbent. The physicochemical properties, including polarity and functional groups, can dictate the extent of interaction and therefore influence whether physisorption or chemisorption is mainly responsible for adsorption. Therefore, the material property-environmental factor interaction is a significant consideration in the understanding of adsorption kinetics. Temperature is also essential in controlling adsorption operations in surface chemistry. Generally, elevated temperatures tend to increase kinetic energy and thereby facilitate more intensified movement towards the surface of the adsorbates. However, this increase is not invariably associated with better adsorption; in the majority of exothermic adsorption reactions, high temperatures can inhibit the likelihood of interaction as the system prefers to proceed to a position of lower energy. Therefore, in processes such as vapor-phase adsorption, determination of an optimum temperature is of utmost significance to obtain maximum adsorbent capacity (Jonathan et al., 2022). Furthermore, the adsorption isotherms, which provide information about how quantities adsorbed change with pressure or concentration at given conditions of temperature, provide valuable information about the thermodynamic favorability of adsorption processes and indicate the necessity for stringent temperature control in numerous applications. A second determinant factor is the presence of competitive species in the surroundings, that can have a significant influence on adsorption efficiency. In multicomponent systems, competition among molecules is possible, thus limiting the number of active sites on the adsorbent that are available. This tends to cause a lower selectivity and adsorption capacity, and hence selectivity coefficients and competitive adsorption isotherms become necessary to take into account to be able to make accurate predictions of adsorption behavior (Di et al., 2020). In addition, surface modification to enhance specific interactions, such as the formation of bespoke functional groups for specific adsorbates, can be employed to repress competitive effects. Therefore, the integrity analysis of the adsorbent surface and the environmental environment is important in the optimization of adsorption processes, especially in systems where there are more than one pollutant present at multiple concentrations simultaneously, such as in indoor air quality management.

## 1.3. SURFACE CHARACTERIZATION TECHNIQUES

The comprehension of the complicated nature of surface chemistry is greatly reliant on surface characterization techniques, which constitute the building block in the determination of material interface properties. Scanning tunneling microscopy (STM) and atomic force microscopy (AFM) are just a few of the techniques that allow scientists to visualize the topography of surfaces at the atomic level, enabling the investigation of surface defects, roughness, and surface atom organization. These pieces of knowledge are very valuable in cases where surface properties govern material performance, such as catalysis and sensor devices. Of particular significance, the advancements in these imaging techniques have made it possible to visualize dynamic processes at surfaces, thereby facilitating real-time analysis of surface phenomena. With continuing development of surface characterization through microscopy and spectroscopic techniques, its contribution toward further insight into surface chemistry becomes increasingly invaluable.

Proper surface characterization is critical in catalysis to allow for optimum performance of the catalyst and identification of active sites. Such characterization is offered by techniques like X-ray photoelectron spectroscopy (XPS) and temperature-programmed desorption (TPD), which offer valuable quantitative and qualitative data regarding the surface composition and chemical states of elements on the surface. For instance, XPS explains the oxidation state of the metal species on support materials and hence informs researchers about the likely reactivity and stability of the catalyst under operation conditions. Secondly, the use of spectroscopic techniques in conjunction with in situ characterization has been useful in monitoring catalytic reactions in real time, giving information crucial for the formation of more efficient materials. The ongoing evolution of such characterization techniques continues to enhance the understanding of how surface properties correlate with catalytic activity. Furthermore, the functionality of oxygen species on carbon materials serves as a strong example of how characterization techniques help to elucidate multifaceted surface effects. Current research highlights the challenges that exist due to inadequacies in current characterization methods, which normally complicate the singular functions and properties of different oxygen functional groups on carbon surfaces (Dang et al., 2025). This insufficiency retards progress in different applications ranging from energy storage to environmental purification. With the use of different in situ characterization techniques, including spectroscopic and theoretical, researchers are now revealing the electronic character of these oxygen species and how they affect the chemical conductivity of carbon materials. Moreover, the results of such studies not only disprove the prevalent myths regarding the homogeneity of oxygen species but also indicate that correct surface characterization is key to constructing surface chemistry knowledge (Bruce, 2024).

### 1.3.1. Scanning Tunneling Microscopy (STM)

The development of STM revolutionized the field of surface chemistry since for the first time in the history of the world, it provided an insight into the atomic-scale structure of surfaces. By sensing the tunneling current that occurs when a sharp metallic tip approaches a conductive surface, STM enables researchers to observe individual molecules and atoms in real time. This capability is of utmost value to the understanding of surface phenomena such as adsorption, molecular rearrangement, and electronic states, as it allows for the probing of nanoscale dynamic processes. This has rendered STM an essential tool in investigating the intricacies of surface chemistry where theoretical predictions and experimental findings converge. The versatility of STM extends to various investigations in material science and catalysis. For instance, STM has been employed by researchers in the study of catalytic reactions on surfaces, which play a pivotal role in designing efficient catalysts.

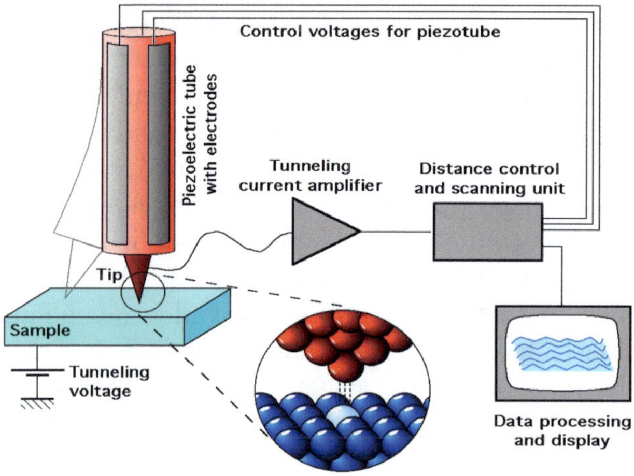


**Figure 1.9.** Scanning tunneling microscope (STM).

*Source:* https://external-content.duckduckgo.com/iu/?u=https% 3A%2F%2 Fscienceinfo.com%2Fwp-content%2Fuploads%2F2023%2F05%2Fimage-7.png&f=1&nofb=1&ipt=c6e32ec9e9d52627b1ba906c41732b1f49ce03efcbf27bd51b92fe031def4192.

By correlating topography information with electronic properties, STM can provide information regarding the active sites of catalysts, including how molecular orientation and electronic structure influence reaction pathways. As surface chemistry plays a fundamental role in catalysis and material properties, STMs ability to probe surfaces at this level enhances understanding of composition-structure-performance relationships in numerous chemical systems and guides the design of next-generation materials and catalysts (Figure 1.9). Moreover, the complementarity of STM with other techniques, such as Tip-Enhanced Raman Spectroscopy (TERS), further augments the acquired information from surface examination. TERS enables molecular specificity to study chemical reactions in combination with STMs spatial specificity. This complementarity enables complex surface reactions, such as on-surface coordination, decomposition, and polymerization, to be studied, revealing vital insight into reaction mechanisms and kinetics, which otherwise would have been concealed. As highlighted by recent advances, the combination of these techniques among themselves creates new research possibilities to explore the mechanistic aspects of surface reactions, thereby underlining the importance of advanced microscopy tools like STM in the general context of surface chemistry (Zhen-Cai et al., 2022; Song et al., 2020).

### 1.3.2. Atomic Force Microscopy (AFM)

Arguably the most significant innovation in surface characterization techniques is AFM, which allows scientists to view surfaces at the nanometer scale. AFM functions by scanning a sharp probe on a spring-mounted cantilever over the surface of a sample, measuring the forces between the probe and the atoms of the material. This approach has gained widespread use within the field of surface chemistry due to its ability to create images of the topography of surfaces, mechanical characteristics, and chemical interactions at previously unmatched resolutions. The addition of AFM to studies has expanded our understanding of many phenomena, including biomolecular interactions and material properties, at the fundamental contributing to technological advancements in many fields, e.g., biochemistry and nanotechnology. In mineral surface studies, AFM techniques have been important in gaining knowledge of the molecular-scale adsorption phenomena. Contemporary studies have employed AFM in combination with theoretical methods like density functional theory (DFT) to investigate the distinctive interactions between biomolecules, for instance, di-glycine, and mineral surfaces, for example, clinochlore. The findings indicated selective adsorption of biomolecules onto hydrophobic regions of the minerals, demonstrating the nuanced role played by surface chemistry in such processes. AFM has the potential to elucidate structural arrangements of biomaterials at high resolution, making it easier for us to understand catalysis and molecular aggregation, which are relevant to imminent advancement in biotechnology and environmental science (Moro et al., 2020). Further, AFMs functionality also covers three-dimensional (3D) imaging and biological sample characterization, which is a revolutionary advancement in biomedical science. Hybrid platforms combining AFM with other fabrication techniques have enabled the mapping of complicated samples in great detail, enabling researchers to collect mechanical and chemical information from complex samples. This is particularly important in the study of massive biological structures such as cells, with their nanolevel architecture playing a decisive role in drug discovery and cancer therapy. For instance, AFM has been instrumental in the identification of vulnerable and resistant bacterial strains, revealing information about their compositional contrast as well as resistance mechanism. These applications emphasize the core significance of AFM in allowing us to understand surface chemistry better and how it applies to biology and material science (Adineh, 2017) (Figure 1.10).

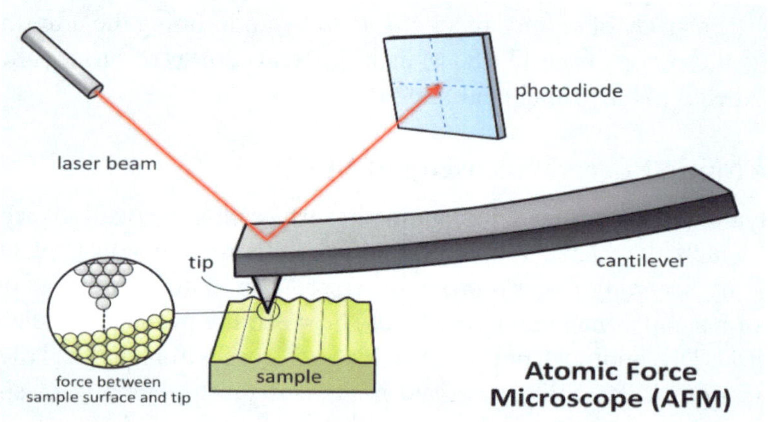


**Figure 1.10.** Atomic force microscopy (AFM).

*Source:* https://external-content.duckduckgo.com/iu/?u=https%3A%2F%2Fwww.nisenet.org%2Fsites%2Fdefault%2Ffiles%2Fcatalog%2Fuploads%2FAFM.jpg&f=1&nofb=1&ipt=dfb2a82d95ebf86e13d690d89ea39d084a2f849d8a693b773e5968a667ccfc62.

## 1.3.3. X-ray Photoelectron Spectroscopy (XPS)

The field of surface chemistry is actually pushed forward by applying the method of XPS, which is a very useful analytical technique utilized for the investigation of surface composition and electronic states of materials. XPS is founded on the principle of photoelectron emission from irradiation of a material by X-rays. The emitted photoelectrons' binding energy informs about the chemical environment of atoms in the surface layer. This capacity to differentiate elemental makeup and oxidation state makes XPS extremely valuable to the study of heterogeneous surfaces, such as in catalysts and semiconducting materials. As an operando technique, XPS is very effective in capturing alterations at the surface in chemical reactions and thus enables scientists to draw meaningful conclusions about surface dynamics, which are crucial in the knowledge of occurrences such as corrosion and catalytic activity, and thus enhancing general knowledge of surface chemistry in practical usage. In particular for battery science, XPS has emerged as a pivotal tool to investigate the intricate surface chemistry at electrode/electrolyte interfaces. Lithium-ion battery surface chemistry, for example, determines their performance and lifetime. The lithiation and delithiation charge transfer processes are intrinsically interconnected with surface states, where XPS provides a unique platform to monitor such changes in situ.

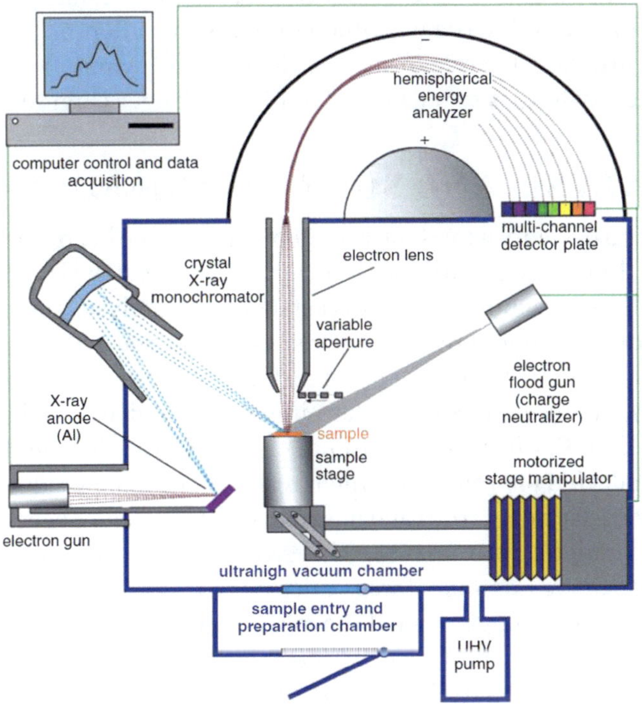


**Figure 1.11.** Schematic diagram of the components of an XPS instrument.

*Source:* https://serc.carleton.edu/msu_nanotech/methods/xps.html.

Recent advances in ambient pressure XPS (APXPS) allow researchers to study these interfaces under conditions of operation and identify effects that go unnoticed with traditional ex-situ methods. Under this process, transient states are now able to be determined, which can be formed during cycling in the battery, giving a better understanding of surface phenomena that dictate the efficiency and capacity of the battery (Liu et al., 2023). These results ultimately lead to the optimization of battery materials and designs. Furthermore, XPS enables wider understanding of surface chemistry beyond batteries and becomes pertinent in fields such as materials science and the preservation of cultural heritage. The ability of XPS in the identification of chemical states and composition has made it an unavoidable tool for detailed characterization at a microscopic level of a wide range of materials, from nanostructured material to ancient relics. For instance, XPS has been utilized to assess degradation processes in art to ensure that preservation methods for culturally significant materials are enhanced (Magdy, 2023). Not only does XPS aid in material characterization by exposing the chemical and elemental properties of surfaces but also fosters information regarding the conservation treatments needed to ensure the preservation of the integrity of cultural heritage. Although the intersection of XPS with other disciplines is growing, the role of surface chemistry becomes more important in developing new solutions to today's scientific challenges (Figure 1.11).

## 1.4. CATALYSIS AND SURFACE CHEMISTRY

The intersection of catalysis and surface chemistry is at the heart of understanding most chemical processes, particularly heterogeneous catalysis. At the heart of catalysis is the interaction between reactants and catalysts surface, which is at the heart of lowering activation energy and enhancing reaction rates. The exclusive characteristics of the catalysts surface, such as composition, structure, and reactivity, are at the heart of determining the efficiency and selectivity of catalytic reaction. For instance, coordination of reactive sites on the surface of a catalyst can decide the routes by which reactants become products. A general knowledge of surface properties and modification is therefore necessary in order to maximize catalytic performance. The evolution of surface characterization techniques, such as atomic force microscopy and X-ray photoelectron spectroscopy, further supplemented the surface chemistry of catalysis, paving the way for rational design of more effective catalytic materials. The evolution of nanotechnology has introduced new dimensions to catalysis, particularly through research on nanomaterials as catalysts. Nanostructured materials such as colloidal nanozymes have great catalytic activity due to the large surface area and exceptional electronic properties. They not only enhance traditional catalytic processes but also open up opportunities in biomedical applications, as evident from their use in the diagnosis and treatment of cancer. The integration of catalytic precision with complex biological systems emphasizes the role of surface chemistry in managing the reactivity and specificity of nanomaterials. The ability to tailor surface chemistry — through methods such as surface modification and interfacial engineering — allows for the development of nanozymes with enhanced biocompatibility and specificity toward tumor-associated reactions. In this case, explanation of the fundamental catalytic mechanisms becomes inevitable to enhance theoretical understanding and applicability (Dheyab et al., 2025). Also, new synthesis techniques for nanomaterials, such as laser techniques, have become more popular for enhancing catalysis and surface chemistry. The techniques offer sophisticated control over the shape, size, and functionalization of carbon-based nanomaterials, e.g., graphene quantum dots (GQDs) and carbon quantum dots (CQDs). The green synthesis route and tunability associated with laser ablation in liquids have proven to be useful in the achievement of these nanomaterials with adjustable properties for various applications, such as catalysis. Systematic tuning of optical and morphological properties of CQDs and GQDs is possible via post-laser treatments, thus influencing their surface interactions and catalytic properties. While. With the scientific community exploring the ultimate potential of these new-age materials, the synergy among catalytic behavior, surface chemistry, and synthesis will necessarily translate into quantum leaps for energy-related and bio-applications, further solidifying the place of catalysis in the grand game of surface chemistry (Rey et al., 2025).

### 1.4.1. Role of Surface in Catalytic Reactions

The role of surface chemistry in catalytic reactions is not to be exaggerated, for surface interactions between reactants and catalysts essentially regulate reaction efficiency and rates. Catalysts operate by developing a surface that the reactants are adsorbed on and under which their conversion into products is promoted while lowering the energy barrier required for the conversion. This is particularly significant in the case of transition metal reactions, when the metal surface itself acts as a site for bond activation and dissociation. The efficacy of palladium catalysts in reactions such as methane oxidation, for instance, shows how the precise fine-tuning of surface properties can facilitate catalysis. Supported palladium catalysts can have high reactivity from the surface structure and composition alterations, which directly influence adsorption states of methane, addressing the key role surface chemistry in catalytic activity (Oh et al., 2023). Further, the design of catalytic systems often involves surface tuning to enhance selectivity and stability. The chemical synthesis of novel materials, such as graphdiyne and its analogues, is one example of this approach, in which the target surface structural characteristics determine reaction pathways and mechanisms. Graphdiyne has a porous structure that not only increases the accessible surface area for reactions but also has a series of asymmetrical charge distributions that can be accountable for enhanced catalytic activities for oxygen evolution and hydrogen evolution reactions. Through the development of catalytic systems that take advantage of new surface chemistries, researchers are able to achieve a delicate balance between high activity and increased stability to enable long-term applications in energy conversion and environmental purification (Liu et al., 2022) (Figure 1.12).

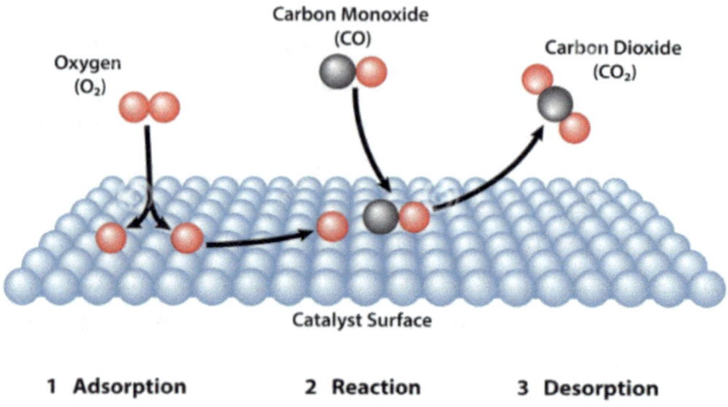


**Figure 1.12.** Catalyst surface with catalytic reaction.

*Source:* https://external-content.duckduckgo.com/iu/?u=https%3A%2F%2Fthumbs.dreamstime.com%2Fz%2Fcatalyst-surface-catalytic-reaction-carbon-monoxide-dioxide-catalysis-oxygen-converter-showing-adsorption-218305638.jpg&f=1&nofb=1&ipt=0d6ad4e8dcf29fd5425de9d584e6e5703e2fee7c0f1a7a83755bf22e6e9906ae.

In addition to accelerating catalytic reaction by surfaces designed, stability of the surface under working conditions is necessary for practical applications. Catalytic surface structure can be greatly changed during the reaction progress to lead to deactivation through processes like sintering or leaching of active metal components. Hence, a correlation between reaction conditions and surface composition has to be established for the design of efficient catalytic systems. The inclusion of stabilizing promoters or protective coatings can minimize such effects, making catalysts long-lasting and efficient in industrial uses. Last, the exploration of surface phenomena provides a glimpse into the enhancement of catalyst design, enabling improvement of catalytic efficiency and sustainability in various industrial processes (Oh et al., 2023).

## 1.4.2. Heterogeneous vs. Homogeneous Catalysis

The distinction between homogeneous and heterogeneous catalysis is at the core of surface chemistry, impacting many industrial processes and academic research. Homogeneous catalysis is when reactants and catalyst are present in the same phase, often dissolved within a liquid solution. This similarity in phases allows for a uniform distribution of the catalyst in the reaction medium, typically leading to enhanced reaction rates and selectivity. But product and catalyst separation after reaction can become complicated, requiring additional purification steps. Heterogeneous catalysis utilizes catalysts of a phase different from that of the reactants, typically solid catalysts with gaseous or liquid reactants. This discrimination facilitates easier recovery of the catalyst, hence heterogeneous catalysis is a viable choice in most industrial applications since it can simplify downstream processing and reduce catalyst separation costs. The choice between heterogeneous and homogeneous catalysis depends on various factors, including reaction conditions, desired product yield, and environmental factors. For instance, in the production of fatty acid methyl esters (FAME) from waste cooking oil (WCO), heterogeneous catalysts have been reported to bring remarkable improvements in reaction performance, particularly when applied in sophisticated reactor systems such as microwave reactors and reactive distillation systems (Aubin et al., 2013). These processes enhance the efficiency of transesterification of the process to achieve higher biodiesel production with less environmental risk from waste disposal. The heterogeneous nature of these catalysts enables them to be employed in combined reaction and separation processes continuous flow reactors, making FAME production quicker and an environmentally sustainable source compared to traditional fuel sources. Furthermore, the environmental footprint of each form of catalysis is being increasingly scrutinized within surface chemistry. Discharge of untreated effluents, often a byproduct of chemical processes, is an important environmental concern, particularly in the case of organic compounds with very high toxicity (Boaventura et al., 2001). Again, heterogeneous catalysis presents an appealing advantage in its efficacy for removal of such effluents and in its ability to work with a wide variety of waste feedstocks. With heterogeneous catalysts, more toxic pollutants can be converted more readily to less toxic or more readily handled products, with implications for both the environmental mitigation and sustainability of chemical manufacture. Thus, the ongoing advancement and optimization of heterogeneously catalyzed processes are a prime intersection of surface science, industrial application, and environmental stewardship.

### 1.4.3. Examples of Catalysts in Industrial Processes

In industrial chemistry, catalysts play the crucial role of enhancing rate and efficiency in reactions. Platinum and palladium in catalytic converters for vehicles are a case in point. The catalysts facilitate the conversion of toxic gases such as carbon monoxide and nitrogen oxides into less harmful substances. The surface characteristics of metals are largely responsible for their effectiveness because they provide active sites to adsorb and react with the reactants. This activity not only helps reduce damaging emissions but also promotes cleaner transport behaviors. Such uses underscore the importance of surface chemistry in catalyst design focused on both environmental and performance needs in industrial applications. In addition to noble metals, researchers are also considering other catalytic materials, including nanomaterials, which can offer similar efficiencies at a lower cost. An example is nanozymes, which are promising catalysts due to their high surface area and tunable properties. Nanomaterial-based artificial enzymes possess the ability to replicate the natural enzyme activity and therefore speed up chemical reactions in various uses, ranging from pharmaceuticals to pollution abatement. The ability to engineer such catalysts at the nanoscale allows increased selectivity and reactivity, making them of interest for utilization in industrial operations. With increased progress in surface chemistry, development of such new material represents an immense shift towards more efficient and sustainable catalytic systems across industries, where material science and catalytic efficiency are prioritized (Jiang et al., 2019). Moreover, hydrogen's catalytic capability is also being utilized in industry, particularly under the banner of decarbonization programs. Hydrogen not only can serve as a clean fuel replacement but as a beneficial reactant for many catalytic reactions like ammonia synthesis and hydrocarbon reforming. Having hydrogen as a participant in these reactions can lead to significant carbon reduction since according to the vision for worldwide sustainability. The present study establishes with certainty that hydrogen technology is no longer a future prospect, but an existing reality that can address the urgent demands for cleaner energy alternatives, particularly in applications such as heat, power, and transport (Staffell et al., 2018). This shows how the principles of catalysis combined with surface chemistry can transform industry processes and pave the way for the world towards sustainability.

## 1.5. SURFACE MODIFICATIONS AND TREATMENTS

For the sake of enhancing the biocompatibility and functionality of nanoparticles, surface modification and treatments are a key element in surface chemistry. These involve intentional alterations of a material's surface properties to enhance interactions with biological systems or induce function in other contexts. As a case in point, surface modification of silica nanoparticles with chosen biomolecules facilitates drug targeting or enhances their optical properties for applications in bioimaging. This is important in fields such as cancer studies, where precision and efficacy in the delivery of drugs matter. Surface modification can further diminish cytotoxicity, improving the safety of nanoparticles, thus making their successful incorporation into clinical practices easier (Gubala et al., 2020). Furthermore, the most critical aspect of surface treatment is engineering the protein corona, i.e., the protein layer that spontaneously adsorts on the nanoparticle surface upon introducing it into biological environments. It is crucial to develop an understanding of the dynamic nanoparticle-protein interactions in order to provide precise predictions of their in vivo behavior. Protein corona modulation can be utilized strategically to promote bioavailability and reduce immune detection, and thereby therapeutic outcome. One of them is superparamagnetic iron oxide nanoparticles (SPIONs), with the potential of site-specific drug delivery that highly relies on their surface chemistry. Through the design of modifications, SPIONs exhibit increased capability to target, which can inflict a big punch on cancer therapy treatment (Palanisamy et al., 2019). Another vital area of surface modification includes the employment of stimuli-sensitive materials, which are able to change vastly in property as a function of particular environmental stimuli such as pH, temperature, or light. These dynamic systems can be employed in nanoparticles to yield controlled drug release systems, thereby optimizing therapeutic effect with minimal systemic side effects. In addition, improved surface modifications enable multifunctional nanocarriers to be established that can simultaneously deliver imaging compounds and therapeutic drugs, facilitating easier cancer treatment protocols. Overall, the study of surface chemistry and modification techniques is pivotal in translating nanoparticles to the complex demands of biomedical applications, bridging the gap between the laboratory and clinic (Gubala et al., 2020).

### 1.5.1. Coating Techniques: Physical and Chemical Methods

The study of the coating techniques, physical and chemical, provides a pivotal foundation in surface chemistry. Physical coating techniques include a variety of methods that are adhesion and surface interaction based and do not alter the chemical composition of concerned materials. Physical coatings like layer-by-layer assembly, dip-coating, and sputter deposition utilize mechanical forces or heat to construct protective films. These methods can be broadly characterized by their ability to deposit thin films that are either heterogeneous or homogeneous, and this ability to deposit such thin films can have a significant influence on the final properties of coated surfaces. The efficacy of the methods is normally based on the conditions under which the epitaxial coatings are deposited, such as environmental conditions like temperature and humidity, along with the material specificity of the substrate and the coating material. On the other hand, chemical coating processes involve the modification of surfaces through chemical reactions that yield a molecular-level change. Such techniques, such as chemical vapor deposition (CVD) and sol-gel processing, allow for the production of coatings with enhanced durability and specialized functionality. For instance, CVD facilitates the deposition of high-purity material that is usually engineered to have specific electrical, optical, or chemical properties. The sol-gel method, where particles of small molecules are converted into a network solid through hydrolysis and polymerization, is unique because of its capability and efficacy in the development of wear-resistance and corrosion-protective coatings (Figure 1.13).

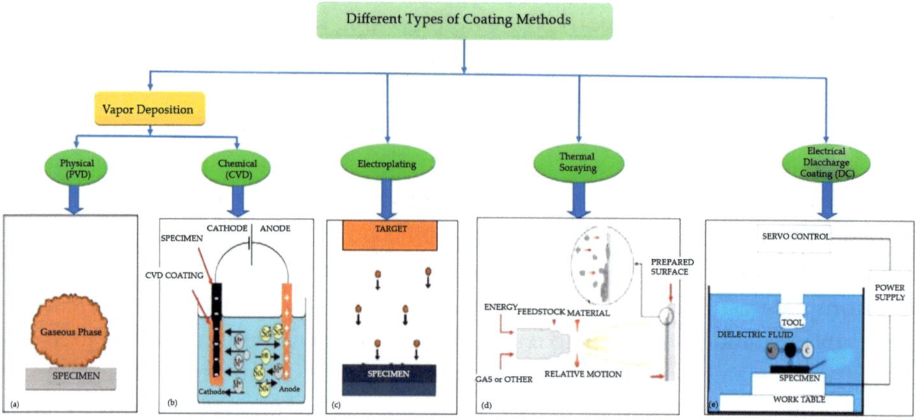


**Figure 1.13.** Different types of coating methods.

*Source:* https://www.researchgate.net/figure/Different-types-of-coating-methods-Reprinted-from-28-Reproduced-from-28-C-The_fig1_368979855.

It is essential to know these chemical processes since they provide general information on how different interfacial interactions can be tailored to produce advanced materials with preferred surface characteristics. These coating techniques are not just used to beautify the surface; they have revolutionary potential in a broad range of fields such as electronics, biotechnology, and materials science. In biomedical applications, for instance, the design of nature-inspired coatings, like those enhancing skin adhesion, illustrates surface chemistry principles translated into real-world applications. The emphasis on wet adhesion properties and biocompatibility in skin adhesive clarifies how the judicious application of physical and chemical strategies can result in innovative solutions for medical devices and wearables.

In addition, application of silver nanoparticle-based coatings accentuates the antimicrobial advantages obtained from a hybrid approach that coalesces two approaches to coating technologies in order to amplify functionality without sacrificing structural integrity. Through breaking boundaries between physical and chemical approaches, researchers are continually uncovering new avenues for performance and use in next-generation coatings and pushing the boundaries of surface chemistry.

## 1.5.2. Self-Assembled Monolayers (SAMs)

The phenomenon of SAMs has significantly helped advance surface chemistry, opening new avenues for molecular-level surface engineering. SAMs are typically formed by spontaneous self-assembly of amphiphilic molecules onto a surface, leading to the formation of a dense and well-packed monolayer. This is achieved through certain interactions, such as van der Waals forces, hydrogen bonding, and covalent bonding, particularly with metal surfaces like gold and silver. Likely the most intriguing aspect of SAMs is their tunability: the chemical composition of the adsorbate can be altered, and surface properties like hydrophobicity, electronic properties, and biocompatibility can be carefully designed by the researcher. This tunable nature presents potential for application in biosensing, microelectronics, and development of novel coatings that can enhance or inhibit surface reactivity. In newer works in this area, N-Heterocyclic Carbenes (NHCs) have been of specific interest as robust building blocks for durable SAMs. Such carbenes present favorable electronic and structural properties that enable the creation of stable monolayers on metal surfaces. Triazolylidene NHCs have been found through experiments to be employed in such a way that their self-assembly on Au and patterned Au/$SiO_2$ substrates resulted in SAMs with high thermal stability compared to conventional NHC SAMs. The work points out the importance of structural and electronic tuning towards making such SAMs more effective, particularly in preventing undesirable metal deposition in IC manufacturing (Figure 1.14).

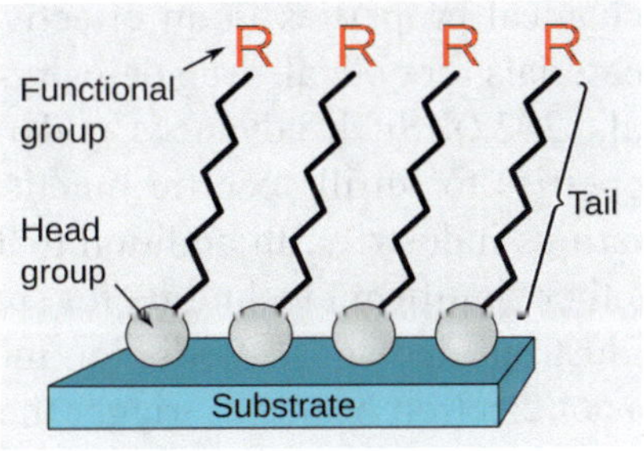


**Figure 1.14.** Representation of a SAM structure.

*Source:* https://upload.wikimedia.org/wikipedia/commons/thumb/9/98/Self-assembled_monolayer.svg/800px-Self-assembled_monolayer.svg.png.

The selective deposition characteristics of Tz SAMs also highlight a new direction for small-molecule inhibitor development that will improve performance in microelectronics fabrication, demonstrating the real-world applicability of these novel materials (Dana et al., 2025). Further, the interaction between adjacent molecules of the SAMs may significantly influence their electrochemical behavior and stability. Current research on Fc-terminated NHC SAMs suggests that these monolayers possess excellent redox kinetics and stability, characteristics desirable for sensors and electrocatalysis. Unlike complex alkylthiol SAMs, with their multivalent redox activity, NHC SAMs provide a more stable electroactive center distribution, yielding more reproducible and consistent electrochemical response. In addition, the ability of $Fc^+$ ions to engage in ion pairing with other anions indicates a high environmental compatibility for such SAMs. The study calls for the revolutionary aspect of NHC SAMs as better alternatives to traditional SAM systems, facilitating their role in the development of surface chemistry concepts and applications (Qi et al., 2024).

### 1.5.3. Surface Functionalization for Specific Applications

The process of surface functionalization plays a crucial role in tailoring material properties for a range of applications in diverse industries, including biomedicine and microfabrication. Researchers can tailor contact between surfaces and their environment by manipulating the surface chemistry of materials to achieve increased functionality. Two-photon polymerization (TPP) techniques, for instance, have expanded the variety of materials available for surface modification. They allow for the creation of intricate 3D microstructures with controllable surface characteristics. In the literature, TPP systems have moved on from typical photoresists to carefully prepared formulations for adjustable hydrophilicity and mechanical properties in an effective manner, unveiling a pathway to customize materials for optical, drug delivery, and tissue engineering applications (Seán et al., 2022). Such advances underscore the requirement to engineer surface properties to fulfill specific functional functions, thereby driving innovation in various industries. In addition to TPP, the application of ionizing radiation is another significant technique for surface functionalization to address various industrial demands. Radiation-induced grafting (RIG) utilizes high-energy ions or electron beams to trigger the creation of functional groups in the surfaces of polymers, expanding their capacity and compatibility with other materials. It is not only a process that changes polymers but also one that facilitates the introduction of reactive sites for future chemical alteration, and therefore broadens the functional horizon of polymers (Ashfaq et al., 2020). The subtleties of radiation chemistry highlight the interconnection between the kinetic nature of irradiation and the surface changes achieved. Parameter conditions such as dose rate and environmental conditions are also controllable, which can induce particular outcomes, which enhance polymers to be more efficient in various applications ranging from medical devices to environmental decontamination. All these approaches illustrate the versatility and nature of adaptability on existing surface functionalization techniques. Applications of surface functionalization are vast and continually expanding, particularly in nanotechnology and biomaterial applications. Through the tailoring of surface properties, scientists have the ability to greatly influence cellular behavior and material biocompatibility, two crucial elements in drug delivery system and tissue scaffold development. For instance, developing hydrogels using radiation-induced polymerization enables control of mesh size and molecular-level interactions, enabling optimized performance in biomedical applications. Recent advancements in functionalization of nanogel platforms suggest promise for the modulated surfaces to interact more effectively with biological systems (Ashfaq et al., 2020). As there is a growing necessity for new materials, proficiency in engineering surfaces at the molecular level becomes even more essential, which is the reason why surface functionalization is crucial in tackling challenges in science and technology at present.

## 1.6. ENVIRONMENTAL AND BIOLOGICAL IMPLICATIONS

Microplastics are a ubiquitous environmental problem, as their diverse compositions and morphologies complicate research into how they affect ecosystems and human health. Entering the environment from many sources—ranging from manufacturing to cosmetics—microplastics are categorized as either primary or secondary microplastics based on the method of formation. Primary microplastics are specifically designed to be small, while secondary microplastics form through the degradation of larger fragments of plastic. The ramifications of such variability are great, for different sizes and morphologies might have the potential to interact differently with natural organisms and systems. Understanding this variability is necessary in order to properly monitor and create environmental policy, since simplified views of microplastics as a monolithic group ignore important factors regulating their environment fate and toxicity in the environment. Therefore, the characterization of microplastics as a class of pollutant is necessary in order to approach their environmental and biological impacts in an integrated way (Chelsea et al., 2019). The persistence of heavy metals in the environment emphasizes their role as important pollutants with important biological impacts. Compared to organic pollutants that can undergo degradation, heavy metals persist in the environment and bioaccumulate in food webs, causing toxic effects on organisms at various trophic levels. When these metals become mobilized by human activities such as mining and industrial processes, they disrupt biogeochemical cycling in nature, thereby leading to elevated concentrations in the terrestrial and aquatic ecosystems. This is of great concern since metals like lead, mercury, and cadmium bioaccumulate in the body tissues of organisms, thereby leading to biomagnification and serious health risks to predators, including humans. The need for intense monitoring of the concentration levels of heavy metals in different ecological compartments is thus imperative since it gives researchers and policy-makers an adequate mechanism to reverse their impact on human health and environmental equilibrium (Ali et al., 2019). The interaction between microplastics and heavy metals is a very critical issue, given the capability of microplastics to adsorb contaminants from their surroundings.

Such action exacerbates the ecological and biological impacts of both pollutants, such that microplastics act as vectors for toxic heavy metals, increasing their bioavailability to terrestrial and marine biota. The combination of microplastics and engaged heavy metals cumulatively poses immediate risks to biodiversity and ecosystem integrity because organisms unintentionally ingest such particles, where there are possible impacts on their wellness. The intricate dynamics among these classes of pollutants necessitate holistic research techniques to adequately understand their interactions within ecosystems. Fending off the twin scourge entails synergistic research, policy formulation, and public awareness strategies that will lay the foundation for out-of-the-box approaches to limiting their environmental footprint as well as protecting ecosystem integrity (Chelsea et al., 2019).

### 1.6.1. Surface Chemistry in Environmental Remediation

The role of surface chemistry in remediation of the environment is instrumental in addressing contamination issues in ecosystems. Surface interactions between substances are pivotal in influencing the efficiency of adsorption processes, which are integral to capturing and purifying contaminants from water and soil. Specifically, the functional groups of the adsorbent surface, such as in the case of activated carbon or metal-organic frameworks, are the most important parameters determining their interaction with the impurities. As an example, modifications of functional groups to form improved functional groups enable enhanced interaction with the intended pollutants, thus improving the overall removal efficiency. This interaction highlights the importance of understanding surface phenomena to build more effective materials for environmental application, where surface optimization can lead to more effective remediation. Recent advances in surface chemistry have been showcased through studies on iron metal-organic frameworks (MOFs) that show the promise of engineered surfaces in environmental remediation. With the application of low-cost and ecofriendly synthesis methods, such as electrochemical processes, scientists have successfully synthesized MOFs for the elimination of organic dye pollutants with their high adsorption and photocatalytic degradation properties. In these instances, modifications like the introduction of polysulfide species into the surface morphology of the MOFs have been seen to significantly increase their photocatalytic activities, giving them exceptional contaminant removal efficiencies, e.g., for methylene blue. Integration of surface chemistry principles in the synthesis of such materials renders them more functional and versatile, thus they can be used to solve different environmental remediation cases, according to findings in recent studies (Brenand et al., 2025). Moreover, principles of green analytical chemistry (GAC) intersect with surface chemistry to promote ecologically compatible methods towards environmental remediation. GAC aims to reduce reagent toxicity and solvent usage while maximizing detection sensitivity and specificity, i.e., through the use of new sorbents like metal-organic frameworks. These innovations not only enhance analytical process efficiency but also contribute to greater environmental goals, fostering greener approaches within the field. With increasing demand for cleaner and more cost-effective remediation technologies, the use of surface chemistry concepts in GAC is a significant way forward. By enabling the development of environmentally friendly analytical methods as well as improving pollutant detection, scientists are better equipped to contribute to the larger goals of environmental and health protection (Mehta et al., 2025) (Figure 1.15).

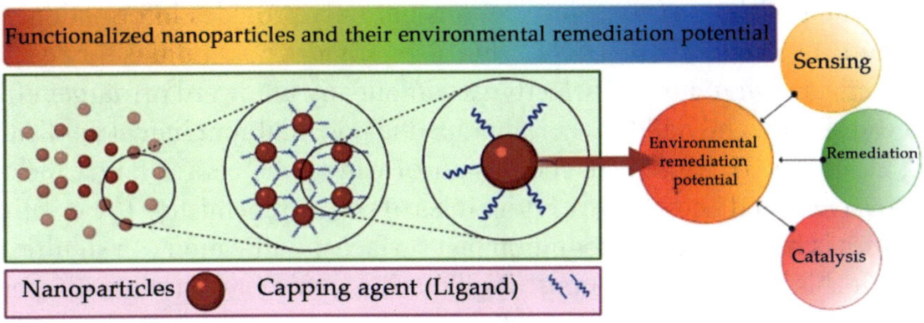


**Figure 1.15.** Functionalized nanoparticles and their environmental remediation potential.

*Source:* https://media.springernature.com/lw685/springer-static/image/art%3A10.1007%2Fs40097-021-00468-9/MediaObjects/40097_2021_468_Figa_HTML.png?as=webp.

### 1.6.2. Interactions of Surfaces with Biological Systems

Surface-biological system interactions lie at the heart of numerous applications in biomedical science, particularly in the design of nanoparticles and other materials for drug and diagnostic purposes. Foreign materials such as colloidal nanozymes, upon entering biological systems, naturally have interactions with an excess of biomolecules to form a protein corona. This surface-engineered layer can significantly alter the physicochemical properties of the nanoparticles and thus their biological activity, biodistribution, and clearance. Surface interactions govern how such materials interact with tissues and cells, a feature significant to their efficacy and safety for drug delivery and imaging and other applications. The subtle modulation of surface chemistry is thus fundamental, deciding not just the biodynamic fate of the particles but also their drug activity, and underpinning thus the necessity for an appreciation of surface-biomolecule interactions within a broad framework of drug design and drug delivery. Looking further into material catalytic activity, reducible metal oxide nanozymes (rNZs) have been identified for enzyme-mimic and catalytic activity. These nanozymes utilize their complex surface topography to carry out reactions akin to biological enzymes, which can be of utmost significance for their application in cancer treatment and biosensing. Significantly, the surface-active sites of rNZs can be highly vulnerable to environmental factors, thus influencing their reactivity along with interaction with biomolecules. The composition and architecture of such surfaces are tuned to facilitate substrate coordination and redox cycling so that processes essential for imaging and for the therapeutic interventions can be modulated. The understanding of these surface interactions not only provides information about the performance and stability of rNZs but also provides possibilities for designing nanomaterials with improved selectivity without any untoward off-target effects, which can significantly add to their therapeutic potential for clinical applications (Craig et al., 2023). Moreover, the design of nanomaterials for target therapies needs to have sound knowledge of principles of surface chemistry. The creation of nanozymes is based on the recognition that surface modification can significantly alter their biological interactions. Voluntarily modifying surface properties, through techniques such as atomic dispersion and ligand coordination, enables the design of materials that are capable of evading immune system detection or inducing targeted cellular uptake. Such innovations are imperative in achieving precision medicine goals, where the therapeutic entities not only possess greater catalytic activity but are also active within the intricate biological context. By critically exploring how surface interactions influence the development of the protein corona and biological identity of such materials, scientists can devise means of designing "designer" nanozymes that possess selective designs towards a specific biological response, thus having future promise in oncology and precision therapeutics (Dheyab et al., 2025).

### 1.6.3. Nanomaterials and Their Surface Properties

Surface properties of nanomaterials have decisive roles in determining their functional applications in various fields, from medicine and catalysis to environmental sciences. As material size is minimized to the nanoscale, the surface area-to-volume ratio increases exponentially, leading to enormously higher reactivity and the ability to interact than with bulk material. This phenomenon occurs particularly in multicomponent magnetic nanoparticles that possess specific structures and functionalities because of their engineered surface chemistries. Recent advances in synthesis procedures have enabled the creation of diverse morphologies, such as core-shell and yolk-shell morphologies, that enhance their performance in applications such as multimodal imaging and nanomedicine. However, to enable mass use, the synthesis procedures must be mass-produced while also being environmentally friendly, mirroring the call for additional innovation in nanomaterials (Garc et al., 2022). Another interesting property of nanomaterials is the ability to be compatible with nature materials, as demonstrated by nanocelluloses (NC). As green biomaterials, NC not only inherits beneficial properties from cellulose but also possesses unique nanomaterial features, such as high surface area and active sites. The versatility of NC can enable it to find broad applications, particularly for water treatment operations where traditional methods prove inadequate. Current research highlights NCs potential use in coagulation, flocculation, and membrane filtration, with emphasis on how a proper understanding of elements like surface chemistry and charge can optimize these technologies. Through the integration of NC in water treatment systems, scientists hope to create cost-effective and sustainable technologies for clean water production, addressing a world imperiled by a large demand for it (Das et al., 2022). The correlation between nanomaterial surface properties and their applications underscores the necessity of studying structure-property relationships in depth. With the increasing demand for nanotechnology, an understanding of these relationships is essential to design novel materials with desired functionalities. Studies on properties such as purity, crystallinity, and morphology could possibly bring to the fore the role and efficiency of nanomaterials in varied applications. As for magnetic nanoparticles and NC, advances in synthesis methodologies and characterization technologies allow the window for innovation. Mapping regions of unawareness and exploring the integration of new materials into existing technologies allows researchers to greatly improve the effectiveness of nanomaterials in practical applications, cementing their role as the driving force behind surface chemistry science (Garc et al., 2022; Das et al., 2022).

## SUMMARY

In synthesizing the overall themes covered under the essay of the Fundamentals of Surface Chemistry, it is important to emphasize the pivotal position that surface interactions occupy in many applications. Surface chemistry is not a mere supporting aspect of material science but the basis that determines the catalytic activity, adsorption effects, and overall functioning of many materials. The significance of hydroxyl groups at surfaces, metal oxides in particular, cannot be overstated as quoted. Their influence on catalyst activity and properties is a prime example of how surface chemistry can be directly applied to performance in real life (Yun et al., 2023). These chemical building blocks are the foundation upon which technological and industrial advancement builds, and therefore more research in this field is essential. Moreover, the intricacy of interactions between etching methods and surface modification dictates the nature in which surface chemistry informs the characteristics of advanced materials. The ability to control etch profiles in 3D NAND technology's manufacturing process using the amorphous carbon layer as an example illustrates the engineering precision required to optimize material properties. The outcomes revealed through simulation not only find practical applications but also relate theoretical concepts to realities of engineering. These insights enhance our knowledge of the underlying mechanisms governing behavior at the surface and also bring the interactive role of surface chemistry and technology development (Zhang et al., 2023). This union of theory and practice is responsible for innovation and accelerating advancement in material design. Last but not least, the deduction from these findings necessitates interdisciplinary investigation of surface chemistry. As indicated by the diverse application context, from catalytic reactions to semiconductor production, it becomes evident that single-headedness is not sufficient. By merging several approaches, including thermogravimetric and plasma simulations, scientists can construct solid baselines for an improved understanding of surface interactions. With advancing technology, future research must quantify and characterize the surface chemical complexities, ultimately leading to the optimization of existing technologies and new material discovery. Continued funding in this area will encourage significant advancements that are the foundation for scientific and technological progress.

Among the most critical aspects of surface chemistry discovery is protein surface interactions and biological resistance, particularly under conditions of extreme stress. Cellular dehydration has been found to be a significant stress where proteins fold and aggregate due to dehydration. The extent to which some proteins endure this dehydration and subsequently regain functionality upon rehydration is largely dependent upon their surface chemistry. Current research shows that such surface topologies are key determinants of tolerance to desiccation, attested by findings that rational modification of protein surface features can enhance their robustness. Such realization underscores the role of surface chemistry in facilitating pivotal cellular processes, particularly where water levels vary, and shows how altering surface characteristics can be an issue of survival or cellular demise (Romero-Pérez et al., 2024). Another important aspect of surface chemistry is electrode material wettability in electrochemical systems. Electrolyte-electrode interface is important in achieving maximum efficiency for electrochemical reactions involved in energy storage devices. Comprehensive understanding of the influence of surface chemistry on electrolyte-wettability can lead to effective electrode design. Surface chemical modification approaches such as polar atom doping, surface functionalization, and in situ surface coatings have been followed to increase wettability, hence optimizing total electrochemical performance. These surface modifications not only improve reaction kinetics but also enable the development of next-generation energy storage devices. With rapidly developing electrochemical systems, ongoing research on surface chemical modification presents a profound promise for technological innovations in energy storage technologies (Zhao et al., 2024). Last but not least, their intersection illustrates a greater tale in the field of surface chemistry, marking its significance in biological and energy systems alike. Surface chemistry in biological systems governs the interactions that allow proteins to function under stress, while in electrochemical systems, surface chemistry governs the efficiency of energy storage devices. As ongoing research uncovers the importance of surface properties, scientists and engineers are better placed to tailor materials to meet specific functional needs, whether that means enhancing the robustness of cells or enhancing electrochemical performance. The continuing investigation of surface chemistry not only makes scientific advancement but also opens the door to new applications in an unusually broad range of areas from biochemistry to materials science and energy technology, thus establishing the essential importance of the field in addressing current scientific challenges.

# REFERENCES

1. Abbatt, J. P. D., Morrison, G., Grassian, V. H., Shiraiwa, M., Weschler, C. J., & Ziemann, P. J. (2022). How should we define an indoor surface? Indoor Air, 32. https://doi.org/10.1111/ina.12955 (accessed on 3rd October 2025).
2. Adineh, V. (2017). Advanced nanofabrication and microscopy methods for multidimensional nanoscale imaging and characterization of biological materials. https://www.semanticscholar.org/paper/3aa67dbc55b3651663e7092b85bb73ab60a1e9d4 (accessed on 3rd October 2025).
3. Ali, H., Khan, E., & Ilahi, I. (2019). Environmental chemistry and ecotoxicology of hazardous heavy metals: Environmental persistence, toxicity, and bioaccumulation. Journal of Chemistry, 2019, 1–14. https://doi.org/10.1155/2019/6730305 (accessed on 3rd October 2025).
4. Ashfaq, A., Clochard, M. C., Coqueret, X., Dispenza, C., Driscoll, M., Ulański, P., & Al-Sheikhly, M. (2020). Polymerization reactions and modifications of polymers by ionizing radiation. Polymers, 12, 2877–2877. https://doi.org/10.3390/polym12122877 (accessed on 3rd October 2025).
5. Aubin, J., Mazubert, A., & Poux, M. (2013). Intensified processes for FAME production from waste cooking oil: A technological review. https://core.ac.uk/download/18540047.pdf (accessed on 3rd October 2025).
6. Biswas, P., Polash, S. A., Dey, D., Kaium, M. A., Mahmud, A. R. R., Yasmin, F., Baral, S., et al. (2023). Advanced implications of nanotechnology in disease control and environmental perspectives. Biomedicine & Pharmacotherapy, 158, 114172–114172. https://doi.org/10.1016/j.biopha.2022.114172 (accessed on 3rd October 2025).
7. Boaventura, R., Faria, J., Figueiredo, J., & Gomes, et al. (2001). Catalytic wet air oxidation of organic pollutants in waste waters. https://core.ac.uk/download/153403454.pdf (accessed on 3rd October 2025).
8. Cai, Z. F., Kumar, N., & Zenobi, R. (2022). Probing on-surface chemistry at the nanoscale using tip-enhanced Raman spectroscopy. CCS Chemistry, 5, 55–71. https://doi.org/10.31635/ccschem.022.202202287 (accessed on 3rd October 2025).
9. Choi, J., Suryanto, B. H. R., Wang, D. B., Du, H. L., Hodgetts, R. Y., Ferrero Vallana, F. M., MacFarlane, D. R., et al. (2020). Identification and elimination of false positives in electrochemical nitrogen reduction studies. Nature Communications, 11. https://doi.org/10.1038/s41467-020-19130-z (accessed on 3rd October 2025).

10. Cortes, F. R. U., Falomir, E., Doñate-Buendía, C., & Mínguez-Vega, G. (2025). A review on pulsed laser-based synthesis of carbon and graphene quantum dots in liquids: From fundamentals, chemistry to bio applications and beyond. The Journal of Physical Chemistry C: Nanomaterials and Interfaces, 129, 10378–10414. https://www.semanticscholar.org/paper/96e9dcc1c5f32145e3a5600a2784c6c725029cac (accessed on 3rd October 2025).
11. Crolla, D. A., Foster, D. E., & Vaughan, N. D. (2015). Encyclopedia of Automotive Engineering. Choice Reviews Online, 53, 53–1072. https://doi.org/10.5860/choice.191861 (accessed on 3rd October 2025).
12. Dang, Y., Liu, Y., Xiang, P., Tan, Z., Tian, Z., Greiner, M., Heumann, S., et al. (2025). Carbon surface chemistry: Benchmark for the analysis of oxygen functionalities on carbon materials. Advanced Materials, 37. https://www.semanticscholar.org/paper/cba2716e9be5a4e108fd7b2a148c69ddd0d038be (accessed on 3rd October 2025).
13. Das, R., Lindström, T., Sharma, P. R., Chi, K., & Hsiao, B. S. (2022). Nanocellulose for sustainable water purification. Chemical Reviews, 122, 8936–9031. https://doi.org/10.1021/acs.chemrev.1c00683 (accessed on 3rd October 2025).
14. Dheyab, M. A., Aziz, A. A., Abdullah, W., Alanezi, S. T., Kasasbeh, W. H., Fohely, F., Khaniabadi, P. M., et al. (2025). Engineering colloidal nanozymes for cancer diagnosis and therapy: From surface chemistry to catalytic mechanisms and precision medicine. ACS Applied Bio Materials. https://www.semanticscholar.org/paper/1ff06bd4079deaa4e62636ddf50722de95bfb6cc (accessed on 3rd October 2025).
15. Di, J., Li, L., Wang, Q., & Yu, J. (2020). Porous membranes with special wettability's: Designed fabrication and emerging application. CCS Chemistry, 3, 2280–2297. https://doi.org/10.31635/ccschem.020.202000457 (accessed on 3rd October 2025).
16. Diercks, C. S., & Yaghi, O. M. (2017). The atom, the molecule, and the covalent organic framework. Science, 355. https://doi.org/10.1126/science.aal1585 (accessed on 3rd October 2025).
17. Dube, E., & Okuthe, G. (2025). Silver nanoparticle-based antimicrobial coatings: Sustainable strategies for microbial contamination control. Microbiology Research. https://www.semanticscholar.org/paper/a960985d0aea504461e22b67c4b1dd574eab4364 (accessed on 3rd October 2025).

18. Feringa, B. L. (2017). The art of building small: From molecular switches to motors (Nobel Lecture). Angewandte Chemie International Edition, 56, 11060–11078. https://doi.org/10.1002/anie.201702979 (accessed on 3rd October 2025).
19. García Díez, A., Rincón-Iglesias, M., Lanceros-Méndez, S., Reguera, J., & Lizundia, E. (2022). Multicomponent magnetic nanoparticle engineering: The role of structure-property relationship in advanced applications. Materials Today Chemistry, 26, 101220–101220. https://doi.org/10.1016/j.mtchem.2022.101220 (accessed on 3rd October 2025).
20. Gates, B. C. (2024). Mononuclear metal complex catalysts on supports: Foundations in organometallic and surface chemistry and insights into structure, reactivity, and catalysis. Chemical Science, 15, 16821–16843. https://www.semanticscholar.org/paper/297b196f51cb64f19cb7933f4b2797cf8e89db04 (accessed on 3rd October 2025).
21. Gubala, V., Giovannini, G., Kunc, F., Monopoli, M. P., & Moore, C. J. (2020). Dye-doped silica nanoparticles: Synthesis, surface chemistry and bio applications. Cancer Nanotechnology, 11. https://doi.org/10.1186/s12645-019-0056-x (accessed on 3rd October 2025).
22. Hall, S. J., Klein, B. P., & Maurer, R. (2022). Characterizing molecule–metal surface chemistry with ab initio simulation of X-ray absorption and photoemission spectra. The Journal of Physical Chemistry C: Nanomaterials and Interfaces, 127, 1870–1880. https://www.semanticscholar.org/paper/f995145d872e5657886d14f851c505506ba5c0e2 (accessed on 3rd October 2025).
23. He, F., & Li, Y. (2022). Advances on theory and experiments of the energy applications in graphing. CCS Chemistry, 5, 72–94. https://doi.org/10.31635/ccschem.022.202202328 (accessed on 3rd October 2025).
24. Hou, L., Liang, Q., & Wang, F. (2020). Mechanisms that control the adsorption–desorption behavior of phosphate on magnetite nanoparticles: The role of particle size and surface chemistry characteristics. RSC Advances, 10, 2378–2388. https://www.semanticscholar.org/paper/fc363f3a05792c4461d0175702f4cbd959e88e24 (accessed on 3rd October 2025).
25. Husain, S., Nandi, A., Simnani, F. Z., Saha, U., Ghosh, A., Sinha, A., Sahay, A., et al. (2023). Emerging trends in advanced translational applications of silver nanoparticles: A progressing dawn of nanotechnology. Journal of Functional Biomaterials, 14, 47–47. https://doi.org/10.3390/jfb14010047 (accessed on 3rd October 2025).

26. Intergovernmental Panel on Climate Change. (2023). Climate change 2022 – Impacts, adaptation and vulnerability. https://doi.org/10.1017/9781009325844 (accessed on 3rd October 2025).
27. Jackman, J. A., Ferhan, A. R., & Cho, N. J. (2019). Surface-based nano plasmonic sensors for bio interfacial science applications. Bulletin of the Chemical Society of Japan, 92, 1404–1412. https://doi.org/10.1246/bcsj.20190112 (accessed on 3rd October 2025).
28. Jiang, D., Ni, D., Rosenkrans, Z. T., Huang, P., Yan, X., & Cai, W. (2019). Nanozyme: New horizons for responsive biomedical applications. Chemical Society Reviews, 48, 3683–3704. https://doi.org/10.1039/c8cs00718g (accessed on 3rd October 2025).
29. Kuśmierek, K., Białek, A., & Świątkowski, A. (2020). Effect of activated carbon surface chemistry on adsorption of phenoxy carboxylic acid herbicides from aqueous solutions. Desalination and Water Treatment. https://www.semanticscholar.org/paper/f8c135e23b884dc05f4206e2dd5473d210339d66 (accessed on 3rd October 2025).
30. Liu, Q., Ericson, T., Kallquist, I., Lindgren, F., & Hahlin, M. (2023). Operando ambient pressure X-ray photoelectron spectroscopy for Li-ion battery electrode/electrolyte interface probing. ECS Meeting Abstracts. https://www.semanticscholar.org/paper/01939d1d9b731d8a884d1b3d54724d878736549c (accessed on 3rd October 2025).
31. Liu, T., Vivek, J. P., Zhao, E. W., Jiang, L., García-Aráez, N., & Grey, C. P. (2020). Current challenges and routes forward for nonaqueous lithium–air batteries. Chemical Reviews, 120, 6558–6625. https://doi.org/10.1021/acs.chemrev.9b00545 (accessed on 3rd October 2025).
32. Liu, Y., Gao, Y., He, F., Xue, Y., & Li, Y. (2022). Controlled growth interface of charge transfer salts of nickel-7,7,8,8-tetracyanoquinodimethane on surface of graphdiyne. CCS Chemistry, 5, 971–981. https://doi.org/10.31635/ccschem.022.202202005 (accessed on 3rd October 2025).
33. Magdy, M. (2023). X-ray techniques dedicated to materials characterization in cultural heritage. ChemistrySelect. https://www.semanticscholar.org/paper/2985d8c963a423d64f594da3efb0cc0f991d4f90 (accessed on 3rd October 2025).
34. Mehta, A., Sharma, A., Sharma, N., Tyagi, L., & Sethi, V. A. (2025). Recent advances and applications of green analytical chemistry in environmental monitoring, food safety, and pharmaceutical analysis. Current Green Chemistry. https://www.semanticscholar.org/paper/acae2131e3906896949a858a2b2c94b45ca8d55b (accessed on 3rd October 2025).

35. Moro, D., Ulian, G., & Valdrè, G. (2020). Nano-atomic scale hydrophobic/philic confinement of peptides on mineral surfaces by cross-correlated SPM and quantum mechanical DFT analysis. Journal of Microscopy, 280. https://www.semanticscholar.org/paper/f4a2a95dc421545ac6540ba59a37df2618ef3eee (accessed on 3rd October 2025).
36. Nanan, D. A. R., Lomax, J. T., Bentley, J., Misener, L., Veinot, A. J., Shiu, W. T., Liu, L., et al. (2025). Self-assembled monolayers of triazolylidenes on gold and mixed gold/dielectric substrates. Journal of the American Chemical Society. https://www.semanticscholar.org/paper/6ae81adf6f4f6816e96eb549fde716dfce141f71 (accessed on 3rd October 2025).
37. Neal, C. J., Kolanthai, E., Wei, F., Coathup, M., & Seal, S. (2023). Surface chemistry of biologically active reducible oxide nanozymes. Advanced Materials, 36. https://www.semanticscholar.org/paper/818906af70872c6448017ddb6ba7e16af1a3b464 (accessed on 3rd October 2025).
38. Nicolau, C., Burguera, S., Piña, M., & Bauzá, A. (2025). Noncovalent binding phenomena in the adsorption of amino acids on Ag/Au surfaces: A theoretical approach. Physical Chemistry Chemical Physics: PCCP. https://www.semanticscholar.org/paper/aad2a98e100b6148a0974e6bcebdc4d4465a08d8 (accessed on 3rd October 2025).
39. Oh, J., Boucly, A., van Bokhoven, J. A., Artiglia, L., & Cargnello, M. (2023). Palladium catalysts for methane oxidation: Old materials, new challenges. Accounts of Chemical Research, 57, 23–36. https://doi.org/10.1021/acs.accounts.3c00454 (accessed on 3rd October 2025).
40. O'Halloran, S., Pandit, A., Heise, A., & Kellett, A. (2022). Two-photon polymerization: Fundamentals, materials, and chemical modification strategies. Advanced Science, 10. https://doi.org/10.1002/advs.202204072 (accessed on 3rd October 2025).
41. Palanisamy, S., & Wang, Y. M. (2019). Superparamagnetic iron oxide nanoparticulate system: Synthesis, targeting, drug delivery and therapy in cancer. Dalton Transactions, 48, 9490–9515. https://doi.org/10.1039/c9dt00459a (accessed on 3rd October 2025).
42. Pruszyński, P., Xu, L., & Hart, P. (2022). Water chemistry challenges in pulping and papermaking–Fundamentals and practical insights: Part 1: Water chemistry fundamentals and pH. June 2022. https://www.semanticscholar.org/paper/2342b2ec318b1af60c47edbcbc11a32c7e0d21b1 (accessed on 3rd October 2025).

43. Qi, L., Mayall, R., Lee, D. S., Smith, C., Woods, A., Narouz, M. R., Hyla, A., et al. (2024). Energetics and redox kinetics of pure ferrocene-terminated N-heterocyclic carbene self-assembled monolayers on gold. Langmuir: The ACS Journal of Surfaces and Colloids. https://www.semanticscholar.org/paper/7b325da2fbaf9beb0c10e3e9ce6c66cc1d4c7df7 (accessed on 3rd October 2025).

44. Rochman, C. M., Brookson, C. B., Bikker, J., Djuric, N., Earn, A., Bucci, K., Athey, S. N., et al. (2019). Rethinking microplastics as a diverse contaminant suite. Environmental Toxicology and Chemistry, 38, 703–711. https://doi.org/10.1002/etc.4371 (accessed on 3rd October 2025).

45. Romero-Pérez, P. S., Moran, H. M., Horani, A., Truong, A., Manriquez-Sandoval, E., Ramirez, J., Martinez, A., et al. (2024). Protein surface chemistry encodes an adaptive tolerance to desiccation. https://www.semanticscholar.org/paper/887b7d719ccf28454e20d5aaa1494e0547063ae4 (accessed on 3rd October 2025).

46. Savéant, J. (2008). Introduction: Molecular and biomolecular electrochemistry. Chemical Reviews, 108(7), 2111–2112. https://www.semanticscholar.org/paper/da11edb77b14f908a7b397b656816a54ccb67598 (accessed on 3rd October 2025).

47. Song, J., Wei, C., Huang, Z. F., Liu, C., Zeng, L., Wang, X., & Xu, Z. J. (2020). A review on fundamentals for designing oxygen evolution electrocatalysts. Chemical Society Reviews, 49, 2196–2214. https://doi.org/10.1039/c9cs00607a (accessed on 3rd October 2025).

48. Souza, B. A. S., De Freitas, I. C., Stavale, F., Guzzo, P. L., Machado, G., & Navarro, M. (2025). Synthesis of a hierarchical FeBTC/FeS structure with enhanced water stability for environmental remediation. ChemistrySelect. https://www.semanticscholar.org/paper/feaee6b2ff2c1f4fe0481b9925119eb530cbd437 (accessed on 3rd October 2025).

49. Staffell, I., Scamman, D., Velazquez Abad, A., Balcombe, P., Dodds, P. E., Ekins, P., Shah, N., et al. (2018). The role of hydrogen and fuel cells in the global energy system. Energy & Environmental Science, 12, 463–491. https://doi.org/10.1039/c8ee01157e (accessed on 3rd October 2025).

50. Steffen, W., Richardson, K., Rockström, J., Cornell, S., Fetzer, I., Bennett, E. M., Biggs, R., et al. (2015). Planetary boundaries: Guiding human development on a changing planet. Science, 347. https://doi.org/10.1126/science.1259855 (accessed on 3rd October 2025).

51. Trepte, K., & Voss, J. (2022). Data-driven and constrained optimization of semi-local exchange and nonlocal correlation functionals for materials and surface chemistry. Journal of Computational Chemistry, 43, 1104. https://www.semanticscholar.org/paper/1cd3f58606e7fd93ad4e4d786b278cd8570a962d (accessed on 3rd October 2025).
52. Wong, V., & Tung, S. (2016). Overview of automotive engine friction and reduction trends–Effects of surface, material, and lubricant-additive technologies. Friction, 4, 1–28. https://www.semanticscholar.org/paper/9877be4f065633c9ebe0c6fd8fd9ff4325ddc59d (accessed on 3rd October 2025).
53. Yang, X., Liu, X., Chau, Y. Y., Qin, X., Zhu, H., Peng, L., Chan, K. W. Y., et al. (2025). Role of chemistry in nature-inspired skin adhesives. Chemical Science, 16, 10665–10690. https://www.semanticscholar.org/paper/de3c189773ed115b29ee17227062823817776264 (accessed on 3rd October 2025).
54. Yun, T., & Chandler, B. (2023). Surface hydroxyl chemistry of titania- and alumina-based supports: Quantitative titration and temperature dependence of surface Brønsted acid-base parameters. ACS Applied Materials & Interfaces. https://www.semanticscholar.org/paper/9802014f8606829d584ffa4947805e1c41f8a269 (accessed on 3rd October 2025).
55. Zhang, D., Chang, S., Luan, P., Kaphle, A., Hisamatsu, T., Shearer, J. C., Park, M., et al. (2023). High-aspect-ratio amorphous carbon mask etch profile control through plasma and surface chemistry optimization. 12499, 1249906–1249906-8. https://www.semanticscholar.org/paper/7ef043aee82bd68b7bba7feefde270d33dda5e3c (accessed on 3rd October 2025).
56. Zhao, L., Peng, Y. Y., Dou, P., Li, Y., He, T., & Ran, F. (2024). Surface chemistry of electrode materials toward improving electrolyte-wettability: A method review. InfoMat. https://www.semanticscholar.org/paper/de1dabc55231d9db1f75fa0d683dab206e720291 (accessed on 3rd October 2025).

CHAPTER 2

# STRUCTURE AND PROPERTIES OF SOLID SURFACES

## Contents



## INTRODUCTION

The development of solid surfaces is a fundamental aspect of materials science because it has a significant effect on the physical and chemical characteristics of materials. The surface of a solid in contact with the surrounding medium, from air to liquids or other solids, is of utmost significance in most applications, from catalysis to electronics. Understanding the connection between the structure of these surfaces and their properties enables researchers to engineer materials for specific functionalities, thereby enhancing their performance and application. Therefore, research into surface phenomena such as defects, atomic structures, and reactivity at surfaces becomes important in advancing materials technology. Such research transcends theoretical advances to practical applications in fields such as nanoengineering, where surface phenomena dictate macroscopic properties. Among the areas of focus in solid surface studies is the role of doping and structural modification, which possess a great capability to alter surface characteristics. For instance, the addition of niobium oxide into lanthanum aluminate via the solid-state reaction has been reported to optimize phase evolution and defect control in $LaNb_xAl_{1-x}O_{3+}\delta$ powders. The research-based observations indicate that the impact of the introduction of $Nb^{5+}$ doping levels affects the microstructural properties, which causes a drastic reduction in reflectance and defect properties. Such enhancement renders these materials feasible for application in high-reflectivity coatings, indicating the critical connection between functional properties and surface structure, as evidenced by a maximum infrared reflectance of over 99.7% in some samples. These findings highlight the necessity of precisely tailoring surface compositions in order to create desired material properties. In addition, structure at the surface and dopant concentration are important parameters that control thin film properties. The use of chemical spray pyrolysis to create Al-doped $\alpha$-$Fe_2O_3$ thin films is a typical example of the relationship between surface properties and functionality. The films deposited are polycrystalline with preferred orientations which affect their optical and gas-sensing properties. Studies reveal that aluminum doping alters the optical band gap and enhances formaldehyde sensing characteristics, with certain concentrations being optimally sensitive at room temperature. Such findings highlight the significance of alterations in surface structure, as they are directly connected with enhancements in chemical reactivity and sensing performance. These figures augment the overall message of how within broader discourse of solid surfaces, the marriage of structure and properties remains a focus point for material design and utilization.

## DEFINITION OF SOLID SURFACES

The phrase solid surfaces is actually grounded in their structural and physical characteristics, which are of special significance in a broad variety of uses in a large scope of sciences. Solid surfaces are normally called the superficial layer of a solid material, with clearly defined characteristics that distinguish them from the bulk material. These surfaces are characterized in terms of features like the atomic arrangement, surface energy, and reactivity that determine the material interactions with the surroundings. Essentially, the solid surface is a significant interface where processes such as adsorption, catalysis, and electrochemical reactions occur. This interface effect has broad ramification in fields such as materials science, nanotechnology, and surface chemistry, where it is crucial to understand the nuances of solid surface properties for designing and optimizing new materials. A significant feature of solid surfaces is their interaction with ions and molecules, which can be investigated using techniques such as physisorption. The importance of physisorption in the characterization of porous solids points to the need for special consideration of solid surface properties, especially in the accurate measurement of surface area and pore size. According to recent advances, conventional procedures are being employed to achieve increased accuracy in physisorption data interpretation and thereby allow researchers to derive information regarding structural properties of solid surfaces (Thommes et al., 2015) (Figure 2.1).

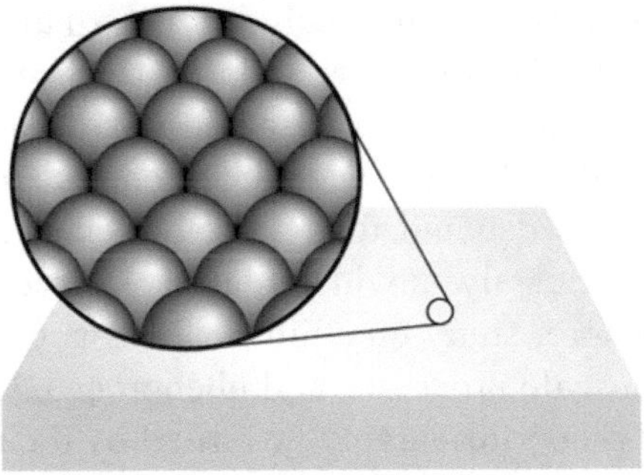

**Figure 2.1.** Solid surfaces.

*Source:* https://slideplayer.com/slide/9523703/.

These methodologies not only expand our understanding of solid surfaces but also help to define the parameters controlling their physical properties. Thus, a comprehensive understanding of solid surface definitions and relative properties can facilitate innovations in various areas of application and allow materials with tailored properties tuned for specific environments. Furthermore, the solid surface theory intersects with cutting-edge fields of research, such as topological insulators, which have peculiar electronic properties because of their surface state. Topological insulators, characterized by gapless surface states resulting from certain symmetry constraints, offer an interesting platform in which to investigate the dynamics of the solid surface (Qi et al., 2011). The treatment of these materials complements classical definitions of solid surfaces with the inclusion of quantum mechanical character of surface interactions. This synergy between historic surface science and recent theoretical advances broadens our understanding of solid surfaces, particularly in the context of their application in electronic and superconducting devices. Thus, the idea of solid surfaces extends beyond mere physical description to encompass a broader set of phenomena influencing material performance in different environments.

## Importance of Studying Solid Surfaces

Solid surface studies are essential to the advancement of numerous scientific and technological disciplines, particularly in materials engineering and science. Material properties are often controlled by surfaces and can affect such phenomena as adhesion, catalysis, and corrosion resistance. In catalysis, for instance, the efficiency of chemical processes, such as the oxygen evolution reaction (OER), is highly influenced by surface interaction. Recent research has highlighted the way that solid surface structures are tuned with methods like co-doping to optimize electrocatalytic performances toward successful and cost-effective catalysts which are able to replace precious metals. It is particularly true in the context of sustainable energy applications, where the study of alternative materials encourages the construction of systems capable of delivering the world's demand for renewable energy solutions, as seen in the design of perovskite catalysts with suitable electrocatalytic properties (Xu et al., 2015). Furthermore, study on solid surfaces is central to advancing our understanding of nanoscale processes and phenomena. Colloidal nanocrystals, for example, possess properties largely governed by their surface structure and chemistry, which can be significantly different from those of the bulk material. The reason is that their high surface-to-volume ratio typical of nanomaterials renders surface properties paramount for functionality. The advance here not only encourages fundamental comprehension of surface science and crystallization but also leads to the prospect of new material synthesis with designed electronic and optical properties. Property engineering is vital in applications in electronics, photovoltaics, and nanocatalysis, where surface tailoring can elevate material performance to record levels. Thus, an extensive analysis of solid surfaces plays an imperative role to the total utilization of future nanotechnology and its multivalent applications (Peng, 2010). Apart from industrial and technological uses, solid surface research has important influences on environmental science and health. Surfaces play a critically active role in processes between materials and their nearby surroundings, governing phenomena such as pollutant adsorption and biocompatibility. An understanding of such interactions can be employed to create materials that will inhibit environmental contamination or enhance the effectiveness of biomedical devices. Surface modifications, for instance, can lead to enhanced hydrophilicity for enhanced biological integration or tailored surface roughness for improved adhesion and reduced biofilm formation. Besides, knowledge by virtue of the study of solid surfaces also affects recycling of materials and sustainability strategy since surface properties play a central role in determining material performance throughout its life cycle. Ongoing research in this area continues to elucidate the complex microstructure-functionality relationships, thus the need for interdisciplinarity in light of contemporary challenges in the majority of domains.

## 2.1. ATOMIC STRUCTURE OF SOLID SURFACES

The atomic arrangement of solid surfaces is a major factor determining their global properties and functioning. At the atomic level, solid surfaces possess distinctive configurations, which can significantly differ from their bulk. This phenomenon, typically referred to as surface reconstructions, arises because of the incomplete coordination of surface atoms, leading to altered bonding interactions and geometry. These structural variations can have colossal impacts on electronic material properties, adhesion, and surface reactivity. For example, alteration in surface atom arrangements can influence catalytic activity by altering accessible sites for reactants, underlining the necessity of atomic configurations for optimal material performance. It is crucial to study the mechanisms of these variations in engineering surface properties in applications from materials engineering to nanotechnology (Figure 2.2).

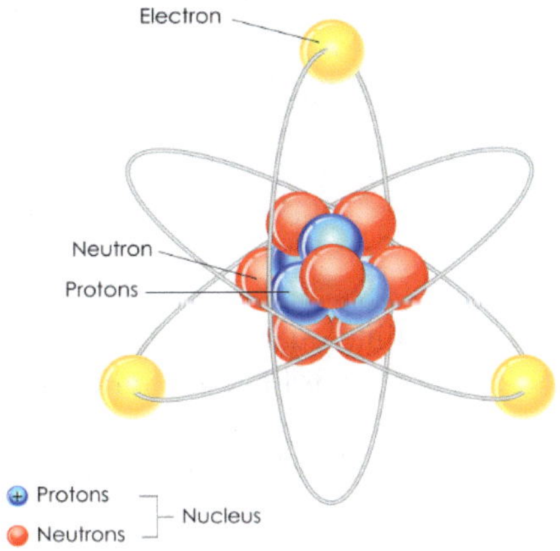


**Figure 2.2.** The structure of atoms.

*Source:* https://external-content.duckduckgo.com/iu/?u=https%3A%2F%2Fkidspressmagazine.com%2Fwp-content%2Fuploads%2F2014%2F04%2Fdreamstime_xl_319762051.jpg&f=1&nofb=1&ipt=cb4757792b562db20b8485e2ff4476fe74923d696a977fca4860e98285abcd59.

Advanced computational techniques are required in an attempt to describe the atomic configuration of solid surfaces because experimental methods may be inadequate to describe the level of complexity involved. The development of models and simulations—particularly those derived from first principles—facilitates a better understanding of atomic configurations under various conditions. Recent advancements in Gaussian Approximation Potentials have been found effective in replicating the density functional theory outputs for silicon and other compounds, unveiling the subtle dynamics between surface imperfections and their influence on material behavior. Such computational methods bring to light physical properties that would otherwise be unfeasible to calculate with traditional electronic structure methods.

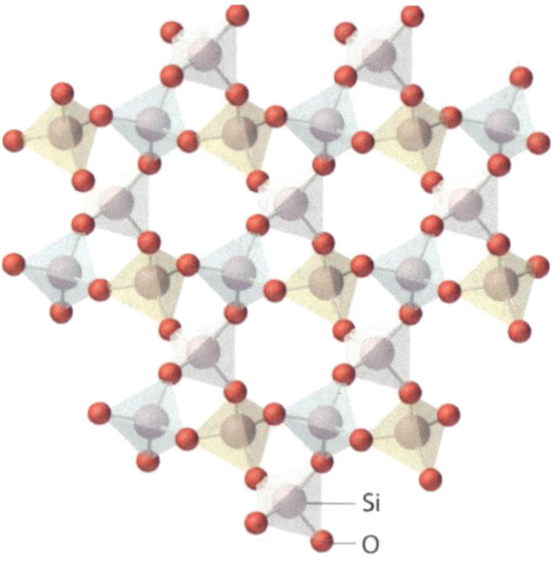


**Figure 2.3.** The lattice of crystalline quartz ($SiO_2$). The atoms form a regular arrangement in a structure that consists of linked tetrahedra.

*Source:* https://files.mtstatic.com/site_4334/61016/0/webview?Expires=1753077531&Signature=hk9Sr3uBTr2xt0K28VwuTP9V1bhD9kANrK2LP8nhZyhMLk19Iy2WLaYdddWqBBHjBl8Eha8wJVEVgJiUGaMehmRf9WfnJMP~FgiGPQ~K8IC9esGxOAGmVyu71QQrpiJlc54q6CwRrdepyK-QRtELDXYsV-GnbXqcwym-fadWKkVw_&Key-Pair-Id=APKAJ5Y6AV4GI7A555NA.

With the integration of machine learning techniques, researchers can predict how alterations of surface atomic structure impact material responses and enhance the design and application of solid surfaces in new technologies (Albert et al., 2018). Study of solid-biomolecule interaction also shows the significance of atomic structure at solid surfaces. Protein-surface interactions are crucial in fields such as biomaterials and nanomedicine, illustrating how atomic structure governs binding modes and stability. Computational models play a complementary role to experiments, revealing the complexity of adsorption processes and specificity determinants in these interactions. The intricacies of molecular dynamics involving very different time and length scales make it problematic, hence multi-scale models are employed to describe the interaction precisely. These involve aspects such as surface reconstruction which may affect the energetics and dynamics of adsorbed proteins, rendering results challenging to interpret and requiring the formulation of robust simulation methodologies. By carefully observing protein-surface systems, scientists can gain knowledge of the underlying atomic interactions that govern such processes and use lessons learned in the design of novel materials (Ozboyaci et al., 2016) (Figure 2.3).

### 2.1.1. Arrangement of Atoms in Solids

The arrangement of atoms in solids is one of the fundamental principles that determine the structural properties and physical attributes of materials. At the atomic level, solids can be defined as crystalline and non-crystalline (or amorphous). Crystalline solids have highly ordered atomic arrangement, which is repeated within a reproducible lattice structure, giving rise to definite geometrical shapes. The regularity facilitates predictable properties such as melting points, hardness, and magnetic properties. On the other hand, non-crystalline solids lack such order, and this leads to a variety of properties that may include reduced mechanical strength and reduced uniformity in thermal and electrical conduction. This fundamental disparity of atomic structure is significant in the understanding of how materials behave under varying conditions and has an ensuing effect in usage in fields such as materials science and nanotechnology. Understanding the importance of atomic structures in bimetallic nanoparticles is reflective of how specific atomic positions can influence material performance. For instance, studies on Au/Pd nanoparticles demonstrate that the unique alignment of palladium atoms between gold atoms results in extraordinary catalytic functions. As revealed by molecular dynamics simulations, the surface structure of such nanoparticles is neither entirely homogeneous nor entirely random; rather, it is dominated by rough surfaces with occasional active sites maximizing catalytic efficiency. Such findings illustrate the extremely significant influence that localized atomic positions may have on the reactivity and efficiency of solid surfaces. Therefore, it is important to know the relationship between atomic structure and functional properties for designing efficient catalysts and other novel materials, which demonstrates the intricacies of atomistic-level organization. In biological membranes too, atomic organization has a command presence in biophysical interactions. Phospholipid bilayer structuring and structural mobility are vital players in cellular functions such as membrane fusion, signaling, and transport (Figure 2.4).

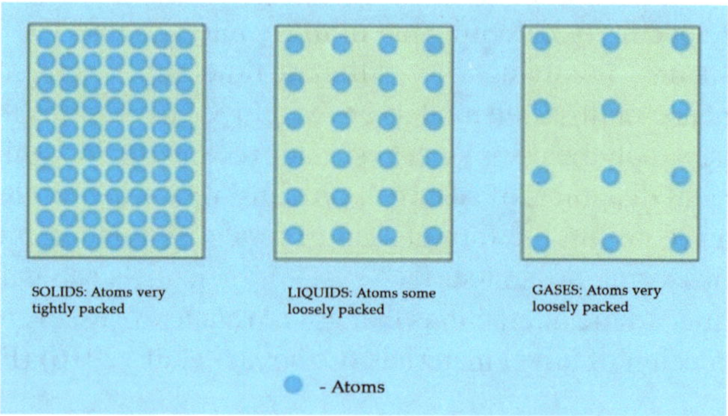


**Figure 2.4.** Arrangement of atoms in solids liquids and gases.

*Source:* https://www.shutterstock.com/image-vector/diagram-show-arrangement-atoms-solids-260nw-1792924147.jpg.

Experimental techniques like atomic force microscopy (AFM) have already shed a little light on such atomic interactions, detailing how lipid composition and packing influence membrane properties. For instance, the presence of some lipid types and their interactions with fusogenic proteins strongly affect adsorption and fusion processes. The novel physical phenomena revealed by these studies highlight the way atomic arrangements dominate biochemical activities under various conditions, thus highlighted the importance of detailed understanding of atomistic structures in synthetic materials as well as biological systems. Investigation of such arrangements further acquires the understanding of the fundamental characteristics of solid surfaces and their diversified applications.

## 2.1.2. Surface vs. Bulk Atomic Structure

The distinction between atomic surface and bulk structures is fundamental to explaining the physical and chemical properties of solid materials. Bulk structures are characterized as having periodic positioning of atoms involving long-range order, while surface structures paint a different scenario as they exhibit significant breaks in this periodicity. In the outermost surface, atoms experience a different environment than those buried in the material, primarily due to the lack of neighboring atoms above. This atom disordering generally causes various electronic and chemical characteristics, including reactivity and catalytic activity, to be influenced. The surface atoms are likely to reside on higher energy levels, which are more chemically active sites crucial to many applications from catalysis to sensorics. The synthesis routes employed to shape solid materials can significantly affect comparative quantities of surface and bulk structures. For instance, procedures like coprecipitation and wet impregnation are very important in deciding the metal catalyst's distribution in solid supports. Research indicates that the synthesis route will have a direct effect on the structural properties and the development of active sites in catalysts (Figure 2.5).

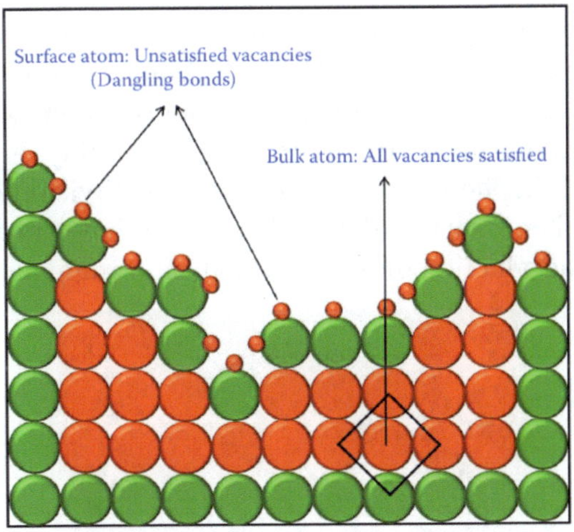


**Figure 2.5.** A schematic representation of the bulk core atom (orange color) satisfied with coordination number, and unsaturated valence atom (green color), refers as dangling bond, responsible for surface energy.

*Source:* https://www.researchgate.net/figure/A-schematic-representation-of-the-bulk-core-atom-orange-color-satisfied-with_fig2_327281815.

One of the studies examined the metal nickel (Ni) on magnesium oxide (MgO) transition metal. Coprecipitated samples were observed to exhibit a more even distribution of Ni cations, which offered greater catalytic activity than wet impregnation-synthesized samples. This demonstrates how variations in preparation processes can yield extreme differences in surface properties and catalytic activity, showing the reliance of synthesis procedures and atomic ordering. The catalytic and atomic structure interaction is also made known through analysis of different elemental substitution in solid solutions. An example is in $Mn_{0.5}Zn_{0.5}Fe_{2-}xY_{x}O_{4}$, where the presence of yttrium (Y) affects the bulk density as well as the surface morphology of the material. Lattice parameters were found in agreement with Vegards rule, which proposes an intrinsic connection between bulk and surface structure atomic ordering, as observed using X-ray diffraction and scanning electron microscopy. In addition, increased content of Y resulted in reductions in average grain sizes, affecting the electromagnetic characteristics of the material. The findings reaffirm the significance of surface-bulk structure interactions in solid materials and the deep implications it has in areas such as magnetic materials and catalysis, where designed surface properties are crucial for optimizing performance.

### 2.1.3. Role of Surface Defects

Surface defects play a key role in determining the properties and behaviors of solids, particularly for electronic and optoelectronic materials. They can come in the form of vacancies, dislocations, or impurities and exert considerable influence on electronic band structures, carrier mobility, and recombination. For instance, in hybrid phototransistors based on methylammonium lead iodide ($MAPbI_3$) and graphene, surface defects have been shown to alter the electronic properties by the introduction of localized states for trapping charge carriers.

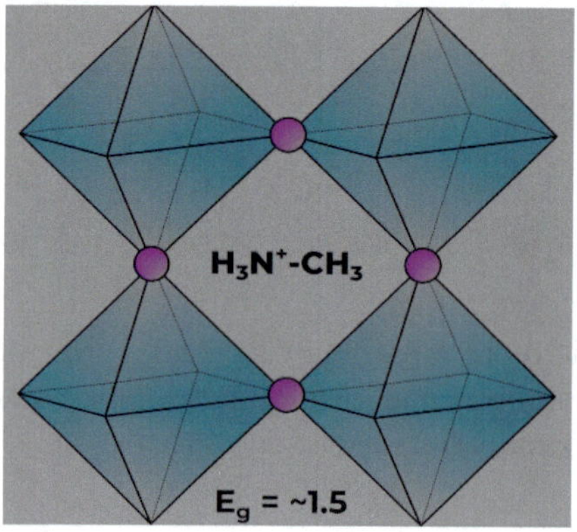


**Figure 2.6.** Methylammonium lead iodide.

*Source:* https://nanoxo.eu/wp-content/uploads/MAPbI3_FIN.png.

Hence, it affects the responsivity and overall performance of the devices. Moreover, as the quality of the $MAPbI_3$ nanowire/graphene interface increases, defects can be minimized, leading to enhanced charge transport and photoconductivity, as shown by devices with responsivities significantly higher than those of traditional solid-state detectors (Spina et al., 2015). Understanding the role played by surface defects is transferred to the development of new materials such as topological insulators, which possess unique surface states protected by topological properties. These materials, with their insulating bulk and surface conductivity, are the exemplars of the interplay between electronic behavior and surface defects. Topological insulator defects can break down the ideal qualities of such systems and produce changes in the dynamics of surface states. For example, the presence of defects or impurities may localize the surface states and hinder the anticipated topological characteristics, hence affecting the materials overall utility in areas such as quantum computing (Qi et al., 2011) (Figure 2.6).

Therefore, the control and treatment of surface defects in these materials are crucial to allow their unique properties to be utilized suitably and to develop new technologies in various applications. Investigations on means to overcome the disadvantageous influence of surface defects have led scholars to pursue novel processing pathways and material combinations. For instance, the combination of carbon nanomaterials like graphene with traditional semiconductors has already been shown to enhance device performance through the enhancement of charge collection efficiency and reducing recombination losses. Leverage both the mechanical strength and electronic properties of carbon materials, scientists are able to design composite structures that can dampen the negative effect of defects. These methodologies are particularly strong in the development of high-efficiency solar cells and photodetectors, where the best interface can make enormous gains in responsivity and device lifespan. More research into the connection between surface defects and methods of material synthesis is necessary for enhancing the development and conception of modern electronic devices, and surface science is central to the development of materials (Spina et al., 2015).

## 2.2. SURFACE ENERGY

The concept of surface energy is the key to understanding the structure and properties of solid surfaces. Surface energy is really the excess energy on the surface of a solid compared to the bulk, caused by unbalanced intermolecular forces. This is a significant effect in numerous applications, including adhesion, wetting, and thin-film formation. The solid surface stability can be affected to a large degree by their surface energy; high-surface-energy materials will exhibit strong adhesion behavior, while low-surface-energy materials could be easily removed or separated. Apart from that, it is also crucial for the enhancement of material performance in coatings, composite, and electronic industries, where the desired interfacial interactions with other materials or environments are typically achieved by the best surface properties. Surface energy not only depends on the intrinsic material properties but also on their microstructure. The existence of defects, grain boundaries, and irregularities alters the effective surface energy, which can affect the mechanical properties of the solid. For instance, surfaces with high porosity or irregularity typically possess lower effective surface energy, influencing their interaction with gases and liquids besides adhesion character. Investigation of intricate structures like topological metamaterials that utilize intricate geometries like mathematically derived minimal surfaces highlights the relationship between surface energy and material performance. It is possible for such structures to be engineered with maximized mechanical properties, with understanding how microstructural features can be tweaked to achieve maximum surface interactions and energy absorbing ability, as in the case of research on solid structures with 3D minimal surfaces (Wang et al., 2023). Moreover, surface energy application can even be observed in the investigation of molecular interaction in crystalline materials, as demonstrated by computational calculations such as Density Functional Theory (DFT). Understanding surface energy at the molecular scale helps scientists predict how various intermolecular forces contribute to the total stability of a solid surface. Using analyses such as Hirshfeld surfaces, researchers can visualize and quantify these interactions and hence comprehend the dynamics of molecular assembly in crystalline systems. This facilitates not just enhanced comprehension of solid properties but also the synthesis of new materials with maximum reactivity and stability. The integration of concepts like energy framework calculations into structural analysis contributes a great deal to the evolution of contemporary material science by clarifying the surface energy-molecular structure-material properties interface (Bakheit et al., 2023) (Figure 2.7).

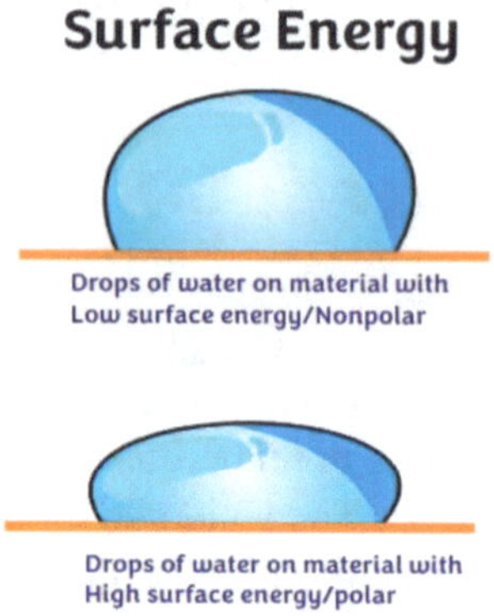


**Figure 2.7.** Surface energy.

*Source:* https://cache.careers360.mobi/media/article_images/2024/11/12/Surface_Energy.png.

## 2.2.1. Definition and Significance of Surface Energy

Surface energy is one of the fundamental concepts of materials science that quantifies the excess surface energy of a solid relative to the bulk. Excess energy is created due to the rupture of intermolecular bonds at the surface since surface atoms have different environments than interior atoms. Thus, surface energy is held directly accountable for governing numerous physical characteristics such as wettability, adhesion, and the formation of thin films. The effect of surface energy is particularly important in those applications ranging from coating to nanotechnology, where the properties of the material are highly dependent on its surface. A profound understanding of surface energy allows the development of materials with special properties designed for particular applications, thereby maximizing functionality in a variety of engineering situations. Surface energy contributes to processes such as adsorption and catalysis, where the surfaces energy landscape fundamentally influences the manner in which particles interact with solid materials. In adsorption, surface energy has the ability to regulate the adsorption isotherm and thus affect the efficiency of energy conversion and storage devices. Moreover, advancements in nanotechnology have emphasized the use of regulation of surface energy to enhance reactivity and maximize catalyst performance. By changing the surface properties through chemical treatments or coatings, researchers can engineer the surface energy to enhance interactions with target reactions. According to various studies, determining the mechanism behind surface energy is essential in developing new applications that effectively harness such properties in drug delivery and sensors (James et al., 2025) (Figure 2.8).

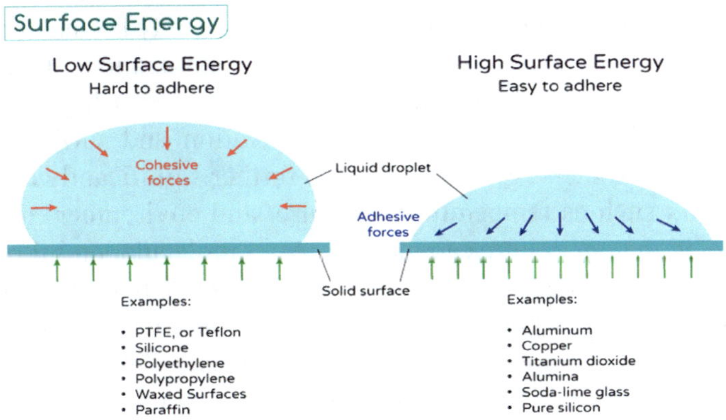


**Figure 2.8.** Significance of surface energy.

*Source:* https://www.sciencefacts.net/wp-content/uploads/2024/10/Surface-Energy.jpg.

Current trends highlight the central role of surface energy in governing material structural dynamics. Phonon modes, or lattice vibrations, couple strongly with surface energy and govern the stability and overall properties of solid surfaces.

The harmonic phonon spectrum presents in a simplified form structural dynamic, which provides insight into the materials' response to perturbations. Notably, hypothetical phonon modes exhibit structural instabilities that may arise due to minimal surface energy; such a realization aids in phase transition determination and material arrangement optimization. The relationship between surface energy and phonon dynamics, in addition to identifying intrinsic material properties, also indicates the importance of surface energy in material innovation and design (Pallikara et al., 2022). From this point of view, surface energy is an essential aspect of material science, driving research and innovation in solid surface structures.

## 2.2.2. Factors Affecting Surface Energy

One of the rudimentary elements of surface energy entails the relationship between chemical structure and structural properties of a solid surface. Alterations within molecular or atomic arrangement can lead to varied surface energy characteristics, affecting the interaction of materials with the environment. For instance, a surface can exhibit different energies based on its crystallographic orientation that can significantly regulate adsorption behavior and wettability. This is additionally influenced by the presence of groups or impurities, which can alter the surface energy by changing the intermolecular forces involved, which is particularly important in areas such as catalysis and materials science. These interactions are essential not only for material response prediction but also for surface design for specific applications, i.e., enhancing electrode efficiency in batteries, which has recently been highlighted in the investigation of the intricate dynamics of solvation and surface interaction (Liu et al., 2025). The energy of a surface is further significantly controlled by external factors such as temperature, pressure, and environmental conditions. For example, high temperature can induce greater molecular mobility that alters surface energy. This is especially important in chemical vapor deposition or when thermal cycling is performed in materials. Pressure variations can also change the way atoms or molecules on the surface react with one another, which can contribute to varying adhesion properties and structural characteristics. An understanding of these influences is essential to optimizing the performance of materials across a variety of applications, from coatings to electronics. Paradigmatic is the case of processing hexagonal boron nitride in which conditions for stripping—pressure and temperature—are shown to be key to regulating surface properties and, by extension, performance characteristics of the resultant material (Wang et al., 2024). Furthermore, the interaction between physical properties and surface energy of a material remains key to understanding its overall performance. Surface crystalline quality, thickness, and roughness are some of the factors that play significant roles in determining both the mechanical strength and thermodynamic stability of solid surfaces. High surface roughness, for example, can enhance interaction of surface area with a solvent, leading to increased adhesion or differential wettability and consequently influencing performance. Moreover, thickness can also decide the material's thermal and mechanical behavior, affecting its use in devices having specific operating conditions. Since these properties affect not only surface interactions but the reactivity and stability of the material as well, they are at the forefront of creating material design approaches for applications in advanced technologies (Wang et al., 2024).

### 2.2.3. Measurement Techniques for Surface Energy

Surface energy measurement techniques are crucial in establishing the physicochemical performance of solid surface characteristics. Surface energy plays an important role in the majority of applications like adhesion, wetting, and material mechanical strength. The most widely used surface energy measurement method is the Wilhelmy plate technique, which involves the use of a thin plate fixed on a balance to study the interaction between liquids and solid surfaces. By measuring the force to pull the plate from a liquid film, researchers can calculate the contact angle and then calculate surface energy values. Furthermore, the sessile drop method, whereby a liquid droplet is placed on a solid surface, provides another simple way to measure contact angles. Such methods, together with accurate interpretation and reproducible measurements, make it possible to provide an extensive description of the surface properties essential for application from coatings to medical devices. Another technique with considerable influence on surface energy determination includes gas adsorption techniques, which have gained considerable significance in material science characterization. With recent advancements, modifications to IUPAC guidelines have made these techniques standard for the measurement of surface area and pore size, reflecting their relevance in physisorption studies. Such analysis can help scientists infer surface properties of a material from observations of the way gases adsorb on or deposit on the solid surface. The resulting adsorption-desorption isotherms from such a measurement can give information regarding the energy heterogeneity of the surface. As such, gas adsorption serves as both an analytical tool and a means of quantifying surface energetics in a systematic manner, thereby assisting in the study of surface-related phenomena across fields ranging from catalysis to material science (Thommes et al., 2015). Last but not least, direct measurement methods remain ever-important, with computational simulations becoming ancillary methods to measure surface energy. Molecular dynamics and Monte Carlo computer simulation allow the scientist to model surface interactions and predict energy values irrespective of large quantities of experimental data. Modelling the interactions at an atomic level allows scientists to investigate considerations such as surface roughness, orientation of molecules, and effects of temperature on energy calculations. These simulations significantly allow us to forecast and customize material properties before physical testing. However, it becomes imperative to validate computational findings with experimental data for accuracy and reliability. Complementing each other, these methods clarify the complex nature of surface energy and how this impacts material performance, ultimately resulting in comprehension of the structure and properties of solid surfaces in modern applications.

## 2.3. PHYSICAL PROPERTIES OF SOLID SURFACES

The solid surface material properties are critical in the description of atomic and molecular-scale interactions, particularly materials engineering and science. Surface properties include cleanliness, roughness, and texture, which have a striking impact on adhesion, wetting, and friction. Surface structure perturbations lead to variations in chemical reactivity, which impacts corrosion resistance and catalytic reactions. For instance, a well-defined surface with minimal roughness will enhance the mechanical strength and resistance of the material and thus render it useful for application in industries such as aerospace or automobile manufacturing.

Regulation of such properties is at the center of developing innovative materials that meet specific performance criteria. Profound knowledge of solid surfaces provides the potential to maximize their application in various technological fields. Additionally, the atomic or molecular structure at a surface determines its electronic and optical properties. Surface states and point defects are particularly relevant concepts as such properties have the capability to contribute to charge transport and light absorption in semiconductor materials. As an example, in solid-state qubits for quantum computation, local lattice disturbances might have a significant impact on the behavior of such materials. Techniques such as stroboscopic scanning X-ray diffraction microscopy allow for mapping of dynamic strains on surfaces, which demonstrate the relationship between lattice distortions and photoluminescence responses of material defects in silicon carbide materials. Understanding the nature of these interactions is important for designing efficient qubits and maximizing coherence time in quantum systems and underlines the importance of extensive studies of surface characteristics and their significance in future technologies (Whiteley et al., 2018) (Figure 2.9).

| Type | Example | Structural unit | Forces | Properties |
|---|---|---|---|---|
| Ionic | NaCl, K2SO4 | (+)/(-) ions | Electrostatic attraction | Hard, brittle, high melting pt. |
| Metallic | Fe, Ag, Cu | Metal atoms [(+) ions surrounded by delocalized electrons.] | Electrostatic attraction | Malleable, ductile, conductive, range of MP & BP |
| Molecular | $H_2$, $O_2$, $CH_4$ | Molecules | van der Waals forces (dispersion, dipoloe-dipole, H-bond) | Low/mod. Melting points, poor conductors |
| Network | Graphite, diamond (C), Quartz, etc. (Si) | Atoms with infinite 2 or 3-D network | Covalent-directional electron pair bonds | Wide range of hardness & MP. Poor conductors of electricity (with some exceptions) |
| Amorphous | Glass, polyethlene (plastics) | Covalently bonded network with no crystalline structure | Covalent-directional electron pair bonds | Noncrystalline wide range mp. Poor conductors. |

**Figure 2.9.** *Physical properties of solid surfaces.*

*Source: https://external-content.duckduckgo.com/iu/?u=https %3A%2F%2 Fimage1.slideserve.com%2F2529097%2Fproperties-of-solids1-l.jpg&f=1&nofb=1&ipt=2e45580d48da985d50a1ac77c075f5c13e2c2166bd24b0b6b083d4bd1f4b2a8b.*

Advances in computational modeling techniques have dramatically enhanced our capacity to investigate the physical properties of solid surfaces. Using methods such as weighted graph representation of lattice structures, researchers can now test complex surface geometries more effectively and precisely. This allows for a more precise simulation of how surface parameters such as stress concentration points and material density distributions influence the total mechanical properties of solids. These models are of particular importance in additive manufacturing, in which mechanical performance is directly tied to physical fidelity. The application of these advanced computational methods not only enables optimal surface design for application in the real world but also enables probing of novel materials with engineered properties. This interplay of computational and experimental methods is required for enhancing our understanding of solid surface structure and properties (Liu et al., 2021).

### 2.3.1. Mechanical Properties (Hardness, Toughness)

Mechanical properties, such as hardness and toughness, play a significant role in determining the performance and durability of solid surfaces. Hardness, which is resistance to localized plastic deformation, is usually defined by way of several indentation tests and defines a material's resistance to wear and indentation. Toughness, on the other hand, is the ability of a material to resist energy absorption and plastic deformation prior to fracture, and is important where impact strength is. The complexity of these properties is further complicated by microstructural characteristics, such as grain size and phase distribution, which have a direct influence on both hardness and toughness. An appreciation of the interrelationship of such mechanical properties is required in optimizing materials for use, with solid surfaces capable of enduring mechanical strain encountered in diverse environments. The creation of coatings, such as titanium aluminum nitride (TiAlN), illustrates the significance of mechanical properties in optimizing surface functionality. According to research, the improvement of the rate of nitrogen flow during the deposition process has significant impact on coatings' microstructure and therefore on their hardness, toughness, and wear resistance. With the increase in the $N_2$ flow rate, the developed nanocrystalline structure has higher solid solution content and undergoes alteration of crystal orientation, which contributes to the major improvements of hardness and toughness. But such enhancement typically reaches a peak and then declines, indicating a highly sensitive equilibrium between microstructural qualities and mechanical performance. Therefore, the coating shows improved corrosion protection and mechanical properties, illustrating how accurate regulation of deposition parameters can produce surfaces designed for optimal robustness and performance in challenging environments (Cao et al., 2025) (Figure 2.10).

**Figure 2.10.** Mechanical properties of materials.

*Source:* https://external-content.duckduckgo.com/iu/?u=https%3A%2F%2Fwww.mechdaily.com%2Fwp-content%2Fuploads%2F2024%2F07%2FMechanical-Properties-Of-Materials.webp&f=1&nofb=1&ipt=8c8c45b26d218486c41cde9fe9a54b8f1c8b9de050eb6317b925e16e63c408cb.

In addition to traditional coatings, novel materials such as segmented polyurethane thermoplastic elastomers are designed to include customized mechanical properties for incorporation in specialized devices. The low surface energy components, which play an important role in greatly reducing ice adhesion—a quality defined by excellent toughness—characterize these materials. The unique 3D microstructure formed during processing causes the films to exhibit enhanced resistance to sheer forces with the resulting solid product not having detrimental additives. These advances in material science reveal a groundbreaking approach to enhancing surface properties at the expense of none of integrity (Nowak et al., 2021). Investigation of the mechanical properties, like hardness and toughness, of these novel coatings provides scientists with a better understanding of how such surfaces behave in applications and can potentially push the limits of technical feasibility in protective surface engineering.

## 2.3.2. Thermal Properties (Conductivity, Expansion)

Thermal properties, notably thermal conductivity and thermal expansion, are central to the description of solid surfaces and define their role in various applications. Thermal conductivity measures the capacity of materials to transfer heat, and this is essential in the context of energy efficiency in devices such as thermal barriers and electronics.

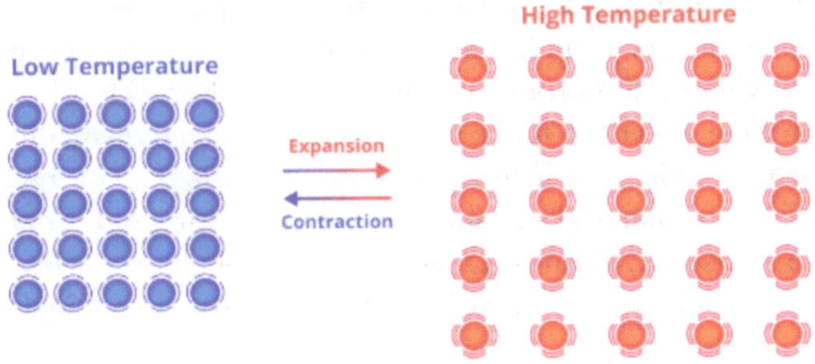


**Figure 2.11.** *Thermal expansion of solids.*

*Source: https://external-content.duckduckgo.com/iu/?u=https%3A%2F%2Fas2.ftcdn.net%2Fv2%2Fjpg%2F05%2F66%2F81%2F95%2F1000_F_566819510_aiCmdRU3RHQTWcIuyov9AgaANuY0pJA9.jpg&f=1&nofb=1&ipt=781825fdaafdc2fac8fdeb84eef78c3c46a5612123bc42316965079a5e263d4b.*

The understanding of the microstructural conductivity parameters has the potential to transform material engineering. For instance, theoretical research into jammed particle system effective thermal conductivity clarifies the dominant influence of packing density and particle shape on heat transfer. By simulating such conditions with granular medium theory, researchers have been able to develop an in-depth understanding of the way structural arrangement determines thermal behavior, which has opened up practical application for materials science and engineering (Asakuma et al., 2012). Of similar importance are thermal expansion properties of solid surface, which describe how materials change dimensions or volume with temperature variations. Particularly important in composite materials and coatings, differential expansion here can lead to stress and potential failure. Doping techniques such as introducing rare earth elements into zirconia have been effective means of managing thermal expansion coefficients. For instance, research has confirmed that 30 mol% $SmO_{1.5}$ and $ErO_{1.5}$-doped zirconia has enhanced insulation together with lower thermal expansion than traditional materials. These developments highlight the importance of combining material compositions to enhance thermal responses in order to improve the functionality and longevity of solid surfaces in differing surroundings (Ansart et al., 2011) (Figure 2.11).

The combination of thermal conductivity and thermal expansion in solid materials governs the integrity of structure and functionality of material. The high thermal conductivity materials are generally prized in uses where heat is easily removed, while low thermal expansion is preferred where there is exposure to thermal changes. Understanding these properties is crucial since scientists and engineers can design materials that are capable of enduring thermal stress without compromising functionality. For example, research in bi-porous cohesive powders finds application in effective medium theories with possible new solutions in thermal barrier coatings. The breakthroughs will further enhance material capabilities beyond their limit, contributing to industries like aerospace, electronics, and energy applications, where reliability in thermal stress is a paramount consideration (Asakuma et al., 2012; Ansart et al., 2011).

## 2.3.3. Optical Properties (Reflectivity, Transparency)

The optical properties of solid surfaces, i.e., reflectivity and transparency, play a very important role in determining their applicability in various technology areas. Reflectivity is a measure of the capability of a surface to reflect light falling on it, which depends on factors such as roughness at the surface, composition of the material, and wavelength of light. Highly reflective surfaces are used in applications such as optical coatings and mirrors, where light transmission efficiency is of utmost importance. Transparency is the ability of a material to allow the passage of light without much absorption or scattering. It is a fundamental optical property of material such as glass and transparent plastics, which are utilized in windows and lenses. Understanding these properties not only sets the form of functional materials, but also drives engineering decisions in a wide range of fields, including photonics and materials science. Correlation between reflectivity and transparency can also have a significant impact on the functional use of solid surfaces. For instance, high reflectivity materials may possess lower transparency, and their use may be limited where both traits are needed. The optical bandgap of the material, determining its reaction to light, is a major factor in this regard. As has been shown through the research on copper oxides, the arrangement of atomic structures differently affects their optical characteristics uniquely. The contrast between the behavior of materials like $CuAlO_2$, which is transparent because of its two-dimensionality, and $Cu_2O$, with color and reflectance owing to three-dimensional Cu···Cu contacts, underscore how finely adjusted structural characteristics condition optical properties. These findings can therefore find application in optimizing the design of new materials with customized optical functionalities (Buljan et al., 2001). Other than reflectivity and transparency, the significance of such optical properties is observed in current applications such as liquid crystal technologies. The modulation of the optical properties within liquid crystals depends on their structure organization, showing the instrumental role played by materials surface properties in deciding their optical behavior. As an example, orientational order changes of liquid crystals induced by electric fields can create reflectivity and transparency modulation, thus enabling dynamic optical functions. Structure-optical response interaction, in particular for complex systems, highlights the material design's role in implementing functionalities. From smart windows that change the degree of transparency to display technology based on light modulation, the optical properties of solid surfaces continue to be the drivers of innovation across most fields, ultimately linking material structure to performance outcomes (Blinov et al., 1995).

## 2.4. APPLICATIONS OF SOLID SURFACE SCIENCE

The field of solid surface science is vast and interdisciplinary and has far reaching implications across a range of fields such as materials science, engineering, and biotechnology. One such important application is in the field of additive manufacturing, where the ability to design surface structure at the micro and nanometer length scale leads to the construction of cellular materials with well-characterized mechanical and functional properties. This is particularly relevant in tissue engineering scaffold design, where engineered surfaces can provide the necessary support for cellular adhesion and proliferation. The advancements in solid surface science enable researchers to tailor these scaffolds with various geometric modeling strategies, making them more tissue biologically compatible and allowing for improved integration of artificial materials into clinical uses. So, solid surface science is a critical tool for the creation of biomaterials, where form is combined with function to meet anticipated tissue engineering and regenerative medicine requirements (Concheri et al., 2018). Superhydrophobic surfaces are another field in which solid surface science is making a significant contribution. Superhydrophobicity is the state of highly textured surfaces with extreme water-repellence. Significant advancements have been achieved in recent years to develop structures capable of repelling water efficiently, and this has wide-reaching implications for applications such as self-cleaning surfaces, anti-fogging coatings, and waterproof textiles. Polymeric materials have recently emerged as key players in this area due to their versatility as well as the possibility of being engineered for particular functional properties. By being more practical use and less theoretical framework-oriented, scientists are beginning to design treatments and coatings that can be tailored to the requirements of specific use-cases, paving the way for integrating advanced surface science into common products (Asmatulu et al., 2011). The integration of solid surface science into technology is not limited to uses in biomedical and materials sciences; it extends into fields such as electronics and environmental science. Nanotechnology-based surface modification has potential to improve the performance of electronic devices by increasing conductivity and minimizing energy loss. Surface treatment can also reverse environmental issues such as corrosion and fouling that pose significant problems in industrial settings. By improving structure-property relationships, researchers and engineers can design surfaces for peak performance with minimal environmental impact. The continued exploration of solid surface interaction and modification will definitely bring forth new vistas for green technology, ultimately deciding the fate of material science and industrial application as never before imagined.

### 2.4.1. Catalysis and Surface Reactions

The intricate relationship between surface reactions and catalysis is the core of heterogeneous catalysis, where the nature of solid surface largely determines the efficiency of reactions. The surface structure, composition, and electronic state of a catalyst often determine the catalytic efficiency. These factors directly influence adsorption processes, which are critical in the catalytic process. Catalytic reactions, for instance, involve the adsorption of reactants onto the surface of the catalyst, followed by subsequent surface reactions where the adsorbed species are converted to product. Therefore, the design and control of catalyst surfaces become unavoidable since geometrical and electronic characteristics determine the courses of the reaction and selectivity, hence affecting yield and efficiency at the industrial level. Knowledge of such mechanisms promotes the development of optimized catalytic materials that improve reaction rates and reduce energy expenditure. Some new advances in nanoparticle synthesis have triggered a big influence on the realization of nanoscale catalysis. The organometallic approach, for example, offers the potential for the synthesis of size-controlled metal nanoparticles with higher catalytic activity due to their large surface-to-volume ratio. By employing ionic liquids as stabilizers and reaction media, researchers have been able to modify the surface characteristics of nickel nanoparticles. This modification leads to varied catalytic functions for specific reactions, such as selective hydrogenation of unsaturated carbonyl compounds. The tailored interaction among the nanocrystals and the ionic liquid modulates the electronic environment of active sites and influences reaction routes, highlighting the complexity of catalysis and surface reactions at the nanoscopic scale (Riisager et al., 2024) (Figure 2.12).

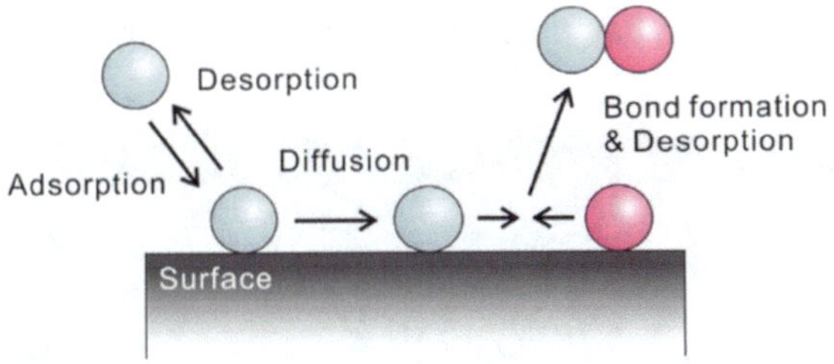


**Figure 2.12.** Elementary steps in a catalytic surface reaction proceeding on a solid surface.

*Source:* https://www.researchgate.net/figure/Elementary-steps-in-a-catalytic-surface-reaction-proceeding-on-a-solid-surface_fig1_272095292.

These innovations represent a significant advancement in rendering chemical processes more efficient and selective across various industries. Moreover, the addition of new materials such as MXenes has opened up new avenues in catalysis, particularly in the improvement of electrochemical reactions. MXenes with their remarkable conductivity and capability of modifying their surface properties have been researched as electrodes in supercapacitors as well as catalysis. By combining MXenes with gold nanoparticles, researchers have created composite materials that support quick ion transport and improved electron transfer, which are necessary to activate surface reactions. The gaps between the layers created by the interaction of MXenes and metal nanoparticles not only improve mass transfer but also increase the density of active sites for reaction. This versatile surface modification approach demonstrates the potential to produce high-performance catalytic materials with the capability to significantly boost reaction kinetics and stability in various electrochemical and catalytic processes (Mustafa et al., 2022). These advancements demonstrate the ongoing improvement in catalyst design with an emphasis on optimizing surface interactions in a bid to achieve more effective chemical conversions.

### 2.4.2. Semiconductor Technology

The structural and surface properties of solid materials dominate in the area of semiconductor technology. As device size decreases, the volume-to-surface ratio is smaller, and therefore the surface effects become important for the operation of the device. Semiconductor surface oxidation becomes a problem, particularly during processing, as surface states significantly impact electronic properties. The relationship between electronic properties and surface structure necessitates a deep understanding of the oxidation processes affecting materials, particularly for III–V semiconductors. Experiments confirmed that the adjustment of thermal oxidation treatments can lead to desirable property changes in the materials that potentially enhance their applications in actual semiconductor devices.

**Figure 2.13.** Semiconductor technology.

*Source:* https://pegglass.com/wp-content/uploads/2020/07/semiconductor-1-1024.png.

Through the resolution of such surface-related problems, semiconductor technology is based at an atomic level, introducing innovations in electronic devices as well as improving overall reliability in their performance (Mäkelä et al., 2019). While studying the inherent causes of semiconductor surfaces, it is crucial to comprehend the role of crystalline oxides. It has been shown in recent studies new phases of crystalline oxides on a variety of III–V semiconductor surfaces, such as GaSb(100) and InSb(111)B. These findings indicate the role of structure in controlling the electronic and chemical behavior of solid surfaces. Employing high-resolution surface science tools in idealized conditions, scientists were able to distinctly characterize these oxide layers and evaluate their passivation effects. For instance, the presence of a stabilizing crystalline InOx interfacial layer beneath $Al_2O_3$ in InGaAs infrared detector structures demonstrates how surface engineering can achieve better device performance. These improvements not only maximize current semiconductor applications but also influence future device development (Mäkelä et al., 2019) (Figure 2.13).

Chemically sensitive semiconductor devices (CSSDs) classification presents yet another critical element to understanding semiconductor technology and underlying principles. Recognition of the various physicochemical effects that impact these devices enables improved design and manufacturing processes, taking into consideration compatibility and enhanced functionality within real applications. An effective CSSDs classification system illustrates how various semiconductor materials respond to external influences and surface modification. Through the careful examination of current CSSDs, scientists can optimize methods of material selection and surface treatment that maximize device sensitivity and overall performance. In the end, knowledge of solid surface structure and properties is not abstract; it lies at the heart of developing semiconductor technology, optimizing device performance, and providing electronic components with reliable operation in a rapidly changing technological environment.

### 2.4.3. Coatings and Surface Treatments

The application of coatings and surface treatments is essential in enhancing the properties of solid surfaces, significantly impacting their function and longevity. By modifying surface characteristics such as hardness, corrosion resistance, and chemical inertness, these treatments can extend the lifespan of materials in various environments. Coatings, including paints, varnishes, and polymer films, form a protective coating that prevents environmental degradation. In addition, surface treatments like anodizing or chemical vapor deposition can alter the surface chemistry, providing specific properties like hydrophobicity or improved adhesion. These advancements provide flexibility to the materials application across industries, ranging from aerospace to biomedical uses, where specific surface characteristics are critical for performance and safety in high-stakes applications. Nanotechnology becomes increasingly important in surface treatments and coatings, particularly in the development of complex materials. Nanostructured coatings provide certain advantages such as improved surface area, reduced weight, and superior mechanical properties. Nanoparticles added to coatings can provide adhesion, improve environmental resistance, and introduce new functions such as antimicrobial or self-cleaning properties. The integration of nanomaterials in surface treatments promotes novel applications, which can revolutionize many fields, including medicine and electronics. Of particular interest are drug delivery system materials, which show how nanotechnology can enhance therapeutic efficacy, as the application of lipid nanoparticles for gene therapies illustrates, revolutionizing conventional methods for treatment and diagnosis in clinical practice, pointing to the confluence of coatings, surface treatments, and novel medical technologies (Patra et al., 2018). Genomic medicines' analysis and their delivery platforms reveal the extensive influence of surface modification and coatings on therapeutic advancement. For instance, lipid nanoparticles (LNPs) offer an excellent non-viral drug delivery platform for genetic drugs, which overcomes the shortcomings of degradation and cellular uptake. The formulation design of LNP is significantly based on the properties imparted by surface treatment, including electrostatic interactions and hydrophilicity, in order to enhance biocompatibility and efficacy of gene silencing therapy. Novel strategies in surface modification are also crucial in offering targeted delivery to specific tissues and enhancing therapeutic effects. Through the success of integrating materials science and therapeutics, innovation in coatings and surface treatments not only makes possible efficient drug delivery but also emphasizes the importance of tailored surface properties to respond to the growing demands of modern medicine (Pieter et al., 2017).

## SUMMARY

In summary, study of the shape and properties of solid surfaces reveals the complex relationship between material properties and functional function. The development and refinement of various surface coatings and modifications affirm the capability to control interaction on the microlevel. Of particular interest are developments in dental materials showing how physicochemical surface behavior can inhibit unwanted biofilm development, shedding insight into prevalent oral health issues. More generally, research demonstrates the ways in which tailored surface properties, such as patterning at the nanoscale or chemical modification, can inhibit bacterial adhesion with remarkable efficiency, ultimately enhancing patient recovery. The applications run further than in dentistry, though, with the suggestion that the principles might be translated across a broad range of applications, from biomedical implantations to environmental engineering, thereby corroborating the premier importance of sustaining surface science research (Sterzenbach et al., 2020). Moreover, the employment of cutting-edge materials, such as mechanically interlocked molecules (MIMs), is a paramount aspect of material science where the nature of bonding dynamics transcends typical chemical interactions. The study of mechanical bonds offers avenues for unprecedented control over molecular assemblies and molecular machines, which may revolutionize various technologies. As well-defined in research, the design and synthesis of such intricate structures emphasize the potential of constructing systems that operate at the nanoscale, where emergent behavior interfaces with conventional chemistry principles. This uncertainty presents interesting prospects for researchers to further investigate the field of supramolecular chemistry, thereby building a story that interdigitates function and design use (Stoddart, 2017). In conclusion, the critical analysis of solid surface structures demands an interdisciplinary approach to bring together chemistry, physics, and engineering knowledge. With the advancing technology, a pressing need exists for continued research of new surface modifications and materials that not only optimize functionality but also operate according to sustainability. Future research should aim at developing biocompatible and sustainable materials to address the growing requirement for sustainable processes. By facilitating inter-disciplinary collaboration, scientists are able to access the collective expertise necessary to conquer the challenges of complex solid interfaces and create opportunities for innovation that delivers better performance in a range of industrial and technological sectors. The relentless pursuit of understanding solid surfaces is central not just to advancing current applications, but to inspiring future innovation in material science (Sterzenbach et al., 2020).

A study of the properties and character of solid surface properties shows that such properties are of crucial significance in determining the behavior of substances in various applications. Surface atomic structure is significantly responsible for phenomena such as corrosion resistance, adhesion, and catalysis. Knowledge about crystallographic orientation and surface defects is needed in order to be able to predict solid surface reactivity. Further, surface-modification methods such as etching and coating can adjust surface properties to meet specific performance requirements in an optimum manner. Surface properties can be engineered to possess enhanced functionality by varying surface structure. This plays a critical role in the development of emerging fields like electronics, nanotechnology, and materials science. The interaction between chemical processes and surface characteristics is thus a focus of work in solid-state physics and chemistry. Moreover, the energy topography of a solid surface is the key to stability and reactivity. Surface energy, a measure of bond disruption at the surface compared to the bulk material, governs processes such as wetting and adhesion. Higher surface energy generally corresponds to greater reactivity and is therefore a critical parameter in applications with catalysts and sensors. In addition, the nature of surface interactions on solid materials, as determined by van der Waals interactions or covalent bonds, impacts material performance under varied conditions. This sequence of interactions gives an indicator that proper understanding of surface energy can lead to improved designs for materials. The study of surface phenomena is also applied to its relevance in phase transitions and thermodynamic stability of different structural phases, emphasizing the interconnectedness of surface properties and overall material behaviors. With ongoing progress in research, there are newly developed methods for atomic-scale characterization of solid surfaces. Advanced microscopy and spectroscopy techniques, including scanning tunneling microscopy and atomic force microscopy, have revolutionized the way scientists study surface structures. These methods allow quantitative characterization of surface topography, electronic states, and chemical composition. Furthermore, computational modeling provides theoretical confirmation of experimental data, insight into the dynamic response of surfaces under varying conditions. The union of experimental and computational methodologies not only facilitates understanding of solid surfaces but also opens the door to engineering advanced materials. Thus, integration of technological advances in surface characterization is needed to realize the full potential of solid surfaces for beneficial uses, and drive further advances in many fields such as nanotechnology and catalysis.

# REFERENCES

1. Aghanim, N., Akrami, Y., Ashdown, M., Aumont, J., Baccigalupi, C., Ballardini, M., Banday, A. J., et al. (2020). Planck 2018 results. Astronomy and Astrophysics, 641, A6–A6. https://doi.org/10.1051/0004-6361/201833910 (accessed on 3rd October 2025).
2. Ansart, F., Barnabé, A., Bonino, J. P., et al. (2011). Sol–gel processing and characterization of (RE-Y)-zirconia powders for thermal barrier coatings. https://core.ac.uk/download/12043062.pdf (accessed on 3rd October 2025).
3. Asakuma, Y., Beaulieu, M., Bershadsky, I., Castellanos, A., Checchetto, D., Da-Wen, L., Flueckiger, J., et al. (2012). Models for metal hydride particle shape, packing, and heat transfer. http://arxiv.org/abs/1205.1073 (accessed on 3rd October 2025).
4. Asmatulu, R., Bai, Y., Balu, R., Barthlott, W., Bayer, I. S., Bhushan, B., Bhushan, M., et al. (2011). The super hydrophobicity of polymer surfaces: Recent developments. https://core.ac.uk/download/4150466.pdf (accessed on 3rd October 2025).
5. Bakheit, A., Attwa, M. W., Kadi, A., & Alkahtani, H. (2023). Structural analysis and reactivity insights of (E)-bromo-4-((4-((1-(4-chlorophenyl) ethylidene) amino)-5-phenyl-4H-1,2,4-triazol-3-yl) thio)-5-((2-isopropylcyclohexyl) oxy) furan-2(5H)-one: A combined approach using single-crystal X-ray diffraction, Hirshfeld surface analysis, and conceptual density fun. Crystals. https://www.semanticscholar.org/paper/e94f53be68e60c80978fc42aa3f7d7378a6c5e46 (accessed on 3rd October 2025).
6. Bartók, A. P., Kermode, J. R., Bernstein, N., & Csányi, G. (2018). Machine learning a general-purpose interatomic potential for silicon. Physical Review X, 8. https://doi.org/10.1103/physrevx.8.041048 (accessed on 3rd October 2025).
7. Bergveld, P., & Saaman, A. A. (1985). A classification of chemically sensitive semiconductor devices. https://core.ac.uk/download/pdf/11473084.pdf (accessed on 3rd October 2025).
8. Blinov, L., Chigrinov, V., & Patel, J. (1995). Electro-optic effects in liquid crystal materials. Physics Today, 48, 85–87. https://www.semanticscholar.org/paper/0a82ec9b7dd87e213a424f8fb4dee2613656b311 (accessed on 3rd October 2025).
9. Buljan, A., Llunell, M., Ruiz, E., & Alemany, P. (2001). Color and conductivity in $Cu_2O$ and $CuAlO_2$: A theoretical analysis of $d^{10}...d^{10}$ interactions in solid-state compounds. Chemistry of

Materials, 13, 338–344. https://www.semanticscholar.org/paper/f4c3454d7fde7fb49df75f76857d9f957b7a56d4 (accessed on 3rd October 2025).

10. Cao, H., Xiao, O., Wu, X., Chen, L., Wu, J., Wu, J., Wang, J., et al. (2025). Mechanical and electrochemical properties of titanium aluminum nitride coatings with different nitrogen flow rates on CrMnSi steel by filter cathode vacuum arc technology. Coatings. https://www.semanticscholar.org/paper/d1eae66b5072f09080073550c49dbdbbbc155ac5 (accessed on 3rd October 2025).
11. Concheri, G., Meneghello, R., Rosso, S., et al. (2018). Geometric modeling of cellular materials for additive manufacturing in biomedical field: A review. https://core.ac.uk/download/154326170.pdf (accessed on 3rd October 2025).
12. Cullis, P. R., & Hope, M. J. (2017). Lipid nanoparticle systems for enabling gene therapies. Molecular Therapy, 25, 1467–1475. https://doi.org/10.1016/j.ymthe.2017.03.013 (accessed on 3rd October 2025).
13. Kumar, S., & Hoh, J. (2000). Direct visualization of vesicle–bilayer complexes by atomic force microscopy. Langmuir, 16, 9936–9940. https://www.semanticscholar.org/paper/71f4fc2082cf4eca94f5a9fcd3b133cac2c729a5 (accessed on 3rd October 2025).
14. Liu, M., Jiang, Z., Wu, X., Liu, F., Li, W., Meng, D., Wei, A., et al. (2025). Reinventing the high-rate energy storage of hard carbon: The order-degree governs the trade-off of desolation–solid electrolyte interphase at interfaces. Angewandte Chemie. https://www.semanticscholar.org/paper/6b12f926aa96867d96b11a23c4909986f9c152bf (accessed on 3rd October 2025).
15. Liu, S., Liu, T., Zou, Q., Wang, W. M., Doubrovski, E., & Wang, C. C. L. (2021). Memory-efficient modeling and slicing of large-scale adaptive lattice structures. J. Comput. Inf. Sci. Eng., 21. https://www.semanticscholar.org/paper/32d3b2414bfc37b970cdf81d5f72a2a771c82383 (accessed on 3rd October 2025).
16. Liu, W., Zou, Y., Chen, Y., Lei, Z., Zhao, L., & Song, L. (2024). Optimizing the structure and optical properties of lanthanum aluminate perovskite through $Nb^{5+}$ doping. Nanomaterials, 14. https://www.semanticscholar.org/paper/5a7bbc2444d69f1608c48cad1102f03fe5d1cd60 (accessed on 3rd October 2025).
17. Mäkelä, J. (2019). Investigation and suppression of semiconductor–oxide related defect states: From surface science to device tests. https://core.ac.uk/download/198017053.pdf (accessed on 3rd October 2025).

18. Matějů, D., Franzmann, T. M., Patel, A., Kopach, A., Boczek, E. E., Maharana, S., Lee, H. O., et al. (2017). An aberrant phase transition of stress granules triggered by misfolded protein and prevented by chaperone function. The EMBO Journal, 36, 1669–1687. https://doi.org/10.15252/embj.201695957 (accessed on 3rd October 2025).
19. Mejía-Rosales, S., Fernández-Navarro, C., Pérez-Tijerina, E., Blom, D., Allard, A., & José-Yacamán, M. (2007). On the structure of Au/Pd bimetallic nanoparticles. Journal of Physical Chemistry C, 111, 1256–1260. https://www.semanticscholar.org/paper/bb52d2c3565fb6f03184b1f7efcd3a7601dae350 (accessed on 3rd October 2025).
20. Mitchell, J. K., Soga, K., & O'Sullivan, C. (2025). Fundamentals of soil behavior. https://doi.org/10.1002/9781119832348 (accessed on 3rd October 2025).
21. Momin, A. A., Munshi, A., Imran, T., Hossain, A., Mimi, A., Gafur, M. A., Hasan, H. R., et al. (2024). Influence of Y substitutions on the structural and magnetic properties of Mn-Zn ferrites. Jagannath University Journal of Science. https://www.semanticscholar.org/paper/00b240c18205ee97deb18c9d8e38a55f54474826 (accessed on 3rd October 2025).
22. Mustafa, B., Lu, W., Wang, Z., Lian, F., Shen, A., Yang, B., Yuan, J., et al. (2022). Ultrahigh energy and power densities of d-MXene-based symmetric supercapacitors. Nanomaterials, 12. https://www.semanticscholar.org/paper/7a141c83927c1a7748654d0144df037e81b2c46d (accessed on 3rd October 2025).
23. Nowak, A., Gross, A. F., Sherman, E., Rodriguez, A. R., Ventuleth, M., Nelson, A. M., Guan, S., et al. (2021). Dual component passive icephobic coatings with micron-scale phase-separated 3D structures. ACS Applied Materials & Interfaces. https://www.semanticscholar.org/paper/d5983e7e2b29a55dcf5481b8dd4e5a1a1a98e7df (accessed on 3rd October 2025).
24. Ozboyaci, M., Kokh, D. B., Corni, S., & Wade, R. C. (2016). Modeling and simulation of protein–surface interactions: Achievements and challenges. Quarterly Reviews of Biophysics, 49. https://doi.org/10.1017/s0033583515000256 (accessed on 3rd October 2025).
25. Pallikara, I., Kayastha, P., Skelton, J. M., & Whalley, L. D. (2022). The physical significance of imaginary phonon modes in crystals. Electronic Structure, 4, 033002–033002. https://doi.org/10.1088/2516-1075/ac78b3 (accessed on 3rd October 2025).

26. Patra, J. K., Das, G., Fraceto, L. F., Campos, E. V. R., Rodriguez-Torres, M. del P., Acosta-Torres, L. S., Díaz-Torres, L. A., et al. (2018). Nano based drug delivery systems: Recent developments and future prospects. Journal of Nanobiotechnology, 16. https://doi.org/10.1186/s12951-018-0392-8 (accessed on 3rd October 2025).

27. Peng, X. (2010). Band gap and composition engineering on a nanocrystal (BCEN) in solution. Accounts of Chemical Research, 43, 1387–1395. https://doi.org/10.1021/ar100025m (accessed on 3rd October 2025).

28. Podili, S., Pothukanuri, N., & Murthy, S. N. (2023). Effect annealing temperature on pure and Al-doped iron oxide thin films for formaldehyde gas sensing application. ECS Journal of Solid-State Science and Technology. https://www.semanticscholar.org/paper/d2900f4a2d824775d49664970bfa693bbb216ede (accessed on 3rd October 2025).

29. Qi, X. L., & Zhang, S. C. (2011). Topological insulators and superconductors. Reviews of Modern Physics, 83, 1057–1110. https://doi.org/10.1103/revmodphys.83.1057 (accessed on 3rd October 2025).

30. Riisager, A., Krishnan, D., Schill, L., Axet, M., & Philippot, K. (2024). Ni nanoparticles in functionalized ionic liquids for chemo selective hydrogenation catalysis. ECS Meeting Abstracts. https://www.semanticscholar.org/paper/d486273fefac6f7d3c1d8f701d9ee097fbb98f8b (accessed on 3rd October 2025).

31. Spina, M., Lehmann, M., Náfrádi, B., Bernard, L., Bonvin, E., Gaál, R., Magrez, A., et al. (2015). Microengineered $CH_3NH_3PbI_3$ nanowire/graphene phototransistor for low-intensity light detection at room temperature. Small, 11, 4824–4828. https://doi.org/10.1002/smll.201501257 (accessed on 3rd October 2025).

32. Sterzenbach, T., Helbig, R., Hannig, C., & Hannig, M. (2020). Bioadhesion in the oral cavity and approaches for biofilm management by surface modifications. Clinical Oral Investigations, 24, 4237–4260. https://doi.org/10.1007/s00784-020-03646-1 (accessed on 3rd October 2025).

33. Stoddart, J. F. (2017). Mechanically interlocked molecules (MIMs)—Molecular shuttles, switches, and machines (Nobel Lecture). Angewandte Chemie International Edition, 56, 11094–11125. https://doi.org/10.1002/anie.201703216 (accessed on 3rd October 2025).

34. Thommes, M., Kaneko, K., Neimark, A. V., Olivier, J. P., Rodríguez-Reinoso, F., Rouquérol, J., & Sing, K. S. W. (2015). Physisorption of gases, with special reference to the evaluation of surface area and pore

size distribution (IUPAC Technical Report). Pure and Applied Chemistry, 87, 1051–1069. https://doi.org/10.1515/pac-2014-1117 (accessed on 3rd October 2025).

35. Ulucan, T. H., Wang, J., Onur, E., Chen, S., Behrens, M., & Weidenthaler, C. (2024). Unveiling the structure–property relationship of MgO-supported Ni ammonia decomposition catalysts from bulk to atomic structure by *in situ*/operando studies. ACS Catalysis, 14, 2828–2841. https://www.semanticscholar.org/paper/c92fcdffa964450d0de6b45068285de188e594c7 (accessed on 3rd October 2025).
36. Wang, H., Li, S., & Wang, C. (2023). Energy absorption characteristics of triply periodic minimal surface lattices. Journal of Physics: Conference Series, 2535. https://www.semanticscholar.org/paper/a960bc6ebf89e8727f8331b622b00b7a5009b6ef (accessed on 3rd October 2025).
37. Wang, Y., He, L., Wang, Y., Chen, J., Ekholm, A., & Liu, J. (2024). A critical review of hexagonal boron nitride manufacturing technology for electronics cooling applications. 2024 IEEE 26th Electronics Packaging Technology Conference (EPTC). https://www.semanticscholar.org/paper/13d673301c235f7ce60f50fbc122ac8dd4a7c145 (accessed on 3rd October 2025).
38. Whiteley, S., Heremans, F. J., Wolfowicz, G., Awschalom, D., University of Chicago, Department of Physics, Institute for Molecular Engineering, et al. (2018). Correlating dynamic strain and photoluminescence of solid-state defects with stroboscopic X-ray diffraction microscopy. Nature Communications, 10. https://www.semanticscholar.org/paper/cfdb3ddfb391774060a2257f6f85d8a9de203420 (accessed on 3rd October 2025).
39. Xu, X., Su, C., Zhou, W., Zhu, Y., Chen, Y., & Shao, Z. (2015). Co-doping strategy for developing perovskite oxides as highly efficient electrocatalysts for oxygen evolution reaction. Advanced Science, 3. https://doi.org/10.1002/advs.201500187 (accessed on 3rd October 2025).

CHAPTER 3

# ADSORPTION PHENOMENA

## Contents



## INTRODUCTION

Adsorption phenomena are inherent processes occurring at phase interfaces with effects across various scientific and technological uses. Adsorption is essentially the adsorption of molecules of the gaseous or liquid phase on solid surfaces, which may dramatically alter the reactivity and surface properties of the solid. Adsorption has essential roles in catalysis, environmental science, and materials construction. By studying molecular interactions with surfaces, researchers can develop innovative ways to enhance the performance of many materials, making them maximize their applications in various sectors. The interplay between molecular characteristics and surface properties accounts for the complexity of adsorption processes, and thus intensive research into this interdisciplinary field is required. Geometry and molecular interactions that govern the manner in which substances adhere to surfaces form a pivotal part of adsorption. For example, it has been discovered only recently that oxygen vacancies introduced onto the surface can significantly influence the adsorption properties of organic molecules. For example, the creation of an oxygen vacancy on the (1010) ZnO nonpolar surface alters the electronic properties, enhances the band gap, and introduces localized energy states for phenol. These modifications make possible the adsorption charge transfer events, since they enhance surface reactivity. Understanding these underlying processes enhances our comprehension of adsorption phenomena and paves the way for surface engineering to achieve maximum interaction in application-specified design, improving efficiency in sensors and catalysis, among other uses. Besides, the nature of the interactions between the adsorbate and substrate is also significant in determining the effectiveness of the process of adsorption. For instance, research on lipid-coated liquid crystal interfaces suggests that the orientation and ordering of the liquid crystals can significantly be affected by the presence of proteins being adsorbed. In particular, the signal protein JM-EGFR produces an ideal change in optical properties of lipid-coated liquid crystals, which signifies successful detection of the protein with this interface. This suggests the potential for using adsorption phenomena in novel sensing technologies which leverage optical changes for analysis. The diversity of adsorption applications—ranging from environmental clean-up to biosensing—is indicative of the very essentiality of understanding adsorption phenomena as a keystone in theoretical and applied science."

## Definition of Adsorption

The adsorption process is the adsorption process by which a compound, usually a gas or liquid, adheres to the surface of a liquid or solid. The process may be defined as the accumulation of particles at the surface between two phases with an altered concentration of the substance at or near the surface. Importantly, adsorption can occur in two general categories: physisorption, on the basis of the weak van der Waals forces, and chemisorption, on the basis of the creation of chemical bonds between the surface and adsorbate. The interplay among these categories is of central importance in a wide range of applications, from catalysis to environmental science, and highlights the importance of surface interactions in determining the efficiency and selectivity of a materials behavior in context. Due to scientific research on novel adsorbents and their applications, researchers continue to discover the intricate process of adsorption mechanisms and their supportive roles across scientific fields (He et al., 2022; Canossa et al., 2020). In essence, the effectiveness of adsorption is related to a variety of factors including surface area, porosity, as well as the nature of the adsorbent and adsorbate (Figure 3.1).

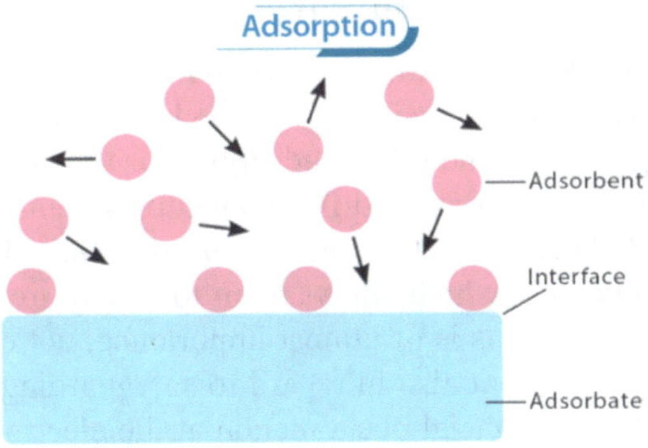


**Figure 3.1.** Adsorption.

*Source:* https://external-content.duckduckgo.com/iu/?u=https%3A%2F%2Fwww.chemistrylearner.com%2Fwp-content%2Fuploads%2F2023%2F12%2FAdsorption.jpg&f=1&nofb=1&ipt=2f35e1065c423eac97858434e457ba83d2285f4267f17803cad969bc401d8429.

Materials of high surface area such as activated carbon and zeolites have high adsorption capacities because increased surface area offers more active sites for interaction. Chemical properties of the partners also come into play; polarity and functional groups, for instance, can affect adhesion and stability. Besides, thermodynamic theory dictates that adsorption tends to be an exothermic process where heat is released as the molecules migrate to the surface, rendering itself a lower-energy system on contact. The process is a manifestation of the underlying physical and chemical processes that dictate material selection for adsorption application both in theoretical research and real applications (He et al., 2022; Canossa et al., 2020). Understanding the definition of adsorption takes into account more than the mere interaction process itself but also the intricate dynamics that govern surface behavior. As an example, adsorption kinetics, which establish the rates at which molecules adhere to surfaces, can provide insight into diffusion processes as well as to hypothetical barriers to adsorption. Besides, the isotherm equations such as Freundlich and Langmuir give the concentration of the adsorbate in the bulk phase against the quantity adsorbed onto the surface. The models aid in predicting behavior in varied conditions of the environment. Finally, the thorough investigation of adsorption phenomena identifies its importance in diverse areas such as material design, environmental treatment, and chemical applications in industry, which will lead to further technological progress and sustainability plans (He et al., 2022; Canossa et al., 2020).

## Importance of Adsorption in Various Fields

The significance of adsorption extends beyond numerous fields and influences development in environmental science, nanotechnology, and biochemistry. In environmental engineering, the role of adsorption is essential in treating water since pollutants are usually removed from aqueous streams through means of adsorption. Activated carbon and zeolites are used frequently due to their high surface areas and porosity, which allows them to adsorb toxic substances. The efficiency of such adsorbents is of utmost importance, not only with regard to meeting regulatory needs but also in regard to safeguarding public health and safety. Understanding interfacial phenomenon at the electrochemical interface (EI) thus becomes critical in overcoming mass transport and heterogeneous reaction problems. The electrical double layer (EDL) behavior in aqueous systems depicts how energy storage and conversion technologies are crucially dependent on adsorption and therefore call for scientific research as well as practical application (Pham et al., 2024). In the field of nanotechnology, the importance of adsorption is further magnified with researchers exploring various materials and their unique characteristics at the nanoscale. Adsorption processes control nanomaterial synthesis and nanomaterial function, determining such factors as stability, dispersibility, and biocompatibility. Manipulation of adsorption characteristics not only facilitates the assembly of novel nanoarchitectures but also impacts drug delivery and biosensing technologies. Supramolecular chirality, wherein high-order organization of molecular subunits leads to chiral structures, demonstrates how weak forces can be exploited in optimizing nanomaterial function. This understanding is coming to the fore for its capacity to create chiral materials to mimic biological performance, thereby bridging the gap between synthetic materials and life organisms. Consequently, certain mechanisms of adsorption are central to moving forward with nanotechnology and its use in various industries (Dou et al., 2020). Biochemical science has also progressively realized the paramount role of adsorption in most biological functions. Biomolecule adsorption onto surfaces or into supramolecular hydrogels strongly regulates cell behavior, i.e., adhesion, proliferation, and differentiation. These interactions are regulated by noncovalent forces such as hydrogen bonding and van der Waals interactions that regulate the overall functionality of the material in biological environments.

For instance, the development of chiral supramolecular hydrogels has been promising for various biomedical applications, including drug delivery and tissue engineering. By regulating the adsorption profiles of such materials, their functionality and biocompatibility may be enhanced, enabling them to facilitate new strategies towards addressing biomedical issues. Studies on the adsorption phenomena are thus still very much pertinent in promoting biotechnology as well as the efficiency of therapeutic interventions (Dou et al., 2020).

## 3.1. TYPES OF ADSORPTION

There are two primary types of adsorption in the case of adsorption phenomena, namely physisorption and chemisorption. Physisorption is weak van der Waals forces that permit molecules to bind onto a substrates surface. This type of adsorption is generally reversible and occurs at low temperatures as it does not involve the breaking of chemical bonds. It is primarily interaction by means of physical forces, hence not so specific as in chemisorption. Physisorption plays a critical role in numerous applications, e.g., gas storage and separation processes, where the stored material can be released easily. This process is required to comprehend if one desires to maximize ad-sorption systems, particularly in applications where desorption has to be rapid, such as in cross-flow microfiltration processes where membrane fouling can practically undermine efficiency (Albasi et al., 2011). Chemisorption, on the other hand, is distinguished by the formation of strong covalent or ionic intermolecular forces between the adsorbent and adsorbate. This tends to be applicable to irreversible states under normal operating conditions because of the vast energy that is required in order to break such bonds. Chemisorption occurs at elevated temperatures and is a more selective interaction between involved molecules. This type of adsorption takes on great importance in fields of catalysis, where the efficiency of chemical reaction relies on the adsorption characteristics of reactants on catalyst surfaces. Moreover, micropollutant research in reverse electrodialysis systems illustrates the different behaviors of different adsorbates because some micropollutants interact differently with ion-exchange membranes, significantly influencing transport behavior and overall system efficiency (Cornelissen et al., 2019). The fact that these two adsorptions occur shows the complexities of adsorption phenomena, influencing many industrial and environmental systems. From the treatment of water to the development of advanced materials, each type has a role in optimizing performance as well as sustainability. For instance, with the realization of processes involved in physisorption and chemisorption in filtration, membranes can be designed that minimize fouling and maximize the rejection of desired compounds. Hence, advancements in membrane technology are often reliant on the control of these adsorption characteristics, thereby enhancing operational effectiveness in various water treatment processes. Because of the importance accorded to the regulation of adsorption processes, more research into the implications of both physisorption and chemisorption will be required for the development of innovative solutions toward addressing contemporary challenges in environmental science and engineering (Figure 3.2).

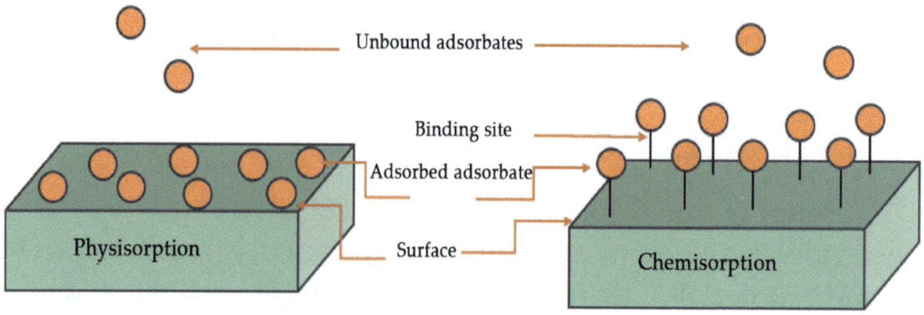


**Figure 3.2.** Type of adsorption (physisorption and chemisorption).

*Source:* https://www.researchgate.net/figure/a-Type-of-adsorption-Physisorption-and-Chemisorption-b-Adsorption-mechanisms_fig3_386259600.

### 3.1.1. Physical Adsorption (Physisorption)

Physical adsorption, otherwise referred to as physisorption, is of great significance in most adsorption processes due to the fact that it is founded upon the fundamental principles of weaker van der Waals forces. Unlike chemisorption where there is the formation of chemical bonds, physisorption involves reversible interaction between the surface of the substrate and adsorbate molecules so that gases or liquids can condense on porous materials without undergoing any change in their chemical structure. This type of adsorption is central to gas storage, catalysis, and environmental remediation purposes because it enables materials to capture and retain an extensive range of substances effectively. Furthermore, the superior ability of physisorption to operate at low temperatures than chemisorption highlights the importance of physisorption in industrial purposes requiring gentler conditions while attaining optimal adsorbate adsorption. The characterization of nanoporous materials, particularly hierarchically ordered ones like mesoporous zeolites, significantly enhances our understanding of physisorption phenomena. These materials possess pore networks that connect micro, meso, and macropore regimes, influencing adsorption capacity and transport characteristics. Detailed inquiries regarding the pore architecture of these materials offer useful information in terms of diffusional rates and selectivity of catalytic reactions. For instance, using advanced physisorption methods, researchers are capable of accurately measuring crucial factors such as surface areas and pore volumes that govern the function of such material in actual applications. This synergy between physisorption and structural properties says a lot about its prime importance in optimizing process in industrial separation and energy storage processes, as highlighted in recent research studies on the topic (Schlumberger et al., 2021) (Figure 3.3).

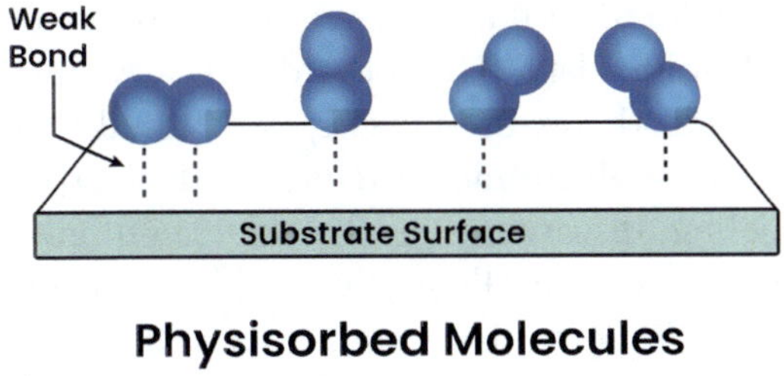


**Figure 3.3.** Physisorbed molecules.

*Source:* https://allen.in/jee/chemistry/physisorption-and-chemisorption.

Moreover, physisorption is a robust technique utilized in the removal of pollutants, particularly wastewater treatment, where it addresses the critical issue of dye contamination. The adsorptivity of various adsorbents allows for selective elimination of toxic dyes and minimizes ecological threats. Physical adsorbents utilize the concept of physisorption to capture dye molecules effectively, thus aiding in the provision of cleaner water sources and diminished levels of toxicity in the environment. Studies foreground the importance of these adsorptive processes, evoking their potential in mitigating ecotoxicological effects of dye contaminants (Dutta et al., 2021). As studies continue, understanding the intricacies of physisorption will pave the way for innovative adsorbents designed for purpose-specific uses, further solidifying its position in environmental science and engineering.

## 3.1.2. Chemical Adsorption (Chemisorption)

Chemisorption, or chemical adsorption, is a fundamental process involving the formation of strong chemical bonds between adsorbent surfaces and adsorbate molecules. In contrast to physisorption, this process is not only characterized by the strength of the interaction, which typically comprises covalent or ionic bonding, but also by its electronic structure dependence of the adsorbent and adsorbate. In catalysis, water treatment, and other technologies, chemisorption plays a vital role. For instance, as highlighted in recent work, the new use of activated carbon from raw sawdust pine has shown good adsorption potentiality for wood dyes. Such carbons' inherent surface characteristics allow for massive dye removal capacity, with processes that are chemisorption kinetics in nature, emphasizing its utility in environmental remediation applications (Pimentel et al., 2023). Investigating chemisorption mechanisms provides us with understanding of the manner in which molecular interactions control adsorption properties. As an example, the transition metal reactivity during chemisorption can be effectively explored in the presence of a Bayesian learning procedure that incorporates ab initio data. This provides a very deep understanding of interactions among adsorbates at specially structured metal sites, unveiling the fine balance between electronic structures and adsorption processes. With this advanced modeling, researchers can calculate configurational reactivity in intermetallics and near-surface alloys, thus bridging theoretical with applied studies on catalyst development. The complexity of these interactions promotes the design of new catalytic materials that utilize the unique electronic features of d-band states to expand the scope of chemisorption to other chemical reactions (Wang et al., 2020) (Figure 3.4).

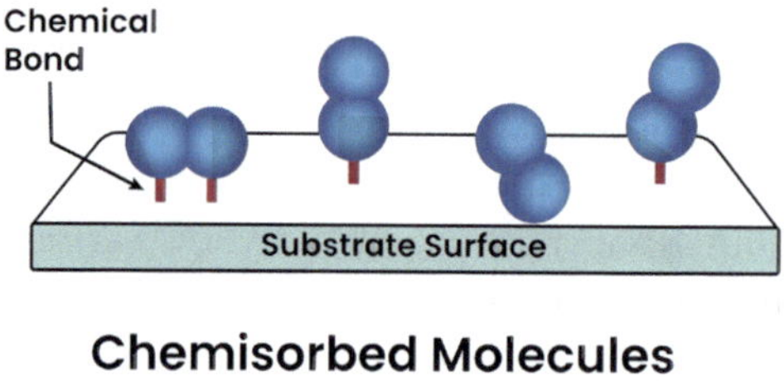


**Figure 3.4.** Chemisorbed molecules.

*Source:* https://allen.in/jee/chemistry/physisorption-and-chemisorption.

Moreover, conditions under which chemisorption occurs—temperature, pressure, and concentration—are closely connected to adsorption efficiency and kinetics. Very effective removals of water contaminants, for instance, have been achieved at optimal pH levels and specific dosages of adsorbents. Literature confirms that equilibrium data fitted to models like Freundlich and Langmuir not only describes the type of adsorption but also determines important parameters toward practical applications. Understanding such dynamics allows chemisorptive processes to be optimized to maximize the elimination of contaminants while minimizing resource consumption. The inferences from such knowledge extend considerably past environmental application to incorporate energy storage and conversion, where chemisorption can be used to improve the functioning of numerous materials and processes that are employed in trapping and converting energy (Pimentel et al., 2023).

### 3.1.3. Comparison Between Physisorption and Chemisorption

To the adsorption phenomena field, it is necessary to know the distinction between physisorption and chemisorption since the two processes are not only different in mechanism but also in effects. Physisorption, characterized by low van der Waals forces, exhibits a comparatively reversible adsorbate-adsorbent interaction.

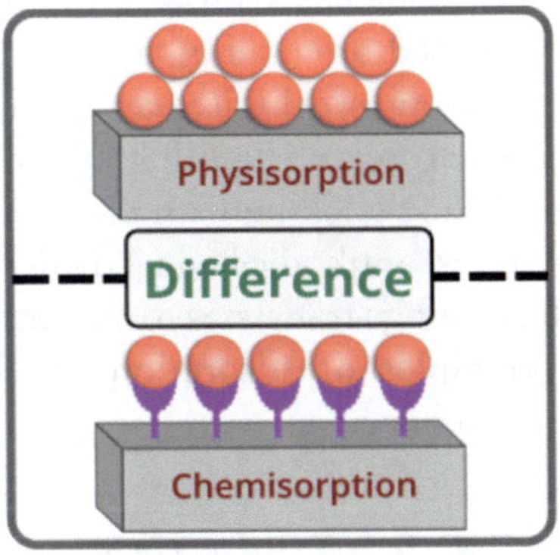


**Figure 3.5.** Differences between chemisorption and physisorption.

*Source:* https://psiberg.com/wp-content/uploads/2022/04/Physisorption-vs.-Chemisorption-The-Two-ways-of-Adsorption-Featured-image-for-differences.svg.

This low interaction is responsible for lower adsorption energies, typically less than 20 kJ/mol, and is responsible for very minimal disruption of the molecular structure of the adsorbates. In contrast, chemisorption involves a tighter, covalent formation of a bond with a resultant drastic alteration in the chemical composition of the adsorbate. The mechanism typically encompasses higher adsorption energies of 40 to a few 100 kJ/mol, which indicates a more permanent change in the adsorption dynamics. These differences not only have implications in fundamental thermodynamics but also dictate useful application in several sectors, particularly catalysis and sensor technology. Use of physisorption and chemisorption is best illustrated when the situation of gas sensors is used, where both types of adsorption play a distinct role in the functionality. For instance, reports have indicated that the sensitivity of palladium-doped titanium dioxide ($TiO_2$) clusters was significantly enhanced towards the detection of chlorine ($Cl_2$). In this case, physisorption precedes chemisorption, essentially reversing the interaction from weak to strong because sensing is occurring. The physisorption is the initial process by which the $Cl_2$ molecules approach the $TiO_2$ surface, and chemisorption creates the permanence of the bonding and enhances the detection efficiencies (Abdulsattar et al., 2023) (Figure 3.5).

This concerted response not only emphasizes the inherent differences between the two processes but also underscores the way these processes can be optimized toward specific applications in industrial and environmental monitoring. Furthermore, the dynamics of antibody positioning in biosensing uncover another compelling axis to the physisorption versus chemisorption contrast. In an APTES monolayer-adsorbed antibody study, it was found that antibodies exhibit varied orientation based on adsorption technique. Physisorbed antibodies possessed a disordered orientation at lower surface concentrations, while chemically bonded antibodies exhibited an ordered, head-on orientation at higher surface densities. The research highlighted how physisorption enables more flexible and faster molecule packing than the rigid stabilization achieved by chemisorption, which fixes antibodies via covalent forces (Gajos et al., 2022). The relevance of the findings translates to biosensor optimization design, demonstrating how the principles of adsorption can influence biomolecular recognition and sensor performance.

## 3.2. MECHANISMS OF ADSORPTION

Knowledge of adsorption mechanisms is of utmost importance in understanding how various substances act at surfaces. Adsorption is a phenomenon whereby the atoms, ions, or molecules of a dissolved solid, liquid, or gas attach to the surface and create a film of molecules. The effectiveness of adsorption processes can vary significantly based on the physicochemical properties of the adsorbate and the adsorbent. For instance, chemical adsorption is typically explained in terms of strong chemical bonds between the surface and the adsorbate, while physical adsorption happens through weaker van der Waals forces. The inherent nature of the surface, such as porosity, hydrophobicity, and charge, also plays significant parts to play in the determination of adsorption efficacy and the specific mechanisms at work. Adsorption mechanisms research illuminates practical applications such as water treatment and the desalination process. Of particular interest is the reverse electrodialysis system, which indicates the process of organic micropollutant interaction with ion-exchange membranes such that varied mechanisms control adsorption behavior. It is revealed in research that on the basis of their charge nature and molecular structure, micropollutants exhibit varied adsorption mechanisms with critical implications for their transport behavior in membranes. The research places into context the intricacies of ion-exchange membrane optimization for enhanced permeability for organic micropollutants and gives a rational platform for the evolution of future systems for effective control of contaminants. The multidisciplinary method of this is imperative to streamline the safety and efficiency of desalination processes where the treated wastewater arrives as a low-salinity stream, emphasizing the use of adsorption processes in the conservation of nature. Further, the research on membrane fouling by other molecular constituents in solutions, for instance, wine or other colloidal fluids, is a critical observation into processes involved in physical adsorption. Identification of processes such as cake layer formation and gel-type layers is a significant consideration in elucidating how variation of nature and concentration impacts membrane fouling during microfiltration operations. For example, tannins, pectins, and mannoproteins in wine differ in their effects on permeate flux that are directly proportional to their action against adsorption on membrane surfaces. An understanding of these processes could spur improved filtration methods and be a significant step towards the optimization of industrial processes wherein fouling is a significant limitation to efficiency. By defining the complex adsorption interactions, researchers can develop ways to reduce fouling, thus enhancing the overall efficiency of filtration technologies in different applications (Figure 3.6).

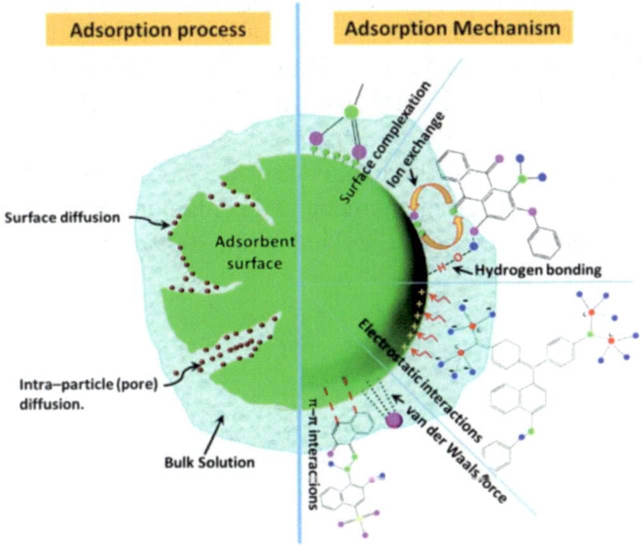


**Figure 3.6.** Mechanism of adsorption process.

*Source:* https://www.researchgate.net/figure/Mechanism-of-adsorption-process-55_fig3_384997347.

### 3.2.1. Van der Waals Forces in Physisorption

In the context of adsorption processes, the Van der Waals forces are significant in the process of physisorption, which is primarily defined by weak intermolecular attractive forces between adsorbate molecules and substrates. These forces are established as a result of temporary fluctuations of electron distribution within molecules, leading to the creation of induced dipoles. Consequently, physisorption has lower binding energies than chemisorption, although it accommodates reversible and elastic interactions that render it significant in a wide variety of applications including catalysis and sensor technology. The significance of such non-covalent interactions is enhanced in aromatic compound systems, where the stacking interactions are extremely strong because of the overlapping π-electron clouds. The gentle nature of physisorption emphasizes the delicate balance of energy landscapes governing molecular assembly on surfaces. Research into the adsorption of molecules over metal surfaces, particularly aromatic hydrocarbons like benzene, illustrates how Van der Waals forces contribute to the overall stability of adsorbate-substrate configurations. More recent research employing newer density functional theory like the application of Van der Waals density functionals (vdW-DF) demonstrated that the addition of these nonlocal correlations leads to significantly disparate calculated adsorption energies. For example, research of benzene adsorption on surfaces such as Cu(111) demonstrates that the addition of these interactions enhances the understanding of binding position shifts and enhances overall binding energy—by up to an order of magnitude, in some cases. This dramatic capacity of vdW forces to direct molecular arrangements on surfaces is testament to their central contribution to physisorption phenomena and challenges classical theories that disregard long-range interactions (Berland et al., 2009) (Figure 3.7).

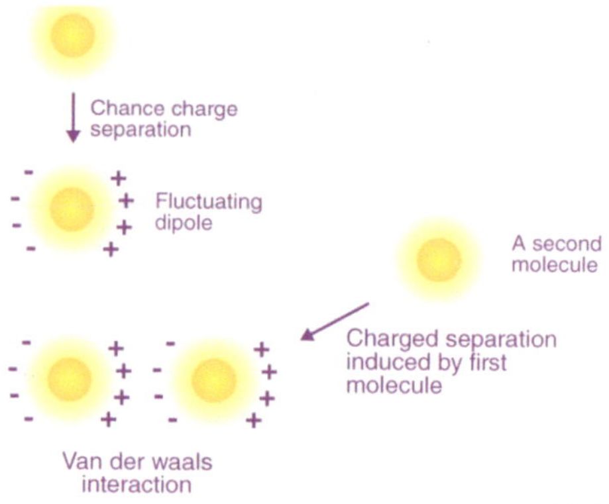


**Figure 3.7.** Van der Waals forces.

*Source:* https://cdn1.byjus.com/wp-content/uploads/2020/10/Van-Der-Waals-Forces-700x538.png.

Furthermore, research into potential alternative theoretical models to include the Van der Waals forces has shed more light on adsorption on closely packed metal surfaces. Recent developments in computational methodology have allowed physisorption on a sequence of metals like Cu, Ag, and Au that engage in weak adsorption alone to be explored. Specifically, the existence of interatomic van der Waals approaches and improved nonlocal functionals allows for accurate treatment of long-range forces dominating adsorption. Examination of adsorption energies and equilibrium positions indicates that accurate treatment of these forces provides close reproduction of experiment, ascertaining the necessity of continuous refinement of computational models and methodology. Together, these experiments verify the governing role of Van der Waals forces in determining the nature of physisorptive interactions across a variety of scenarios (Tkatchenko et al., 2014).

### 3.2.2. Chemical Bonding in Chemisorption

The understanding of chemical bonding in chemisorption is crucial in the elucidation of the mechanisms of adsorption processes. Chemisorption, which is founded on the formation of strong chemical bonding between adsorbate and adsorbent, typically involves electron transfer and the formation of covalent or ionic bonds. The process involves the liberation of a large quantity of energy, distinguishing chemisorption from weaker physical adsorption methods, which are founded on van der Waals forces. The nature of the chemical bonds that have formed contributes a huge factor in the adsorbate's stability and reactivity, and therefore these interactions must be accurately measured. Sophisticated computational tools such as mean-field density functional theory (DFT) enable researchers to model these interactions at the most basic level and make predictions about the adsorption capacity and energetics of various materials, particularly those containing transition metals such as copper, which are part of catalytic and separation processes in the process of carbon capture and sequestration (Pourrahmani, 2024). Theoretical models such as the d-band reactivity theory help to greatly explain chemisorption bonding. According to this model, the electronic structure of metal surfaces, and specifically the Fermi level density of states, plays an imperative role in dictating how adsorbates interact with the substrate. Through the use of Bayesian learning strategies, scientists are able to investigate chemisorption processes at atomically designed metal sites. Most significantly, Bayesian models on ab initio data allow site reactivity prediction in a wide variety of intermetallics and near-surface alloys. These theoretical advance bridges the gap on the complex issue of adsorption site chemical bonding, translating theoretical inputs into viable applications to predict and design novel catalytic materials (Wang et al., 2020). Thus, these models enable us to understand the orbital-wise nature of chemical bonding, illuminating us in terms of how various d-band characteristics influence the substrate-adsorbate interactions. Experimental verification of the theoretical models is essential to enhance our understanding of chemisorption. Electron microscopy and spectroscopy are two of the methods utilized in investigating the chemical bonding nature of adsorbates on metal surfaces. For instance, the Nion UltraSTEM microscope can image metal site coordination environments and surface defects with unprecedented resolution, allowing researchers to correlate these structural characteristics with adsorption behavior. In addition, methods like Electron Energy Loss Spectroscopy (EELS) can furnish details about the oxidation states and electronic configurations of adsorbate interactions. Computational methods complement such characterization to create a more in-depth knowledge base for the involved complex phenomena of chemisorption. As the research continues, the combination of theory development with experimental information will result in the rational design of optimized adsorption-capable materials, ultimately enhancing application areas like catalysis and environmental remediation.

### 3.2.3. Role of Surface Area and Porosity

The porosity-surface area relationship is also critical in the process of adsorption, thus having a major impact on the effectiveness of materials being utilized this way. The higher the surface area, the more the adsorption capacity, since there will be more active sites where molecules can adhere. This correlation is particularly evident for carbon-based materials, where changes in particle size result in differences in both surface area and resultant adsorption properties. For example, experiments show that the smaller microparticles of carbon have greater adsorption capacities due to their larger surface area and enhanced availability of sites for the adsorbate, thus increasing the general efficiency of adsorption.

That pore architecture and surface characteristics define the accessibility of such active sites is an indication of the importance of optimizing these parameters in the design of effective adsorbent materials to apply in a range of applications. Porosity is also as vital in determining the extent to which adsorbates can permeate and fill a material. High porosity allows for gases and liquids' freedom to move, enabling easy transport to adsorptive surfaces. When materials have a specific porous structure, adsorbate-internal surface interaction increases, which could enhance further adsorption sites which might be absent in less porous media. While considering carbon microparticles' properties, it is clear that those produced from organic waste tend to have a typical absence of pores, which restricts their efficiency in some adsorption situations. Nonetheless, the development of bigger pores through new carbon synthesis methods unmasks how the alteration of porosity can make adsorbents capable of trapping more diverse ranges of target adsorbates with high efficiency, making them more useful in environmental remediation and catalysis (Figure 3.8).

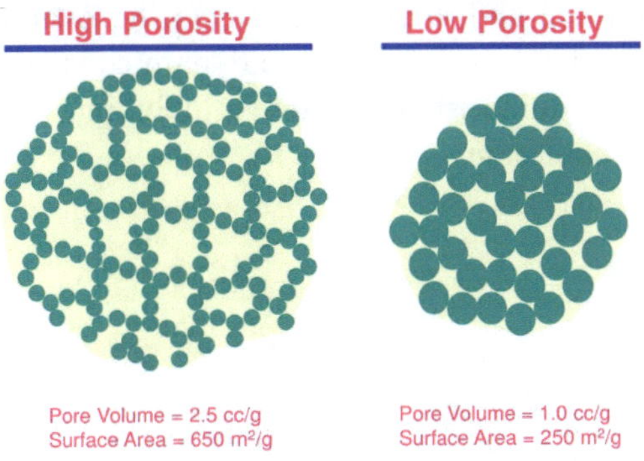


**Figure 3.8.** The influence of catalyst porosity on catalyst.

*Source:* https://external-content.duckduckgo.com/iu/?u=https%3A% 2F%2 Fimage3.slideserve.com%2F6753230%2Fslide2-l.jpg&f=1&nofb=1&ipt=a58f86ec9f1f6b2e1eca6590c9f3efbafef14c6bef970c20220416f9b3054219.

In addition, the structure at the nanoscale also influences surface area and porosity in significant manners. For instance, studies of the synthesis of magnesium oxide (MgO) nanostructures illustrate how external conditions like humidity can reshape nanoparticles into sheet structures with enhanced surface characteristics. This process displays how water incorporation leads to nanosheets and nanorods that have higher surface area, which can undergo constructive surface reconstructions. These conversions contribute to the overall effectiveness of the material in adsorption processes by making active sites more efficient and accessible since water molecules facilitate the conversion. This comprehensive information regarding the importance of surface area and porosity in materials highlights their significance in designing and utilizing effective adsorbents and catalysts in different industries (Nandiyanto et al., 2021; Thomele et al., 2021).

## 3.3. FACTORS AFFECTING ADSORPTION

The adsorption process is highly influenced by a host of factors that regulate the interaction between the adsorbents and adsorbates. Among them is the pH of the solution, which alters the charge on the adsorbent substance and thus its affinity for specific ions. For instance, higher acidity would enhance the adsorption of anionic dyes by higher electrostatic attraction, while the cationic dyes would require higher pH for optimal holding. This is a condition that must be controlled with great precision under the regulation of pH in adsorption operations because it influences the charge balance and the chemical stability of the adsorbent interface and, therefore, affects the overall effectiveness of the adsorption process. Understanding these dynamics is critical to optimizing conditions for adsorption in most applications, including wastewater treatment and environmental remediation.

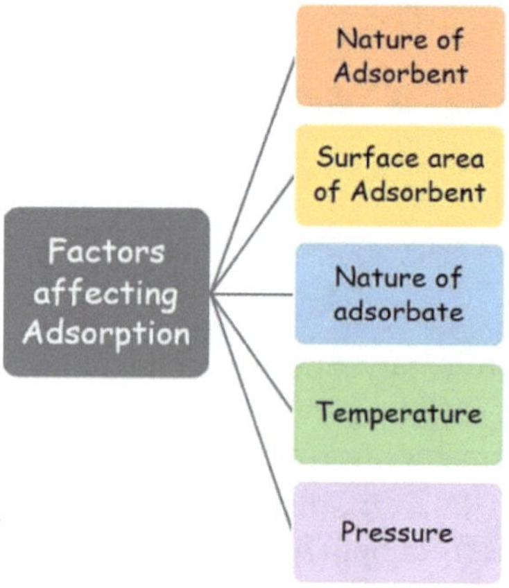


**Figure 3.9.** Factors affecting adsorption capacity.

*Source:* https://external-content.duckduckgo.com/iu/?u=https%3A%2F%2Fbiodifferences.com%2Fwp-content%2Fuploads%2F2018%2F03%2FFactors-affecting-adsorption.jpg&f=1&nofb=1&ipt=109eb45ca378cdf948a4e74bafe4f5a1ee80f611c3238d8ba7cdc11b10a393c7.

The dosage of the involved adsorbent material is another critical variable that influences adsorption. Adsorption capacity can be enhanced with increased dosage of the adsorbent material because more sites become available for analyte attachment. But beyond a certain point, the ratio is one of diminishing returns, in which more adsorbent no longer means proportionate rises in adsorption capacity. This is due to the fact that once saturation of available binding sites is apparent, more than a sufficient amount of adsorbent can result in clumping of particles or inefficient dispersion in the solution, impeding meaningful contact between adsorbent and contaminants. The determination of adsorbent dosage is thus very important in the design of efficient adsorption systems because it maximizes operating cost and process efficiency (N'diaye et al., 2022). Contact time is also a very important parameter influencing the adsorption process. The contact duration between an adsorbent and an adsorbate determines the extent to which impurities are removed from a solution. Initial instant capacity to adsorb typically occurs due to a high gradient of concentration, with increasing deceleration as fewer binding sites are available. Optimal time of contact to achieve effective adsorption is influenced by a number of factors, including type of adsorbate, temperature, and physical properties of the adsorbent. Timing of interactions can further prevent the process of adsorption and result in suboptimal removal efficiencies of desired pollutants. Thus, it is important to exhaustively examine the best contact time to ensure high adsorption efficiency in applications to make it a significant factor together with pH and dosage in the overall framework of adsorption phenomena (Abed et al., 2023) (Figure 3.9).

### 3.3.1. Temperature and Its Impact

Temperature is a crucial parameter in the adsorption processes, influencing the kinetic as well as the thermodynamic conditions of the process. On raising the temperature, the kinetic energy of adsorbate molecules also increases, leading to higher molecular motion. Such enhanced motion will enhance the interaction between the adsorbent and adsorbate surfaces and, in general, lead to higher rates of adsorption. But temperature rising also tests adsorption capacity since the majority of adsorption processes are endothermic in nature, in which more thermal energy can develop strength over the adhesive forces that keep the adsorbent onto the adsorbent.

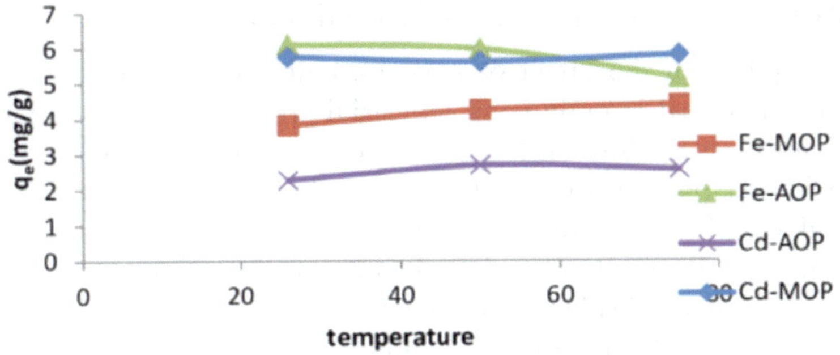


**Figure 3.10.** Effect of temperature on adsorption process.

*Source:* https://www.researchgate.net/figure/Effect-of-temperature-on-adsorption-process_fig1_310581791.

Hence, the proper understanding of the temperature dependency of adsorption plays a central role in achieving the maximum efficiency of most applications, including pollutant removal and gas storage. Additionally, the effects of temperature on adsorption extend to advanced material synthesis and design like that of metal-organic frameworks (MOFs) and covalent organic frameworks (COFs). These new materials possess unique characteristics that can be tuned for applications through alteration of their structural properties and chemical compositions. Freund et al. (2021) highlights how the reticular strategy design within this field presents the potential for precise control over diverse properties with potential to synthesize materials that can perform well across diverse temperature regimes. Since such frameworks can be controlled, researchers are able to design adsorbents with certain performance abilities for fluctuating thermal conditions, widening the applications of MOFs and COFs in industrial contexts (Figure 3.10).

Lastly, the impact of temperatures on adsorption is crucial in the assessment of environmental effects of effluents, particularly in the case of dye pollutants. The efficiency of adsorbents in removing such pollutants can be regulated by alterations in temperature levels within water ecosystems. For instance, elevated temperatures may enhance the desorption process of adsorbed pollutants, leading to increased toxicity levels within ecosystems. Dutta et al. (2021) supports the screening of adsorptive removal processes based on specific thermal conditions so that interventions can be effective against ecotoxicological effects. Since temperature continues to fluctuate due to climate change, studies in this direction are crucial to mitigate any potential environmental hazard of dye pollutants and their removal through adsorption processes.

### 3.3.2. Pressure and Concentration Effects

An understanding of the effect of pressure and concentration on adsorption phenomena is critical in optimizing industrial processes such as catalysis and purification of gases. With an increase in pressure of a gas, the surface area available for adsorption is saturated more rapidly since the increased frequency of molecular collisions with the solid surface leads to greater intensity of adsorption. This effect not only raises the rate of adsorption but may influence the quantities and species being adsorbed. Similarly, the concentration of the surrounding phase adsorbate plays an important role in adsorption kinetics. Higher concentrations tend to bring a greater number of available molecules to occupy the sites of adsorption, hence higher efficiency of the adsorption process overall. This coupling between pressure and concentration reinforces the fine balance possible for determining the efficacy of adsorption under various environmental conditions. The process of adsorption is further complicated by the introduction of noise and concentration fluctuation, which can cause transitions among different states of chemical reactivity. In catalysis situations, as specified in current studies, the introduction of controlled noise by different concentrations can introduce fascinating transitions to control the outcome of reactions. There is an experimental observation that the addition of minute concentration fluctuations can stabilize quicker reaction rates, while higher fluctuations may cause reversible transitions among different adsorption states. These findings imply that pressures and concentrations not only set the initial adsorption characteristics but also influence the stability and dynamics of the reaction pathway, fashioning the broader context of chemical reactions in a subtle way (Cisternas et al., 2018). Pressure and concentration effects are pivotal in membrane reactors, particularly in those that are utilized in hydrogen production. The functioning of such reactors is often based on the precise regulation of gaseous reactants within the system. From experiments, it has been observed that the change in reactant concentrations may significantly alter the transport mechanisms and adsorption phenomena at the membrane interface. Moreover, the change in pressures can enhance or inhibit the efficiency of separation according to the membranes selective permeability towards different gases. Understanding these interdependencies is crucial as any simplification of modeling the adsorption process may have consequences in conceptual discord for reactor design and operation (Annesini et al., 2018). Accordingly, ongoing research into the complex interaction of pressure and concentration in adsorption processes remains of utmost importance in the advancement of theoretical models and practical applications.

### 3.3.3. Nature of the Adsorbate and Adsorbent

The physical and chemical nature of the adsorbent and adsorbate is important in determining the mechanisms of adsorption processes. The adsorbate, or the substance which is to be adsorbed, can vary significantly in its physical and chemical nature, including its size, charge, and polarity. These conditions decide how well an adsorbate would be able to interact with the adsorbent, which would typically be a solid surface. For example, polar adsorbates would be more readily removed by polar adsorbents because there is similar intermolecular interaction, while nonpolar adsorbates would be better placed with nonpolar adsorbents. In addition, size compatibility is also a significant consideration: large molecules of the adsorbate might face steric hindrance that prevents them from reaching the surface of the adsorbent and would therefore affect the efficiency of the adsorption process. In contrast to adsorbates, the properties of adsorbents also have significant effects on the adsorption properties. The surface area, pore size distribution, and surface charge of the adsorbent can facilitate or inhibit the adsorption process. A higher surface area provides more active sites for adsorption, thus the potential to adsorb adsorbates. The chemical composition of the adsorbent also influences its selectivity towards various adsorbates. Adsorbents such as activated carbon, silica gel, and zeolites have unique structures that allow selective adsorption based on the characteristic of the adsorbates. Recent research has shown that thermal treatment and pH level can alter the surface characteristics and morphology of adsorbents to lead to variations in their adsorption capacity (Matijevic et al., 2025). The nature of adsorbent and adsorbate interaction encompasses the dynamics of adsorption processes too. Temperature, pressure, and presence of competitive species affect adsorption behavior to a considerable extent. For instance, an increase in temperature will enhance the kinetic energy of adsorbate molecules, which may enhance the rate of adsorption to a saturation level. Apart from interactions among different adsorbates—competitive or cooperative—adsorption efficiency may also be altered. Such sophistication means that neither the adsorbent nor the adsorbate should be looked at separately but as part of a concerted system that determines the overall effectiveness of adsorption. These relationships need to be understood to ensure optimization in applications across various areas like environmental remediation and industrial operations (Koumpouras et al., 2008).

## 3.4. ADSORPTION ISOTHERMS

The study of adsorption isotherms forms a part of the knowledge of the adsorption phenomena, offering details about the interaction between adsorbate and adsorbent. Adsorption isotherms express the relation between mass adsorbed on the surface of a solid and its concentration in the liquid phase in equilibrium. Some of the models like Langmuir and Freundlich isotherms are conceptual models to examine such relations. For instance, Langmuir isotherm works in terms of equal adsorption sites without interaction among adsorbate molecules, and hence it is best for systems where monolayer adsorption occurs. The Freundlich isotherm is compatible with heterogeneous surfaces involving higher complexity in adsorption interactions compared to the Langmuir isotherm. Comprehension of these models not only increases knowledge about adsorption mechanism but also directs formulation of efficient methods for various applications like waste treatment and biopharmaceutical separation processes. The effectiveness of an adsorbent is integrated with the adsorption characteristics in a complex manner and relies on a number of operating parameters. These involve pH, adsorbate initial concentration, dose of adsorbent, and contact time, all of which play the key part in the effectiveness of the adsorption processes. Research has shown, for instance, that the adsorption of dyes from aqueous solution is greatly affected by pH, and lower pH levels tend to favor anionic dye adsorption. In addition, the rising adsorbent dosage typically contributes to increased adsorption capacity due to greater availability of active sites for interaction with the dye molecules. Contact time also influences the adsorption kinetics such that increased contact time allows for stronger interactions between the adsorbent and adsorbate surfaces. Careful analysis of the aforementioned factors is crucial for optimizing adsorption isotherms and their usefulness in environmental remediation and industrial processes like those explored in N'diaye et al. (2022). In more advanced applications, multi-component adsorption isotherms are crucial in describing the complex dynamics of multi-adsorbate systems. Such systems tend to be prevalent in biopharmaceutical applications where several proteins need to be separated simultaneously. The multi-faceted nature of mass transfer and adsorption under such circumstances demands novel methods for precise modeling. Novel technologies, such as high-throughput screening approaches integrated with capillary electrophoresis microfluidics, have revolutionized multi-component adsorption isotherm determination in recent years. These technologies perform flawlessly with low quantities of samples and reagents and give extremely high accuracy to measure component concentration levels. Their application in protein studies like bovine serum albumin is a pointer to the role of sophisticated models in the explanation of adsorption at an industrial level, as set by Lietta et al. (2021). Understanding multi-component adsorption is vital in the optimization of separation processes as well as the enhancement of efficiency in purification steps (Figure 3.11).

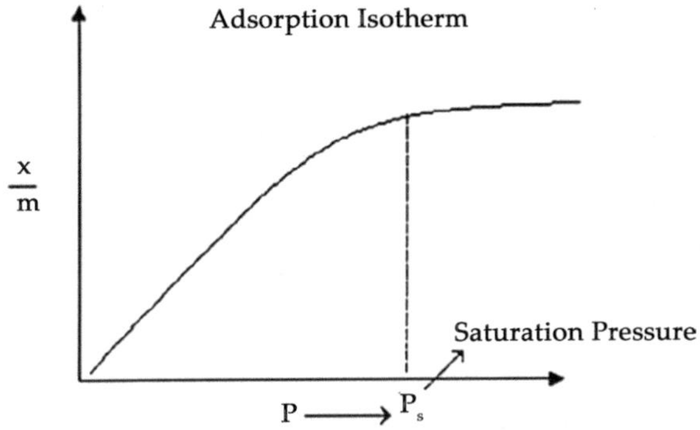


**Figure 3.11.** Adsorption isotherm.

Source: https://external-content.duckduckgo.com/iu/?u=https%3A%2F%2Fchemistnotes.com%2Fwp-content%2Fuploads%2F2022%2F04%2Fbasic-adsorption-isotherm.png&f=1&nofb=1&ipt=438280d1ab86c1bb17b07ffaf81957e4e603517d4924ea4956ad4f3204bf3420.

## 3.4.1. Definition and Significance of Isotherms

The concept of isotherms is fundamental to adsorption phenomena as the amount of adsorbate adsorbed on a solid adsorbent against the concentration in the surrounding phase at a specified temperature. Isotherms assist in illustrating the way adsorbate molecules distribute themselves on the solid surface and the fluid phase, thus providing a quantitative estimate of adsorption capacity. Various isotherm models, such as Langmuir and Freundlich, describe this relationship with different assumptions and interpretations. Isotherms go beyond theoretical study since they are at the core of practical applications across industries from environmental engineering to chemical manufacturing. By correctly modeling such relationships, scientists can predict the performance of adsorbents in processes such as filtration, separation, and pollutant removal, thus contributing to designing more efficient systems for industrial processes. In the context of research on adsorption, isotherms provide crucial information regarding mechanisms that take place during adsorption. For instance, they are able to offer insights into the nature of interactions between adsorbents and target molecules and aid in determining adsorption sites and energies.

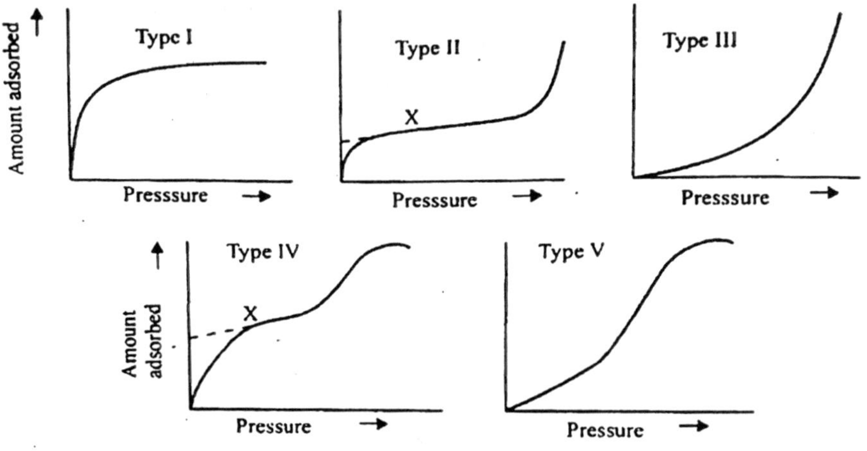


**Figure 3.12.** The adsorption phenomenon involves five different forms of adsorption isotherm.

*Source:* https://chemistnotes.com/physical/adsorption-isotherm-and-its-types/.

This information is critical in optimizing the selection of materials used in various applications; for example, adsorption of p-Nitrophenol (PNP) on activated carbons plays a significant role in wastewater treatment, where performance can be modelled and optimized using accurately defined isotherm equations such as the Double-Langmuir model and the Dubinin–Radushkevich equation. By correct fitting of experimental data to such models, researchers are able to gain valuable information about not only the equilibrium concentration, but adsorption kinetics as well, ultimately improving the efficiency of treatment processes (Ledesma et al., 2023). Isotherm analysis also plays important implications for the establishment of theoretical frameworks and practical procedures in the science of adsorption. The quantity of quantitative information provided by isotherms allows one to assess the adsorption capacity under varied conditions, leading to greater optimization of adsorption systems. The research into gas adsorption, for example, methane and nitrogen in MIL-101, demonstrates the fine balance of molecular interactions and structural organization controlling the behavior of adsorbates. High-resolution investigations, including Monte Carlo and molecular dynamics simulations, have been able to capture such complexity by illustrating how structural features influence selectivity and diffusion within adsorbent structures (Shao et al., 2024). Lastly, the value of isotherms is their ability to occupy the gap between theoretical predictions and practical application, making them inevitable resources for knowledge development and innovation in adsorption technologies (Figure 3.12).

### 3.4.2. Langmuir Isotherm Model

The Langmuir isotherm model is pivotal in the study of adsorption phenomena, particularly the manner in which adsorbates adsorb onto the surface of a solid adsorbents. The model is premised on the knowledge that adsorption occurs on specific homogeneous locations in the adsorbent where the same location can hold one molecule of an adsorbate.

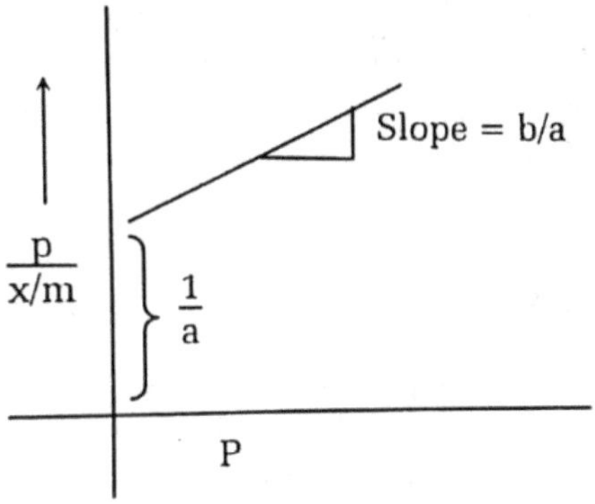


**Figure 3.13.** *Langmuir adsorption isotherm graph.*

*Source: https://chemistnotes.com/physical/langmuir-adsorption-isotherm-assumptions-derivation-and-limitations/.*

When such positions are occupied, further adsorption does not occur. This model is applicable to all kinds of systems in the analysis of many various systems and hence is crucial for forecasting the effectiveness of adsorbents in actual applications. Recently, in recent studies, superiority of the Langmuir isotherm in the modeling of adsorption processes has been proven using empirical data. For instance, one of the biosorption studies used walnut husk and was indicating that the iron sorption potential was increasing with temperature as predicted by the Langmuir model quite well. Under optimal conditions, this study indicated that the maximum removal rate was an impressive 93.28%. Regression coefficients based on experimental data were generating strong support for the Langmuir model over other isotherm models such as Freundlich. These findings demonstrate the capability of the Langmuir isotherm to describe adsorption behavior well, especially in systems with low active sites and homogeneous energy distribution on the surface of the adsorbent, like in iron adsorption using walnut husks (Ihab et al., 2025) (Figure 3.13).

In addition, the Langmuir isotherm model has been used successfully in the instance of novel adsorbents, hence reinforcing its relevance in the adsorption studies. For example, recent research on the use of a composite material—nano-$Fe_3O_4$/pomegranate peel biochar/alginate hydrogel beads—demonstrated the efficacy of the model in describing the adsorption of Cr(VI) ions from solutions. The experiment displayed a highest adsorption capacity of 303.03 mg/g for Cr(VI), indicating the applicability of the models to different materials and contaminants. The studies also demonstrated kinetic modeling, in particular the pseudo-second-order model, to be in accordance with the adsorption kinetics encountered. These facts not only confirm the validity of the usage of the Langmuir isotherm in contemporary adsorption science but also guide the design of new materials that can remove contaminants from water bodies more efficiently (Parlayıcı et al., 2025).

### 3.4.3. Freundlich Isotherm Model

The Freundlich isotherm equation is one of the core equations employed to describe adsorption processes, particularly when surfaces are heterogeneous. The model is empirical, and it considers the fact that the adsorption process occurs on surfaces of different affinities towards the adsorbate, hence making it a better representation than simpler models such as the Langmuir model. Freundlich isotherm is typically expressed in logarithmic form and indicates that the relationship between adsorbate concentration and adsorption capacity is nonlinear. This nonlinearity is particularly useful when used for actual applications, wherein adsorption capacity could differ independently over varying concentrations. The model has been thoroughly tested by analytical studies, which have recognized its strength in addressing the complexity involved in the adsorption of varied impurities, such as organic impurities in wastewater treatment systems, thus setting its applicability in environmental engineering contexts. In practical applications, Freundlich model has also been instrumental in determining the adsorption capacity of different materials, including those used in biological treatment systems. For instance, studies in Adsorbable Organic Halide (AOX) adsorption on activated sludge have used Freundlich isotherm to explain batch adsorption data. Though the model will not consistently provide best fit for all fraction of contaminant, it is valuable for assessment of adsorption trends over a broad range of molecular weights. Research indicates growing AOX adsorption affinity with higher molecular weights, which demonstrates the models capability to suggest dynamic competition for adsorption sites between various contaminants (Figure 3.14).

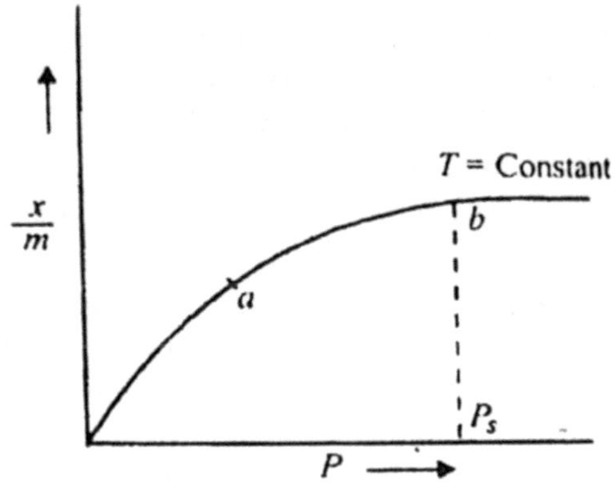


**Figure 3.14.** Freundlich adsorption isotherm.

*Source:* https://chemistnotes.com/physical/freundlich-adsorption-isotherm-equation-and-limitations/.

These findings emphasize the effectiveness of the Freundlich isotherm in the optimization of treatment strategies and the acquisition of a clearer understanding of the influence of specific characteristics on adsorption behavior in multifaceted waste streams, exemplified by investigations of different bleaching sequences on the removal of AOX. While the Freundlich isotherm model effectively depicts numerous of the processes that occur during adsorption, it is far from perfect. For example, it does not justify the saturation effects that may occur at high concentrations, where the Langmuir isotherm would be more descriptive. The most effective application of the Freundlich model typically justifies low concentration conditions when adsorption sites are favorably available. In addition, competitive adsorption is common in different systems, such as concurrent adsorption of phenol and p-hydroxybenzoic acid on activated carbon. Such a situation, although the Freundlich model could be of assistance, complex models considering competitive adsorption kinetics would be more accurate, according to research investigating the detail of adsorption interactions. Consequently, although the Freundlich isotherm remains a part of the understanding of adsorption phenomena, coupled model research is essential to advance the field and achieve effective resolution to environmental problems.

## 3.5. APPLICATIONS OF ADSORPTION

Uses of adsorption are extensive and unavoidable in various scientific and industrial fields, particularly environmental decontamination. One of the significant ones is water purification, where heavy metal ions are a huge risk to human life and the ecosystem. The majority of the traditional purifications of water are not efficient enough to remove the impurities, and therefore newer technologies using adsorption have gained prominence. Nanomaterials and other advanced materials are established effective adsorbents owing to the fact that they possess high surface areas that enhance their interaction with impurities. The effectiveness of iron oxide nanoparticles, for example, illustrates that one of the adsorption processes can be employed to eliminate heavy metals from water and thereby present a potential solution to water quality (Barbero et al., 2024). Apart from water purification, adsorption also plays an important role in other technological processes, such as gas separation and catalysis. In such processes, the selective adsorption process enables the separation of desired substances from mixtures, a process which is core to chemical manufacturing and petroleum refining industries. The adsorbent structure tailored for specific systems matters; materials can be optimized at the molecular level to ensure that their surface properties are optimized for specific applications. A case in point is the use of fractional calculus in the analysis of gas molecule diffusion, which clarifies the connection between adsorption kinetics and material properties, and eventually to a better understanding of the physical phenomena involved (Barbero et al., 2024). This mathematical model not only helps optimize existing technologies but also gives directions for designing new ones. Moreover, the application of adsorption transcends technological limits, to pharmaceuticals and veterinary medicine, where drug delivery and performance depend upon it. Adsorption impacts drugs' interactions with living systems and influence their absorption, distribution, and overall therapeutic actions. The ongoing advancements in veterinary pharmacology highlight the significant potential of adsorption in formulating the bioavailability of administered drugs to animals. By applying the mechanisms of adsorption, researchers can develop methods towards the formulation of drugs that ensure active ingredients to be adequately absorbed within biological tissue. In addition, adsorption in drug delivery systems offers the potential for targeted treatment, maximizing therapeutic effect while minimizing side effects, thereby reflecting the versatility and importance of the phenomenon across a number of disciplines (Prasad et al., 2024).

### 3.5.1. Environmental Remediation

Environmental remediation is an essential method of addressing soil and water pollution and involves sound means of offsetting environmental risk. Solidification/stabilization (S/S) is one well-established method used during the process that successfully transforms pollutants into less mobile forms. The technique depends on the application of cementitious materials that chemically transform pollutant composition, and therefore reduce their solubility and mobility. Besides limiting contaminants, the S/S process also facilitates recycling of treated products to be reused in construction, thus promoting sustainability. Furthermore, the cost-effectiveness and simplicity of application of S/S technology account for its popularity in contaminated site management, which further enhances its significance in environmental remediation in general (Goh et al., 2008).

Another innovative use in environmental restoration is the application of electrokinetic (EK) techniques, which take advantage of electrical fields to mobilize contaminants for removal. The technique has been viewed with interest for its ability to address heavy metal contamination in sediments by optimizing the treatment process through real-time monitoring. Recent advancements with applications of frequency-domain time-lapse measurements allow researchers to track the electrokinetic transportation of contaminants, yielding data about physical changes within the environment during treatment. By taking measurements of such parameters as resistivity, researchers can ascertain the effectiveness of various conditioning agents and modify the remedial approach for optimum results.

According to research, the interaction between the resistivity variation and parameters such as pH and ionic concentration is of utmost significance, indicative of the prospect of EK techniques in enhancing environmental remediation processes (Cereato et al., 2020). The complex interaction among different remediation strategies identifies the significance of managing environmental pollution effectively. S/S and EK techniques indicate the prospect of technology advancements to optimize environmental remediation processes monitoring and application. The addition of techniques such as adsorption phenomena also further enhances our understanding of contaminant behavior and response to treatment. As the science progresses, however, examination of novel materials and new technologies will be essential to further improving remediation outcomes. Ultimately, through leveraging a range of techniques for environmental remediation, we can work toward the security and sustainability of our ecosystems with minimized effects of legacy contamination on human and ecological health.

### 3.5.2. Catalysis in Chemical Reactions

The catalytic role in chemical reactions is fundamentally related to the process of adsorption. Catalysis provides a surface that facilitates the adsorption of reactants and thereby lowers the energy of activation for a reaction to occur. Such interaction is vital as the efficiency of the catalytic process is dependent on the strength and nature of the bonds being formed between the catalyst surface and reactants. With adsorption, the reactants undergo a process usually characterized by a reorganization of electronic structures, which can be optimized for increased catalytic activity. Manipulation of orbital alignment and adsorption kinetics has been shown to be an important factor in controlling the reaction yield, with a strong correlation found between surface chemistry and catalysis (Khurshid et al., 2024).

Thus, understanding of adsorption dynamics is essential to create more effective mechanisms for catalysis. Adsorption phenomena in catalysis are also underscored by the recent findings on hot electrons (HEs) which facilitate catalytic mediation of reactions through mechanisms beyond conventional approaches. HEs could accelerate reaction rates by accessing inaccessibility otherwise, but their effective utilization remains an area of research. Current studies show that parameter tuning such as electronic orbital alignment can be linked to spectacular improvement in catalysis, five-fold increases in some reactions (Khurshid et al., 2024). This highlights the intricate relationship between electronic action and adsorption kinematics. These developments not only illuminate the intrinsic mechanisms that govern catalysis but also open up new avenues for catalyst design with a multi-purposing, environmentally friendly emphasis towards chemical applications, ultimately towards more energy-conserving processes.

Furthermore, recent studies of surface Van Hove singularities (SVHS) indicate that these exotic electronic states have significant effects on catalytic activity by optimizing the adsorption characteristics of a broad class of reactants. For example, the graphene/$Ca_2N$ heterojunction and $Pt_2HgSe_3$ (001) surface both show how SVHS may act as an electron bath to facilitate the adsorption of oxygen and subsequent reaction processes with very low energy barriers. These active electronic states not only promote CO oxidation but also assist in hydrogen evolution reactions, which indicates their universal catalytic ability. Furthermore, the tunability of SVHS allows for fine-tuning of chemical reactivity, thus bridging the gap between computational investigations of adsorption processes and practical catalytic applications. Unlocking and tapping into such unique electronic properties can thus revolutionize the catalyst design and development field.

### 3.5.3. Separation Processes in Industry

Industrial separation operations are of basic significance to guarantee the efficiency and selectivity of various operations, particularly those with adsorption processes involved. In a majority of industrial procedures, information regarding the interaction between solid substrates and adsorbates is of main significance to further improve separation processes. Above all, the polar and dispersive character of solid materials can irrevocably influence their adsorption ability. For instance, the performance of adsorption-based separation processes can be enhanced through modification of surface properties of adsorbent materials.

As outlined in research, evaluation of polar surface free energy and Lewis acid-base character assists in acquiring a more comprehensive understanding of adsorption phenomena and facilitates forward progress in sectors such as coatings, adhesion, and filtration systems (Hamieh, 2024). Physico-chemical characterization of solid substances by chromatographic methods represents the intricate relationship between adsorption phenomena and physico-chemical properties. Employing various chromatographic methods in the last four decades has succeeded in satisfactorily interpreting adsorbent retention times, thus allowing for evaluation of variable specific surface variables like dispersive and polar characteristics. These observations are vital to the optimization of adsorption processes since variations in retention time can bring about varying results on different solid substrates.

The application of a novel method based on London dispersion forces enhances significantly the precision of the measurement of polar free energy and adsorption constants for components such as silica and alumina (Hamieh, 2024). This advance in methodology has an immediate effect on product design and quality assurance in industries involving separation process-based manufacturing. Lastly, the constant pursuit of better separation processes in industry shows the continuance of developing knowledge regarding adsorption processes. The application of new characterization techniques allows for greater control and comprehension of solid surfaces and results in enhanced performance in separation processes. Systematic study of the enthalpy and entropy of common adsorbates provides valuable information to industries that aim to optimize processes such as catalysis, waste treatment, and resource recovery. With more efficient processes, industries become more efficient and selective, which are essential requirements for competitiveness in today's rapid-changing market. Thus, the field of adsorption research significantly contributes to separation technologies and emphasizes ongoing research in the field (Hamieh, 2024).

## SUMMARY

The study of adsorption processes emphasizes the delicate interactions that are the cause of how molecules adhere to the surface and has applications in a broad range of sciences. In examining the specific interactions of amino acids with metallic nanoparticles, the research emphasizes the relevance of molecular recognition processes that form the foundation of bioinorganic and supramolecular chemistry. The computational research unraveled the weak but significant interactions underlying the stability of the amino acid complexes with gold or silver nanoparticles, offering a window for future investigation of functionalized materials and conditions for stability (Nicolau et al., 2025). The fragility of the delicate balance of these interactions showcases the benefit of surface engineering for specified applications in biotechnology, nanotechnology, and material science. Moreover, the outcome of fluid dynamics in semiconductor manufacturing is to emphasize further the relevance of adsorption phenomena in practical applications. As demonstrated through the spinning liquid film experiments, wave dynamics and turbulence contribute to molecular rates of transport between the wafer surface and the liquid, which in turn influence the quality of the devices. It is significant to comprehend such behaviors as advection and diffusion processes regulate cleaning and etching methods, which are fundamental in semiconductor manufacturing. Enhancing these fluid dynamics in combination with adsorption effects enables researchers to enhance manufacturing processes to the maximum, thus enhancing precision and efficiency in this essential technological sector (Belmiloud, 2024). In conclusion, the interface of adsorption phenomena intersecting molecular interactions along with fluid dynamics provides a multidimensional context required to create different scientific fields. As research continuously dissects the intricacies of these phenomena, particularly within amino acid and nanoparticle interactions and fluid dynamics in industrial application, there lies a potential for tremendous technological advancement. This ongoing research not only brings theoretical understanding forward but also spurs innovative techniques in material synthesis and effective manufacturing procedures. The results of such research serve as the basis for future advancement that is certain to further efficiency, quality, and performance in various applications in science and engineering.

Understanding adsorption processes requires a careful examination of the underlying principles that direct interaction between adsorbates and adsorbents. Essentially, adsorption is a situation where molecules from a gas, liquid, or dissolved solid stick onto the surface, creating a concentration gradient. Adsorption can be described in terms of two main categories: physisorption that relies on weak van der Waals forces and chemisorption that relies on chemical bonds. Effectiveness is also based on a variety of factors including surface area, temperature, and the nature of adsorbent material. As investigation in the field advances, the intensive findings are yet to emerge that unveil the complexities involved in adsorptive interactions and their applications in industry across domains. Coupled with the theoretical concepts, the nature of adsorption phenomena is conclusively determined by a sequence of theoretical models. Of these, the Freundlich and Langmuir isotherms are of key importance in describing the correlation between the concentration of the adsorbate and the amount of substance adsorbed on the surface. Langmuir theory relies on monolayer capacity of adsorption, while Freundlich theory accounts for multilayer adsorption and is more adaptable to empirical use. An understanding of these models is useful in predicting the behaviors of different materials under adsorption, which is crucial for the optimization of processes in fields such as environmental science and materials engineering. Recent studies have indicated the progress in the development of these models as they grow to accommodate newer intricacies, thereby reflecting a continued growth of the knowledge base regarding the dynamics of adsorption into a widening scientific framework, as postulated by Atasoy (2024). Also, the engineering applications of adsorption effects are enormously vast, pointing to its importance in industrial processes and environmental clean-ups. Adsorption techniques have immense uses in wastewater treatment, air purification, and even catalysis, where molecule selectivity plays a crucial role. Progress in material science, e.g., the design of sophisticated adsorbents with tailored qualities, keeps boosting the efficiency of adsorption processes. These advances call for a reevaluation of current practices and challenge researchers to explore new avenues of inquiry. The recognition of challenges posed by these phenomena, particularly in terms of considerations such as surface heterogeneity and kinetics, is still fundamental to ensuing work. Overcoming these challenges is paramount to ensuring the further evolution of adsorption technology and applying it to solving global environmental problems, which aligns with the findings described in Ducker (2024).

# REFERENCES


1. Abdulsattar, M. A., Khalil, S., & Mahmood, T. H. (2023). Adsorption of $Cl_2$ on pristine and Pd doped $TiO_2$ clusters: A transition state theory study. *Passer Journal of Basic and Applied Sciences*. https://www.semanticscholar.org/paper/33e4fab522868e944f6bdaa88d9cdafa690391da (accessed on 3rd October 2025).
2. Abed, A., Faisal, A., & Hussein, A. (2023). Modeling and simulation of copper removal from the contaminated soil by a combination of adsorption and electro-kinetic remediation. *Journal of Engineering*. https://www.semanticscholar.org/paper/0e3db46290c7f8fa9e82915297aeacb77b7f7f91 (accessed on 3rd October 2025).
3. Albasi, C., Bacchin, P., Devatine, A., et al. (2011). Cross-flow microfiltration applied to oenology: A review. https://core.ac.uk/download/12042631.pdf (accessed on 3rd October 2025).
4. Albasi, C., Bacchin, P., Devatine, A., et al. (2012). Analysis of membrane fouling during cross-flow microfiltration of wine. https://core.ac.uk/download/12043964.pdf (accessed on 3rd October 2025).
5. Altameemi, I. A., & Sadaa, A. M. (2025). Thermodynamic analysis of activated walnut husks in the adsorption of iron (III) ions from aqueous solutions: An assessment of feasibility and sustainability. *Academia Open*. https://www.semanticscholar.org/paper/96474a62cd15cd0c60fc32d8c45dc79884819453 (accessed on 3rd October 2025).
6. Andriantsiferana, C., Creanga-Manole, C., Delmas, H., et al. (2013). Competitive adsorption of *p*-hydroxybenzoic acid and phenol on activated carbon: Experimental study and modelling. https://core.ac.uk/download/12043979.pdf (accessed on 3rd October 2025).
7. Annesini, M. C., Cerbelli, S., & Murmura, M. A. (2018). Modeling fixed bed membrane reactors for hydrogen production through steam reforming reactions: A critical analysis. https://core.ac.uk/download/159268736.pdf (accessed on 3rd October 2025).
8. Ascoli, A., Menzel, S., Rana, V., Kempen, T., Messaris, I., Demirkol, A. S., Schulten, M., et al. (2022). A deep study of resistance switching phenomena in TaOx ReRAM cells: System-theoretic dynamic route map analysis and experimental verification. *Advanced Electronic Materials, 8*. https://www.semanticscholar.org/paper/43177852f2e37affe65d647e098c16ec0782c1be (accessed on 3rd October 2025).
9. Atasoy, A. (2024). Sociology and tourism: A systematic literature review. *Journal of Tourismology*. https://www.semanticscholar.org/

paper/ef31ae83623a1e3831980bc193bd77361451e776 (accessed on 3rd October 2025).

10. Barbero, L. R., Evangelista, R. S., Zola, E., Lenzi, A., & Scarfone, A. (2024). A brief review of fractional calculus as a tool for applications in physics: Adsorption phenomena and electrical impedance in complex fluids. *Fractal and Fractional*. https://www.semanticscholar.org/paper/9405c9ac19f82fefa945780e70b48fcbce901397 (accessed on 3rd October 2025).
11. Belmiloud, N. (2024). Characterizing turbulence and wave dynamics under rotation: Enhancing semiconductor processes through CFD analysis and high-speed imaging. *ECS Meeting Abstracts*. https://www.semanticscholar.org/paper/36dacca4f686cd758707fd6d7c558f05390b382a (accessed on 3rd October 2025).
12. Berland, K., Einstein, T. L., & Hyldgaard, P. (2009). Rings sliding on a honeycomb network: Adsorption contours, interactions, and assembly of benzene on Cu (111). http://arxiv.org/abs/0907.4291 (accessed on 3rd October 2025).
13. Canossa, S., & Wuttke, S. (2020). Functionalization chemistry of porous materials. *Advanced Functional Materials, 30*. https://doi.org/10.1002/adfm.202003875 (accessed on 3rd October 2025).
14. Cercato, M., & De Donno, G. (2020). Time-lapse monitoring of an electrokinetic soil remediation process through frequency-domain electrical measurements. https://core.ac.uk/download/296244746.pdf (accessed on 3rd October 2025).
15. Cisternas, J., & Wehner, S. (2018). Detailed analysis of transitions in the CO oxidation on Palladium (111) under noisy conditions. http://arxiv.org/abs/1807.07219 (accessed on 3rd October 2025).
16. Cornelissen, E., Gutierrez, L., Post, J. W., et al. (2019). Key physicochemical characteristics governing organic micropollutant adsorption and transport in ion-exchange membranes during reverse electrodialysis. https://core.ac.uk/download/237011091.pdf (accessed on 3rd October 2025).
17. Delgado-López, J. M., Drouet, C., Gómez-Morales, J., et al. (2013). Progress on the preparation of nanocrystalline apatites and surface characterization: Overview of fundamental and applied aspects. https://core.ac.uk/download/12044395.pdf (accessed on 3rd October 2025).
18. Dengo, N., Gross, S., & Vittadini, A. (2017). Functionalization of colloidal transition metal sulphides nanocrystals: A fascinating and challenging playground for the chemist. https://core.ac.uk/download/141699569.pdf (accessed on 3rd October 2025).

19. Devi, M., Verma, I., & Pal, S. (2021). Distinct interfacial ordering of liquid crystals observed by protein-lipid interactions that enabled the label-free sensing of cytoplasmic protein at the liquid crystal-aqueous interface. *The Analyst*. https://www.semanticscholar.org/paper/43f97937d0b712ec8572581831d1f4675c05b46a (accessed on 3rd October 2025).
20. Dou, X., Mehwish, N., Zhao, C., Liu, J., Xing, C., & Feng, C. (2020). Supramolecular hydrogels with tunable chirality for promising biomedical applications. *Accounts of Chemical Research*. https://www.semanticscholar.org/paper/a960288367e70060f63a1ee10ac68f9234539fb5 (accessed on 3rd October 2025).
21. Ducker, N. (2024). Capturing the situated, dynamic nature of EMI-lecture listening comprehension in real time. *Journal of English-Medium Instruction*. https://www.semanticscholar.org/paper/ce044d0816153dd320f8ec24c58ffd32efd4c972 (accessed on 3rd October 2025).
22. Dutta, S., Gupta, B., Srivastava, S. K., & Gupta, A. K. (2021). Recent advances on the removal of dyes from wastewater using various adsorbents: A critical review. *Materials Advances, 2*, 4497–4531. https://doi.org/10.1039/d1ma00354b (accessed on 3rd October 2025).
23. Freund, R., Zaremba, O., Arnauts, G., Ameloot, R., Skorupskii, G., Dincă, M., Bavykina, A., et al. (2021). The current status of MOF and COF applications. *Angewandte Chemie International Edition, 60*, 23975–24001. https://doi.org/10.1002/anie.202106259 (accessed on 3rd October 2025).
24. Gajos, P., Petrou, A., & Budkowski, A. (2022). Comparison of physical adsorption and covalent coupling methods for surface density-dependent orientation of antibody on silicon. *Molecules, 27*. https://www.semanticscholar.org/paper/39d3bc152f93ea68a4ff6b8bec5d3f40ee649ac4 (accessed on 3rd October 2025).
25. Gapes, D. J. (1994). *The Adsorption of Adsorbable Organic Halide onto Biological Solids: A Thesis Submitted in Partial Fulfilment of The Requirements for The Degree of Master of Technology in Environmental Engineering*. https://core.ac.uk/download/148643381.pdf (accessed on 3rd October 2025).
26. Goh, B. M., & Yin, C. Y. (2008). Solidification/stabilization of soil contaminated with metal: A review. https://core.ac.uk/download/11235912.pdf (accessed on 3rd October 2025).
27. Hamad, M. T., & Ibrahim, S. (2024). Effective fabrication and characterization of eco-friendly nanoparticles composite for adsorption

Cd(II) and Cu(II) ions from aqueous solutions using modelling studies. *Scientific Reports, 14*. https://www.semanticscholar.org/paper/f937f03fc3ae8d8680cbf6a38d9376358a2658bd (accessed on 3rd October 2025).

28. Hamieh, T. (2024). New progress on London dispersive energy, polar surface interactions, and Lewis's acid–base properties of solid surfaces. *Molecules, 29*, 949–949. https://doi.org/10.3390/molecules29050949 (accessed on 3rd October 2025).
29. Hamieh, T. (2024). New progress on London dispersive energy, polar surface interactions and Lewis's acid-base properties of solid surfaces. https://doi.org/10.20944/preprints202401.0638.v1 (accessed on 3rd October 2025).
30. He, F., & Li, Y. (2022). Advances on theory and experiments of the energy applications in graphdiyne. *CCS Chemistry, 5*, 72–94. https://doi.org/10.31635/ccschem.022.202202328 (accessed on 3rd October 2025).
31. Khurshid, F., Muthu, J., Wang, Y. Y., Wang, Y. W., Shih, M. C., Chen, D. R., Lu, Y., et al. (2024). The importance of catalytic effects in hot-electron-driven chemical reactions. *ACS Nano, 18*, 34332–34340. https://www.semanticscholar.org/paper/12159f5162bb8a1087e973b7aa6ce91d46835405 (accessed on 3rd October 2025).
32. Koumpouras, G., & Koumpouras, G. (2008). *Mathematical Modelling of a Low-Temperature Hydrogen Production Process with In Situ $CO_2$ Capture*.
33. Ledesma, E., Sabio, C. M., González-García, S., Román, S., Fernández, M. E., Bonelli, P., & Cukierman, A. (2023). Batch and continuous column adsorption of *p*-nitrophenol onto activated carbons with different particle sizes. *Processes*. https://www.semanticscholar.org/paper/1786a6765fa96111e9c43b7f39c6f79b12bc6aa1 (accessed on 3rd October 2025).
34. Lietta, A., Pieri, A., Innocenti, E., Sparavigna, A. C., Vanni, M., & Barresi, A. (2021). Use of microfluidic capillary electrophoresis for the determination of multi-component protein adsorption isotherms: Application to high-throughput analysis for hydrophobic interaction chromatography. *Pharmaceutics, 13*. https://www.semanticscholar.org/paper/86e2e1a2db6eb3f3bd8a81bef2fcda299a39319c (accessed on 3rd October 2025).
35. Liu, L., Wang, C., Zhang, L., Liu, C., Niu, C., Zeng, Z., Ma, D., et al. (2022). Surface Van Hove singularity enabled efficient catalysis in low-dimensional systems: CO oxidation and hydrogen evolution reactions. *The Journal of Physical Chemistry Letters*. https://www.semanticscholar.

org/paper/c2111bc5795555fde5258e0ebc664271b7c6dafd (accessed on 3rd October 2025).

36. Maldonado, A., Villamagua, L., & Rivera, R. (2019). DFT analysis of the adsorption of phenol on the nonpolar (100) ZnO surface. *The Journal of Physical Chemistry C*. https://www.semanticscholar.org/paper/64e5a11458bf64a12f6ad42628107ee46a7967b0 (accessed on 3rd October 2025).

37. Matijevic, E., & Sapieszko, R. S. (2025). Formation of metallic and metal hydrous oxide dispersions. https://core.ac.uk/download/pdf/42855426.pdf (accessed on 3rd October 2025).

38. N'diaye, A., Kankou, M., Hammouti, B., Nandiyanto, A., & Al Husaeni, D. F. (2022). A review of biomaterial as an adsorbent: From the bibliometric literature review, the definition of dyes and adsorbent, the adsorption phenomena and isotherm models, factors affecting the adsorption process, to the use of *Typha* species waste as adsorbent. *Communications in Science and Technology*. https://www.semanticscholar.org/paper/ba750022b124b869f2b257d75641f483ec415934 (accessed on 3rd October 2025).

39. Nandiyanto, A., Hofifah, S. N., Inayah, H. T., Putri, S. R., Apriliani, S. S., Anggraeni, S., Usdiyana, D., et al. (2021). Adsorption isotherm of carbon microparticles prepared from pumpkin (*Cucurbita maxima*) seeds for dye removal. *Iraqi Journal of Science*. https://www.semanticscholar.org/paper/6c29a3b4abff3ea3ec61e01ed03cb82979b5f5ef (accessed on 3rd October 2025).

40. Nicolau, C., Burguera, S., Piña, M., & Bauzá, A. (2025). Noncovalent binding phenomena in the adsorption of amino acids on Ag/Au surfaces: A theoretical approach. *Physical Chemistry Chemical Physics: PCCP*. https://www.semanticscholar.org/paper/aad2a98e100b6148a0974e6bcebdc4d4465a08d8 (accessed on 3rd October 2025).

41. Parlayıcı, Ş., & Pehlivan, E. (2025). Highly efficient hexavalent chromium removal using nano-$Fe_3O_4$/pomegranate peel biochar/alginate composite as an advanced biosorbent. *Turkish Journal of Analytical Chemistry*. https://www.semanticscholar.org/paper/02181d9277a92fefeadb228e10e23f53fab66d35 (accessed on 3rd October 2025).

42. Perveen, S., Rasheed, M. A., Sana, S., Mumtaz, I., & Qamar, S. (2023). Theoretical analysis of a nonequilibrium transport model of two-dimensional nonisothermal reactive chromatography accounting for Bi-Langmuir isotherm. *ACS Omega, 8*, 3057–3077. https://www.semanticscholar.org/paper/4876d577b82147c8fc12728946c34b79719505e0 (accessed on 3rd October 2025).

43. Pham, B. L., Ranjbari, A., Tadjeddine, A., Dalstein, L., & Humbert, C. (2024). Sum-frequency generation spectroscopy at aqueous electrochemical interfaces. *Symmetry*. https://www.semanticscholar.org/paper/d9e03e6182f839b76d1c945a51c2fe94a169777d (accessed on 3rd October 2025).

44. Pimentel, A. H., Freire, M., Gómez-Díaz, D., & González-Álvarez, J. (2023). Preparation of activated carbon from pine (*Pinus radiata*) sawdust by chemical activation with zinc chloride for wood dye adsorption. *Biomass Conversion and Biorefinery*. https://www.semanticscholar.org/paper/1cd44387f7c6e14ea3b593dca468990286053b82 (accessed on 3rd October 2025).

45. Pourrahmani. (2024). Analyzing the roles of copper in carbon capture and sequestration using mean field density functional theory and scanning transmission electron microscopy. *ECS Meeting Abstracts*. https://www.semanticscholar.org/paper/8043073e92f132c07acefdca37fc37a73fcdc6a3 (accessed on 3rd October 2025).

46. Prasad, R., Prasad, N. R., Prasad, N., Prasad, R. S., Prasad, S., Shrivastav, O. P., Prasad, R. B., et al. (2024). A review on phenomenon of adsorption of inorganic materials: Applications in veterinary pharmacology and material sciences. *Advances in Analytic Science*. https://www.semanticscholar.org/paper/c681927ed096c649a26f58b459ed77391372c9c2 (accessed on 3rd October 2025).

47. Schlumberger, C., & Thommes, M. (2021). Characterization of hierarchically ordered porous materials by physisorption and mercury porosimeter—A tutorial review. *Advanced Materials Interfaces, 8*. https://doi.org/10.1002/admi.202002181 (accessed on 3rd October 2025).

48. Shao, Y., Wang, S., Huang, L., Ju, S., Fan, X., & Li, W. (2024). Adsorption and diffusion of $CH_4$, $N_2$, and their mixture in MIL-101(Cr): A molecular simulation study. *Journal of Chemical and Engineering Data, 69*, 4466–4482. https://www.semanticscholar.org/paper/34c238ed2ed12e17d4cf6d4e6808e81c4bec4dd1 (accessed on 3rd October 2025).

49. Sharma, R., Kar, P. K., & Dash, S. (2022). Correlating solvation-adsorption phenomena of some styrylpyridinium dyes on SDS modified silica surface in organic media. *Journal of Dispersion Science and Technology, 45*, 548–558. https://www.semanticscholar.org/paper/75f8ae596e18c4acd2d7079d6d8c9717af59a9bc (accessed on 3rd October 2025).

50. Straw, W. (2008). Principles of chromatography and separative techniques – Adsorption and partition chromatography. *Journal of the Society of Dyers and Colorists, 101*, 409–416. https://www.semanticscholar.org/pa

per/15fc4c88d673898476251327a5d5dd4d3291e8a3 (accessed on 3rd October 2025).

51. Thomele, D., Baumann, S. O., Schneider, J., Sternig, A. K., Shulda, S., Richards, R., Schwab, T., et al. (2021). Cubes to cubes: Organization of MgO particles into one-dimensional and two-dimensional nanostructures. *Crystal Growth & Design, 21*, 4674–4682. https://www.semanticscholar.org/paper/9c0b05abb02ec44294c47859c451019351ae75cd (accessed on 3rd October 2025).

52. Tkatchenko, A., Michaelides, A., Bruch, L. W., Carrasco, J., Stone, A. J., & Liu, W. (2014). Insight into the description of van der Waals forces for benzene adsorption on transition metal (111) surfaces. https://core.ac.uk/download/31202763.pdf (accessed on 3rd October 2025).

53. Wang, S., Pillai, H., & Xin, H. (2020). Bayesian learning of chemisorption for bridging the complexity of electronic descriptors. *Nature Communications, 11*. https://www.semanticscholar.org/paper/e76a8e3cc46539bef9866e6f788e18b6f14b9c9f (accessed on 3rd October 2025).

54. Wijayanti, E. E., & Solfarina, S. (2023). Comparison of gravimetric method with UV-Vis spectrophotometry for determination of reaction order in natural dyes adsorption process. *J-PEK (Jurnal Pembelajaran Kimia)*. https://www.semanticscholar.org/paper/1b6bc9d1d5b9d04ce7ffd4a6bfc33050d914e029 (accessed on 3rd October 2025).

CHAPTER 4

# SURFACE THERMODYNAMICS AND KINETICS

## Contents



## INTRODUCTION

Surface kinetics and thermodynamics are fundamental researches that describe the behavior of materials on the micro-scale, particularly in the use of materials for energy conversion and storage. On the surface of a material, complex interactions are taking place which influence processes such as adsorption, reaction kinetics, and phase transition. These surface reactions have a tendency to determine the efficacy and efficiency of energy devices, for example, batteries, fuel cells, and catalysts. Under these conditions, it is particularly relevant to study how thermodynamic stability and kinetic pathways can be optimized in order to improve performance. Combining thermodynamic aspects with the modeling of kinetics forms a strong basis for analyzing how surface properties affect the overall properties of materials under various operating conditions. Among the basic features of surface thermodynamics is that of non-equilibrium states, which are capable of affecting the morphology and efficiency of interfaces. For example, electron transfer reactions have been deemed crucial in controlling the interfacial morphology, which has a direct correlation with the efficiency of energy harvesting and storage devices. With the help of non-equilibrium thermodynamic stability theory, one would be able to predict under what circumstances these morphologies are changed and optimized. Furthermore, with the aid of models such as the Butler-Volmer equation, one is able to represent electrochemical reactions as coupled ion-electron transfer kinematics. This coherent approach not only enriches our understanding of surface processes but also suggests the prospect for the design of tailor-made interfaces for specific applications, as noted by Bazant et al. (2020). At the same time, the study of diffusion processes is important to comprehend the kinetics of multicomponent systems. Diffusion, while intuitively straightforward, becomes increasingly complex in systems with more than one interacting component. The nonlinear nature of multicomponent diffusion calls for specific analytical technique compatible with principles of chemical kinetics. By the mechanism-based approach to nonlinear diffusion, it is also possible to build systematically equations that reflect the strong interactions characteristic of multicomponent mixtures. Not only does this follow from principles of thermodynamics, but it also opens the way to conditions necessary for reaction kinetics in such systems. To model and comprehend such processes is of utmost significance to the progress of surface-interaction-based technologies like electrocatalysis and photocatalysis, thereby emphasizing the central synergy between kinetics and thermodynamics in research and application as discussed in Gorban et al. (2011).

## *Definition of Surface Thermodynamics*

The discipline of surface thermodynamics is central to understanding the intricacies of material behavior at interfaces. Surface thermodynamics is an expansion of classical thermodynamics, which specifically refers to the energetic and entropic terms generated in the vicinity of the interface between two phases, i.e., gas-solid or liquid-solid interactions. It comes into focus in discussing how surface characteristics influence phenomena like adsorption, catalysis, and phase transitions. By the investigation of these interfacial properties, surface thermodynamics provides a basis for making predictions as to how materials would interact with varying conditions, and thus it enhances our ability to engineer surfaces for specific purposes. The philosophies underlying that have been formed with the help of surface thermodynamics find themselves assuming a central role in technological advances in areas of industry, particularly catalysis and materials science. Perhaps the most significant part of surface thermodynamics is the Gibbs adsorption isotherm, which describes the relationship between surface excess and chemical potential. What the relationship does is illustrate how molecules react to encountering a surface, effectively bridging molecular forces and macroscopic indication. The significance of such can be evidenced in applications such as plasma catalysis, where surface dynamics are crucial for the maximization of reactions and minimizing inefficiency. As noted, plasma catalysis holds stimulating potential for gas conversion like $CO_2$ to valuable products. Nevertheless, to enjoy these advantages to the fullest, it is important to grasp the underlying processes regulated by surface thermodynamics that control the catalysts performance and reaction kinetics in general (Bogaerts et al., 2020) (Figure 4.1).

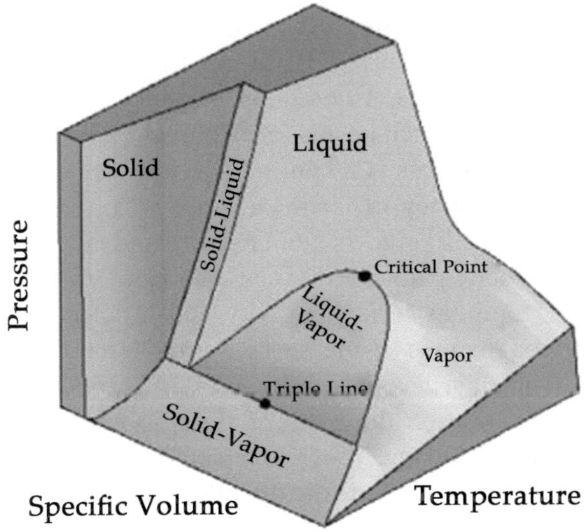


**Figure 4.1.** Thermodynamic surface.

*Source:* https://ars.els-cdn.com/content/image/3-s2.0-B9780123706102500076-f02-05-9780123706102.jpg.

This emphasizes that understanding interfacial phenomena is not merely theoretical but has real-world implications for improving catalytic systems. In addition, developments in computational techniques, especially based on machine learning, have transformed our strategy for surface thermodynamics and kinetics. Machine learning algorithms provide the prediction of intricate molecular interactions and energetics with unprecedented precision, enabling the acceleration of new material and catalyst discovery (No Fé et al., 2020). By applying these techniques within surface thermodynamics, researchers can effectively model kinetic effects and equilibrium states that exist at surfaces to gain a better insight into mechanisms of adsorption and reaction routes. In the face of continuously evolving challenges for catalyst design, the application of machine learning algorithms presents an exciting route to surmounting the limitations intrinsic to traditional modeling approaches. Here, the interplay between surface thermodynamics and computational software highlights the importance of creative solutions to drive the quest for more efficient catalysts and the optimization of surface interactions in a range of chemical processes.

## *Importance of Kinetics in Surface Reactions*

In the chemistry of surface reactions, kinetic consideration is of foremost significance to determine the rates of such reactions and therefore directly impact their application. Kinetics offers insight into how quickly reactants are converted to products at surfaces and is particularly critical in catalysis, corrosion, and sensor science. Conditions can be optimized for enhancing the rate of reactions if mechanisms are known. For instance, surfactant adsorption on various surfaces is a compelling case study, where the intrinsic kinetics govern the efficiency of chemical processes, ranging from flotation and separation processes. Engineers can create better catalytic systems if they are aware of these kinetic parameters since they can create interactions at the molecular level for optimal surface activity and efficiency. The interface of kinetics and thermodynamics at the surface provides a cohesive framework for the investigation of surface processes. While thermodynamics determines the probability of reaction on the basis of shifts in energy, it is kinetics that determines the pathway and speed of movement toward equilibrium. In most cases, reactions that are thermodynamically allowed may be sluggish, at speeds too sluggish to be realized without the help of catalysts or alteration of surfaces in some special way. For example, benzohydroxamic acid (BHA) adsorption on metal-activated surfaces is a well-known instance of such interaction, wherein metal ion availability significantly reduces the adsorption energy barrier (He et al., 2021). This proves the extent to which surface kinetics contribute to making otherwise difficult reactions occur, thus enhancing opportunities for various industrial applications. The assessment of surface reactions commonly relies on the establishment of protective films, such as the solid electrolyte interphase (SEI), that affect both thermodynamic stability and kinetic response of surfaces.

SEI formation is the process that describes the interaction between reactive and protective mechanisms in which kinetics plays a critical role. Understanding of SEI growth mechanisms and structural properties is also important for the optimization of electrochemical device performance, particularly in lithium-ion batteries (Wang et al., 2018). By using advanced computational techniques, researchers have simulated SEI dynamics, which has enlightened understanding of how kinetic considerations influence its efficiency in protecting electrode surfaces while allowing lithium ion transport. The insight gained from such work reveals the central importance of kinetics both in assisting surface reactivity and optimizing system performance in many technologies.

## 4.1. FUNDAMENTAL CONCEPTS OF THERMODYNAMICS

Knowledge of fundamental concepts in thermodynamics is a requirement for the study of surface thermodynamics and kinetics as these formulate energy transfer at the molecular level in various processes. Thermodynamics is a science that gives a framework to characterize how heat, Work, temperature, and energy relate to one another and form a central aspect of evaluating surface interactions in materials. The second and first laws of thermodynamics form the context for the discussion of phase transitions, spontaneous direction of processes, and equilibrium states. For instance, the Gibbs free energy parameter integrates both enthalpy and entropy in a manner that enables one to predict stability and spontaneity in processes at the interface. This thermodynamic basis for energy changes is essential in applications such as catalysis, electrochemistry, and thin film deposition, where surface interactions directly influence resulting material properties and function. Surface effects are thus inherently tied to thermodynamic concepts since they dictate how atoms and molecules act in interfaces. Adsorption is an archetypal example where kinetics and thermodynamics intersect; it is a function of factors such as temperature, pressure, and surface topography. The topology of adsorbed energy can be described by Gibbs isotherms and Langmuir isotherms, which define the manner in which molecules occupy surface sites. Kinetics of such a process are also significant, as they determine how quickly adsorption and desorption occur. In addition, the surface crystallization process discussed at length in literature, is governed by thermodynamic stability and kinetic barriers, pointing towards the complex interaction between the two disciplines. A sound comprehension of these interconnected dynamics is extremely critical for the best possible utilization of materials in research as well as industrial applications. Moreover, the interdisciplinarity of thermodynamics in surface science enables it to further understand nucleation and growth within crystal materials, something that becomes increasingly relevant with the advancement of nanotechnology. Nucleation, as the initiation of the development of crystals, is all about overcoming energy barriers that are heavily related to the thermodynamic variables. The findings of the classical theory of nucleation illustrate how temperature and supersaturation control the likelihood of nucleation events. In addition, kinetic aspects of crystal growth highlight how nucleation at the onset determines the general structural characteristics of the material formed. As scientists probe deeper into the nuances of crystal growth and surface interactions, fundamental thermodynamic concepts still play a dominant role in building credible models and predictions. With the rapidly growing area of nanoscience, knowledge of these fundamental principles will enable us to better understand material behavior at the nanoscale, fueling further innovation and use across a number of disciplines.

### 4.1.1. Laws of Thermodynamics

The basic concepts of thermodynamics are crucial in explaining surface thermodynamics and kinetics, particularly in relation to the first and second laws of thermodynamics. The first law, founded upon the law of conservation of energy, hypothesizes that energy cannot be created or destroyed but can be converted from one state to another. For surface phenomena, this hypothesis highlights how energy transfer dictates reaction paths, adsorption characteristics, as well as desorption. As materials interact within surfaces, energy conversions dominate the decision-making of how molecules align, change, and react. Not only does such knowledge further our comprehension of chemical reactivity at surfaces but also it forms the basis for enhancing industrial processes such as catalysis and material synthesis by aligning energy inputs with wanted reactions, thereby governing beneficial kinetics and thermodynamic stability. Except for the first law, the second law of thermodynamics gives the concept of entropy, which plays a key role in assessing surface interaction feasibility. Entropy increase in closed systems has a tendency to be associated with spontaneous processes, e.g., diffusion and reaction kinetics at surfaces. In adsorption phenomenon studies, for instance, entropy changes in molecular arrangement and phase transitions directly influence surface coverage and binding energies. Significantly, the principle explains how certain reactions happen at a fast pace under some circumstances but are slow in other situations. Incorporating the second law in surface thermodynamics enables scientists to make predictions about whether or not a system will equilibrate and how efficiently it will convert reactants into products, thus shedding light on the kinetics that drive these conversion processes, especially in heterogeneous catalysis (Roque-Malherbe, 2009) (Figure 4.2).

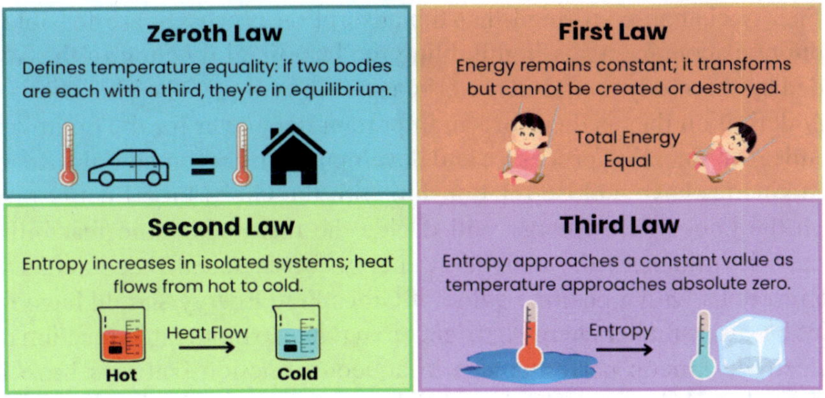


**Figure 4.2.** Four of laws in thermodynamics.

*Source:* https://images.examples.com/wp-content/uploads/2024/04/Laws-of-Thermodynamics.png.

Lastly, thermodynamic concepts are extended past traditional interfaces to influence more sophisticated regions such as nanomaterials and porous materials. Nanoscale material behavior deviates from bulk behavior because it has a greater surface area and altered energy landscapes. The laws of thermodynamics provide a stage on which to frame phenomena characteristic of nanomaterials, such as augmented adsorption and discriminatory reactivity. In addition, thermodynamic interactions in nanoporous materials can guide the development of novel catalysts or energy storage devices, indicating the applicability of the laws in cutting-edge research and application. As scientists learn how energy conservation and entropy govern material behaviors on surfaces, they can apply the laws to engineer responsive materials for utilization in sustainable energy, environmental remediation, and advanced manufacturing (Chesworth, 2008).

## 4.1.2. Gibbs Free Energy and Its Role

In surface kinetics and thermodynamics, Gibbs Free Energy plays a key role in determining the spontaneity and feasibility of chemical reactions and adsorption processes. As the largest quantity of reversible work that can be obtained from a thermodynamic system at constant temperature and pressure, Gibbs Free Energy (G) gives scientists the means to forecast the likelihood that a reaction will take place under specified conditions. A negative Gibbs Free Energy change ($\Delta G$) indicates that a process is able to go on spontaneously, something which is extremely important in surface adsorption phenomena. For instance, during the adsorption of various metal ions, the negative Gibbs

Free Energy change confirmed that the adsorption processes are feasible and can continue spontaneously, highlighting the beneficial function of the insight provided by thermodynamics in material use (Jab et al., 2025). This Gibbs Free Energy definition thus is therefore an important parameter for the optimization of chemical engineering processes and development of new materials. Reaction surface kinetics have a close relationship with the Gibbs Free Energy profile. Though the Gibbs Free Energy will decide the thermodynamic feasibility of a process, it influences the activation energy for chemical reactions as well. Those reactions with a positive value of Gibbs Free Energy should have faster kinetics because smaller energy barriers often relate to more rapid reaction rates. The electrode kinetic performances in aqueous zinc-ion batteries have been investigated, and the control of potential energy surfaces by ion intercalation modifies essentially the Gibbs Free Energy associated with the process of ion migration. The modifications of the Gibbs Free Energy landscape not only influence the rate of reactions but also stabilize the structures upon cycling, combating capacity loss (Chen et al., 2024) (Figure 4.3).

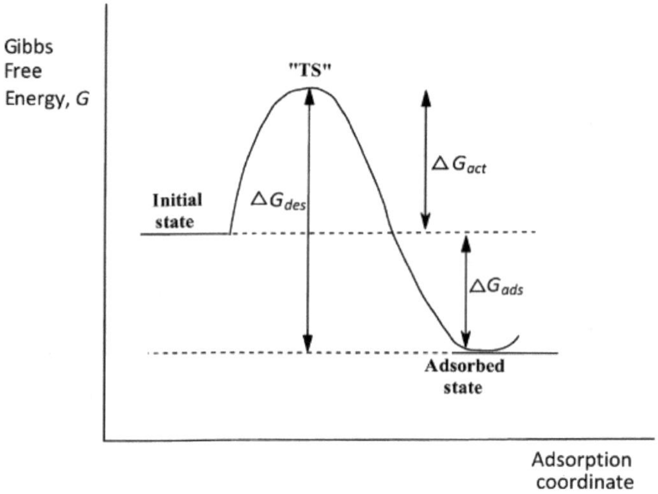


**Figure 4.3.** A two-dimensional Gibbs free energy diagram in which an adsorption process is treated as a chemical reaction, showing the energy barriers for adsorption and desorption with a hypothetical transition state "TS."

*Source:* https://www.researchgate.net/figure/A-two-dimensional-Gibbs-free-energy-diagram-in-which-an-adsorption-process-is-treated-as_fig5_361675840.

Therefore, knowledge of Gibbs Free Energy is essential both in regard to optimizing reaction rates and optimizing material stability for energy storage applications. Beyond its importance in particular reactions and adsorption processes, Gibbs Free Energy also proves to be a critical parameter in the broader understanding of surface science thermodynamic cycles. It enables determining the interdependencies between pressure, temperature, and concentration, so one becomes aware of the conditions of equilibrium and the paths of reactions. Through Gibbs Free Energy changes in adsorption and desorption reactions, researchers can design more efficient catalysis and pollution abatement materials. For example, activated carbon produced from waste tobacco stalks had a large surface area through which metal ion adsorption was possible— an adsorption process whose Gibbs Free Energy change under optimum conditions indicated positive and spontaneous adsorption (Jab et al., 2025). Hence, Gibbs Free Energy not only defines the feasibility of reactions but also determines strategic planning in experimental design and material development in surface thermodynamics and kinetics.

### 4.1.3. Phase Equilibria and Surface Energy

Phase equilibria knowledge is essential in surface thermodynamics and kinetics, particularly in understanding how different states of matter interact at material surfaces. Phase equilibria refers to the equilibrium of different phases, such as solid, liquid, and gas, and their interaction at interfaces. Phase equilibria is significant in ascertaining the thermodynamic stability of materials and material transformation.

For instance, Gibbs phase rule is a concept that provides a framework toward predicting the number of phases of a system in varied circumstances. Its use spans various industries from materials science to biophysics, where surface phase behavior control can significantly enhance performance and efficiency in catalysis and energy storage, to mention a few. The understanding of such interactions emphasizes the necessity to develop predictive models that can illustrate the manner in which molecular dynamics influence macroscopic properties.

Surface energy is another highly impacted aspect of material science. It is the excess energy at the surface compared to the bulk of a material. This phenomenon lies at the core of various processes such as wetting, adhesion, and colloidal system stability. For example, successful water surface potential, which is crucial for determining hydration phenomena, illuminates interactions at various interfaces. As validated in advanced studies, including those using quasi-chemical theory, this potential plays a key role in the characterization of ion behavior in solution and interaction with surfaces, highlighting the complex interrelation between surface energy and phase behavior that dictates thermodynamic processes in liquid systems (Shi et al., 2020) (Figure 4.4).

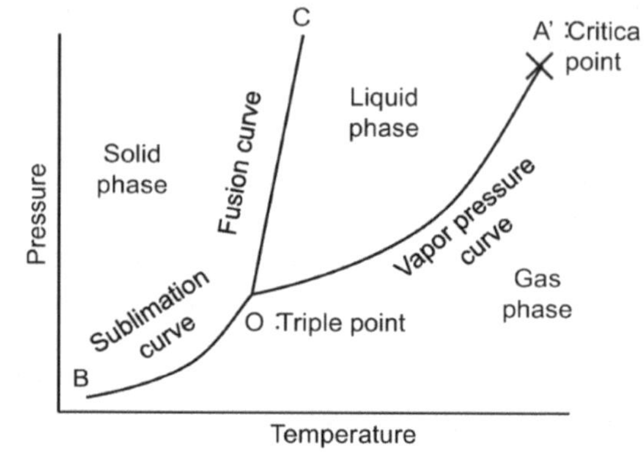


**Figure 4.4.** Phase equilibrium.

*Source:* https://media.springernature.com/lw685/springer-static/image/chp%3A10.1007%2F978-981-13-9935-0_2/MediaObjects/429848_1_En_2_Fig2_HTML.png.

Surface energy manipulation by techniques like surface modification has significant impacts on altering interfacial interaction and phase equilibria behavior. In adsorption kinetics, phase equilibria and surface energy interaction becomes critical in determining system behavior towards equilibrium. Mass transfer kinetics has a significant role in determining how rapidly a system approaches adsorption equilibrium, particularly in porous materials like carbon molecular sieves. These substances, characterized by their unique microporous morphologies, exhibit specific adsorption characteristics that are significantly influenced by surface energy characteristics as well as thermodynamics governing phase states of the adsorbates. A precise understanding of immersion enthalpies' correlation with adsorption isotherms enables scientists to create systems that will optimize separation and purification processes (Hähnel et al., 2017). With observation of such behaviors, potential for predictive modeling is increased, opening the way to develop innovative applications in industrial processes where phase transitions need to be efficient and surface interactions need to be enhanced.

## 4.2. SURFACE PROPERTIES AND CHARACTERIZATION

Knowledge of surface characteristics and properties is highly pertinent in surface thermodynamics and surface kinetics. Materials interface is characterized by distinct features that determine a materials physical as well as chemical behavior. For instance, the surface energy occupies a central role in regulating phenomena such as wetting, adhesion, and surface reactivity. It is at these interfaces that molecular interactions are largely occurring, so characterization surface methods are needed for the analysis of the behavior of material under conditions that are varied. As Mejuto (2020) says, this emphasis on surfaces has been an area of interest for applied chemistry fields through materials science, particularly with the advances in characterization equipment such that the surface effects could be better simulated. Consequently, acquisition of a comprehensive understanding of surface properties is not an academic exercise but is vital to the design and optimization of a vast array of applications from catalysts to biomaterials. Surface characterization techniques such as microscopic methods like scanning electron microscopy and atomic force microscopy allow analysis of structural, mechanical, and chemical properties at the nanoscale. These techniques extend far beyond the qualitative identification of surface morphologies; they provide access to surface composition, defect microstructures, and phase transformations that can influence macroscopic material properties. Also, investigations of phenomena like ice crystal growth and related molecular kinetics with surface attachment kinetics have the potential to reveal controlling mechanisms of material performance in practical applications. The research knowledge acquired through studies like that quoted in Libbrecht (2019), which examines ice crystal dynamics, underscores the dynamic interaction between surface properties and material science. This intersection underscores the importance of academics adopting transdisciplinary approaches, integrating various scientific ideas to address complex surface interactions in various fields. As we proceed with surface property considerations, we can see that thermodynamic principles are behind much of what we observe. For the sake of material prediction, it is useful to know how temperature, pressure, and composition affect surface thermodynamics. Features like the Gibbs adsorption isotherm help us understand how interactions between molecules vary with environmental conditions and hence can illuminate stability and reactivity at surfaces. In addition, the kinetic models that account for the mechanisms of desorption and adsorption are central to describing the rates of these surface reactions. To effectively realize the potential of materials in practical applications, it is important not only to accurately characterize surface properties but also to understand the related thermodynamic and kinetic principles that govern these interactions, thus bridging the gap between theoretical know-how and practical implementation as emphasized in the unifying themes of surface thermodynamics and kinetics.

### 4.2.1. Surface Tension and Its Measurement

Surface tension is a property inherent in liquids as a consequence of the intermolecular cohesive forces that occur at the free surface of a liquid. Surface tension has a critical role in many areas of science, from materials science and biology, where the behavior of liquids can significantly affect structure and function. Surface tension is usually quantified by applying a variety of methods such as the drop weight method, the capillary rise, and the Wilhelmy plate method. Each technique is grounded on a distinct principle of physics; for instance, the Wilhelmy plate technique is the force measurement on a thin partially submerged plate in liquid, which is the effect of surface tensions on the ability of liquids to wet the surface. Knowledge of these techniques is responsible for quantitatively describing surface behavior across a variety of applications, from detergent and surfactant performance to emulsion stability in complex fluid systems. Besides the ordinary measuring techniques, the interaction of surface tension with thermodynamics gives qualitative details of molecular interaction that controls material behavior. The majority of systems are governed by the laws of non-equilibrium thermodynamics that take into account the molecular level fluctuation of heat and mass transfer. For instance, recent work has shown how single interactions in systems can induce liquid-liquid phase separation, with a severe impact on surface tension and interfacial stability. This is particularly the case in biological systems where proteins and nucleic acids can assemble into membranelles condensates, such as nucleoli, by means of these interactions. Such findings suggest that the underlying thermodynamic mechanisms not only account for empirical observations but can also allow one to predict dynamics in the surface interactions in different contexts and bridge gaps between theory and laboratory measurements (Qi et al., 2021) (Figure 4.5).

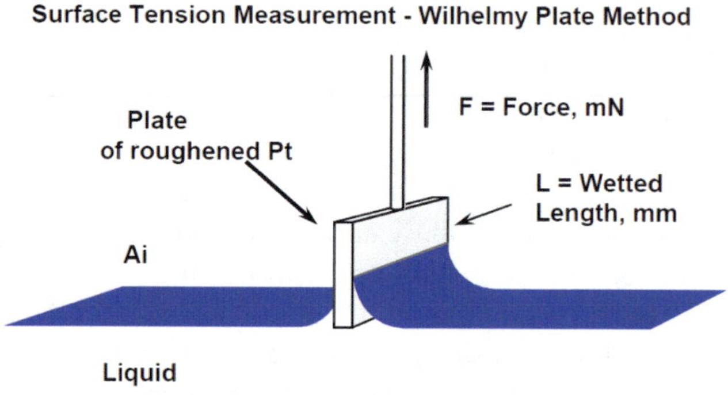


**Figure 4.5.** Measure surface tension.

*Source:* https://external-content.duckduckgo.com/iu/?u=https%3A%2F%2Fclearsolutionsusa.com%2Fwp-content%2Fuploads%2F2021%2F01%2FWilhelmy-Plate-Method.png&f=1&nofb=1&ipt=6efba93844bc7209bc775c4489ec48aa92aaadadc4e013af5215b642e2eeb09c.

Also, being aware of surface tension in the context of energetics and kinetics provides additional context in understanding how materials behave under other conditions. Taking the example of catalytic reactions, surface temperature may be of great importance while changing reaction rates by the manner in which reactants interact in the interface. Such a relationship is fundamental in heterogeneous catalysis, where phenomena at the surface can influence total reaction efficiency. Through the foundations of non-equilibrium thermodynamics, researchers can uncover the nuances of how surface tension contributes to kinetic behaviors and make strides in material design and chemical engineering. Such interdisciplinary work emphasizes surface tension as not only a constitutive property but as a unifying feature within the broader context of surface thermodynamics and kinetics (Cruz et al., 2019).

## 4.2.2. Adsorption Isotherms

The adsorption isotherm theory is a cornerstone for the explanation of the phenomenon of adsorbent-adsorbate interaction, which is becoming more and more influential in designing various industrial processes, particularly wastewater treatment. Adsorption isotherms are utilized to establish the relationship between the amount adsorbed over the surface of an adsorbent and the bulk concentration of an adsorbate at a given temperature. General models, the Freundlich and Langmuir isotherms, provide models for such relationships that describe solute adsorption in terms of concentration. Examples are studies where biosorbents such as kapok husk for indigo carmine adsorption have been found to follow the Freundlich isotherm model for multilayer adsorption on heterogeneous surfaces. The strength of this model lies in its capability to explain the non-uniform distribution of the adsorption energies on the adsorbent surface, and therefore shed some light on how adsorbent utilization can be maximized under actual applications (Zein et al., 2025). Besides providing models for explaining adsorption mechanisms, isotherms are also important in evaluating the kinetic characteristics of adsorption processes. Kinetic measurements are typically accompanied by isotherm measurements to allow scientists to determine the rate at which adsorption occurs, as is relevant for effective design and operation in real-world applications. For example, the aforementioned research on kapok husk determined that the indigo carmine adsorption kinetics closely described a pseudo-second-order model, which indicated that the adsorption rate is a function of accessible active sites and not concentration. This combination of kinetic and isotherm data points to the significance of appreciating adsorption not only as a static equilibrating phenomenon but as a dynamic one responsive to various factors of influence including contact time and temperature (Zein et al., 2025) (Figure 4.6).

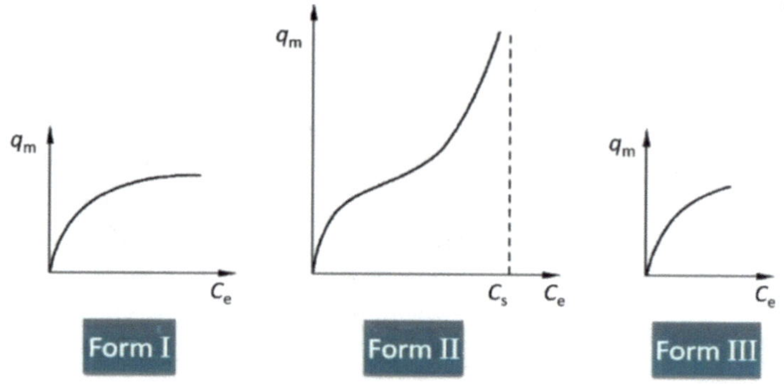


**Figure 4.6.** Relation curve between adsorption equilibrium capacity and equilibrium concentration.

*Source:* https://external-content.duckduckgo.com/iu/?u=https%3A%2F%2Fi.ytimg.com%2Fvi%2FwXN6WFqAdd0%2Fmaxresdefault.jpg&f=1&nofb=1&ipt=663ee7682aea54e8d4ae2a3be493d4bb2a9bed87ca5a6f2e04ed71de0a18fa72.

Furthermore, thermal analysis techniques complement isotherm studies by giving researchers an insight into the energy changes in adsorption processes. The adsorption thermodynamic parameter of adsorption conveys to scientists spontaneity and heat transfer news during the adsorption process. Specifically, in the MM-MDH nanoclay research, they found that dye removal exhibited both exothermic and endothermic tendencies, which shows the complicacies of adsorption behaviors based on unique solute-solvent interactions. These findings in thermodynamics, as well as isotherm and kinetic data, allow for a comprehensive image of adsorption processes. Such a broad consideration increases the requirement for consideration of the principles of thermodynamics and adsorption isotherms, ultimately driving the developments of material choice and optimization of the process in surface thermodynamics and kinetics.

### 4.2.3. Techniques for Surface Characterization (e.g., SEM, AFM)

The ultimate understanding of surface kinetics and thermodynamics demands the application of advanced surface characterization techniques, such as Scanning Electron Microscopy (SEM) and Atomic Force Microscopy (AFM). SEM enables the imaging of surface topographies at high resolution, providing information about the morphology of materials at the microscale and nanoscale. With electron beams, SEM yields images of high resolution that reveal surface defects, grain boundaries, and phase distributions that are critical to interpreting how surface characteristics regulate thermodynamic behaviors. The capability of analyzing sample compositions using Energy Dispersive X-ray Spectroscopy (EDX) also raises the efficiency of SEM so that researchers can relate structural properties to compositional data, hence enriching the analysis of kinetic processes at the surface of oxidation or catalytic reactions.

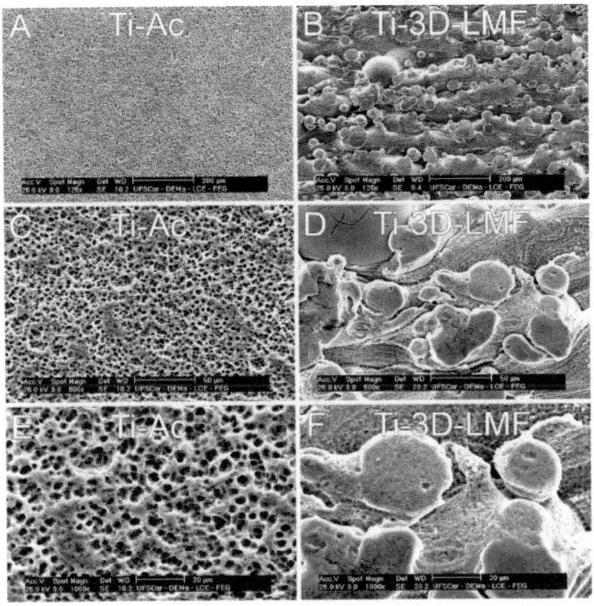


**Figure 4.7.** Surface characterization by scanning electron microscope (SEM).

*Source:* https://www.researchgate.net/figure/Surface-characterization-by-Scanning-Electron-Microscope-SEM-The-columns-represent-the_fig1_378675327.

The finer details of SEM underlie theoretical accounts of surface kinetics, affirming the relationship between surface phenomena and directing thermodynamic forces. At the same time, AFM presents another feature of surface characterization employing a laser and a cantilever to sense the forces between a sharp probe tip and the target surface. This technique is particularly valuable for the description of nanoscale properties such as surface roughness and mechanics, which are especially relevant for nanoelectronics and materials science technologies. AFM excels in providing three-dimensional reconstructions of surface topographies, enabling one to observe atomic-level interactions. The quantitative data acquired via AFM measurements can be correlated with surface thermodynamic properties, allowing for a better understanding of how structural defects influence kinetic routes in processes like diffusion and adsorption. Here, AFM complements SEM, allowing for multi-faceted surface characterization, hence facilitating a better understanding of thermodynamic behavior associated with various materials under diverse environmental conditions (Figure 4.7).

The complementarity between AFM and SEM in surface characterization is responsible for the unending development of nanotechnology, particularly in agriculture and environmental remediation. With these techniques, researchers can design nanomaterials that are able to enhance the fertility of soil and optimize the efficacy of nutrient uptake, thereby addressing serious agricultural issues. As an illustration, the application of nanotechnology in fertilizer production is reflective of how better nutrient delivery systems with less environmental footprint are emerging as critical, as seen in the work on soil-clay interaction at the nanoscale (Inderpreet et al., 2009). Further, surface characterization methods assist in finding the consequences of material composition modifications, like in the situation of Ti-Al based intermetallics where oxidation behavior and scaling traits were modified by alumina formers (Tan et al., 2008). Basically, these advanced characterization methods not only serve a fundamental function in scholarly research but also in practical applications that generate contributions to sustainability and efficiency in various industries.

## 4.3. ADSORPTION PHENOMENA

The adsorption process is the focal point of surface thermodynamics and kinetics since it regulates the interactions between molecules and solid surfaces. Adsorption happens either as a physical adsorption, meaning by weak van der Waals forces, or as a chemical adsorption, where there is the formation of stronger bonds. This discrimination is essential in all types of applications, such as catalysis and sensors, where the quality of the adsorption process is paramount to overall functionality. The thermodynamic forces that regulate these interactions, for example, entropy and enthalpy change, are essential in determining extent and nature of adsorption. With researchers exploring novel materials, such as nanoparticles, their unique surface character leads to better adsorption behavior, which is rather dissimilar from bulk materials. It is imperative to understand these differences to best enhance applications in fields such as environmental science, materials science, and nanotechnology (Joudeh et al., 2022). Surface kinetics also plays a role in regulating the kinetics of the processes of adsorption and the effectiveness of various industrial and environmental applications. The amount by which molecules are absorbed to surfaces is directly related to the availability of active sites, surface roughness, and temperature. Adsorption kinetics in most cases follows Langmuir and Freundlich isotherms, which define the capacity and affinity of surfaces to adsorb, respectively. These models allow for prediction of behavior under various conditions, providing useful information to researchers and engineers looking to optimize catalytic processes or remove pollutants from the environment. Moreover, recent progress in plasma catalysis highlights the interplay between surface kinetics and adsorption. This dynamic approach to catalyst synthesis and treatment emphasizes the significance of adsorption phenomena in recent research, especially when designing affordable and effective catalytic systems (Bogaerts et al., 2020). The significance of adsorption phenomena extends beyond interdisciplinary research, bridging material science, biology, and environmental engineering. Progress in biogenic nanoparticle synthesis is a case in point regarding the significance of adsorption in applications as diverse as drug delivery systems and waste treatment processes. By taking advantage of the unique character of these nanoparticles, scientists can maximize adsorption performance, either increasing the therapeutic effect or pollutant removal. Adsorption research also enables the study of biological processes, e.g., the adsorption of nutrients by plants and microbial interactions within soil systems. With advancing knowledge in surface thermodynamics and kinetics, the delicate control of adsorption response will progressively offer fresh research and application opportunities. This comprehensive outlook highlights the need to unify studies on adsorption in different fields, further establishing its basis in science and technology (Joudeh et al., 2022; Bogaerts et al., 2020) (Figure 4.8).

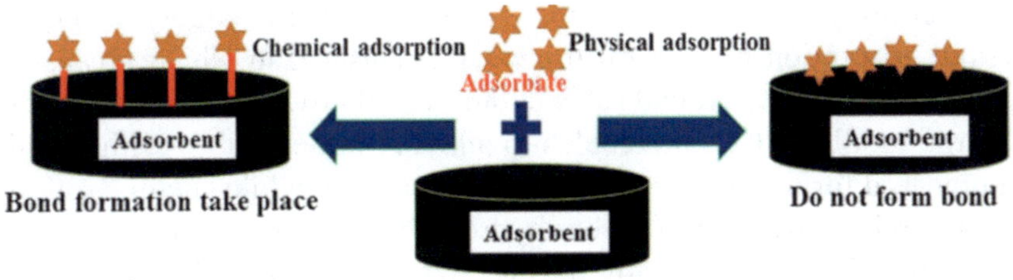


**Figure 4.8.** Chemical and physical adsorption phenomenon.

*Source:* https://www.researchgate.net/publication/361740859/figure/fig1/AS:1143 1281122896791@1677585236862/Chemical-and-physical-adsorption-phenomenon.jpg.

### 4.3.1. Types of Adsorption: Physisorption vs. Chemisorption

Adsorption is a significant phenomenon in numerous applications of surface thermodynamics and kinetics, particularly catalysis, cleaning of environmental pollution, and storage of gases. Physisorption and chemisorption are two broad categories of types of adsorption. Physisorption, with weak van der Waals forces, is a process where molecules adsorb on a surface by nonchemical bonding. This two-way interaction is normally at lower temperatures with less energy change and thus much flexibility of capacity for adsorption. Chemisorption, however, entails the formation of covalent or ionic bonds between the adsorbate and the adsorbent. This is normally irreversible and at higher temperatures where the high energy input is conducive to bond formation. These differences are important to optimizing surface interactions in various industrial technologies. The technological use of physisorption and chemisorption extends beyond basic science and affects developing and running adsorbent materials in carbon capture and photocatalytic technologies, among others. For instance, advances in capturing carbon dioxide ($CO_2$) utilize both types of adsorption, each for a distinct function in enhancing efficiency and sustainability. Advanced materials such as metal-organic frameworks (MOFs) and carbon-based composites call attention to how optimized adsorbent characteristics can achieve maximum capture efficiency through optimized physisorption and chemisorption interactions. In addition, research on these materials is a major contributing area to solutions for carbon emissions and climate change mitigation on the global scale. In the spirit of net-zero emissions, adsorption mechanisms can unleash breakthroughs in efficient $CO_2$ solid adsorbents like those discovered recently, emphasizing the need for systematic investigations into such properties (Dziejarski et al., 2023). Also, the landscape of developing photocatalytic $CO_2$ conversion emphasizes another side of adsorption's role in surface thermodynamics and kinetics. The mechanism is largely dependent on successful adsorption of $CO_2$ on the catalyst surface, in which physisorption and chemisorption play crucial roles in the overall efficiency of the conversion. Physisorption allows $CO_2$ molecules to adhere on the surface with the aim of being prepared for the succeeding chemical reaction, while chemisorption may allow $CO_2$ to be converted to higher hydrocarbon molecules. Such interactions are very important in determining catalytic activity and stability, which are paramount in sustainable energy generation and environmental benefit. Synthesis of advanced photocatalytic materials therefore must be directed towards maximizing adsorption features to attain optimized $CO_2$ conversion processes, as underscored in ongoing research activities to open up a cleaner energy future (Gong et al., 2021).

### 4.3.2. Factors Affecting Adsorption Rates

The performance of adsorption operations is influenced by several significant factors that determine adsorbate and adsorbent surface interaction. Of these factors, pH levels are crucial, since alterations in them may alter the charge and consequently the binding capacity of the adsorbent. For instance, the adsorption property of a hydrogel of sodium alginate shows that pH fluctuation can enhance or diminish adsorption of the pesticide chlorpyrifos and phosalone by modifying binding at the optimum pH of 7 for full effectiveness (Samanta et al., 2024). The time provided for adsorption is also important as equilibrium must be reached to maximize interaction between surfaces. This interaction between solution pH, contact time, and adsorption efficacy highlights the intricacy of adsorption kinetics in real applications. The surface area of the adsorbent material is also another critical factor that affects adsorption rates. Higher surface area offers more active sites for adsorbate molecules, which can effectively improve adsorption capacity.

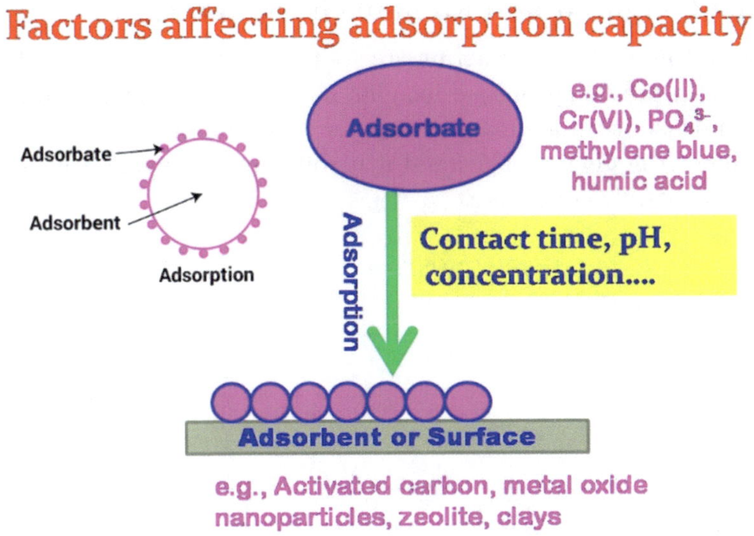


**Figure 4.9.** Factors affecting adsorption capacity.

*Source:* https://external-content.duckduckgo.com/iu/?u=https%3A%2F%2Fi.ytimg.com%2Fvi%2FBE45w0M3QhU%2Fmaxresdefault.jpg&f=1&nofb=1&ipt=e74338bdd73891eda6309d50f11a784214f0036b7dfd15771d124a46514f9edd.

For instance, research on $H_2TiO_3$ lithium ion sieve nanoparticles shows that their high specific surface area leads to greater adsorption capacities of lithium ions in solution (Liu et al., 2024). Particle uniformity also helps in improved dispersibility, hence exposing more adsorption sites and accelerating interaction kinetics. This correlation between surface characteristics and the effectiveness of adsorption highlights the need to optimize adsorbent design to achieve the desired performance for many environmental processes. Finally, the process of adsorption itself, physisorption or chemisorption, is the sole determining factor in setting the rate of adsorption. Chemisorption will typically be coupled with a higher level of interaction, often resulting in a lower rate due to bond establishment. The study of sodium alginate hydrogel illustrates the phenomenon where adhesion is controlled by an endothermic process, implying that the energy dynamics render it more difficult to adsorb (Samanta et al., 2024). Similarly, lithium-ion adsorption in $H_2TiO_3$ has pseudo-second-order kinetics, implying that the adsorption reactions are more controlled by the adsorption energy than mass transfer constraints (Liu et al., 2024). An understanding of these mechanisms is significant for the optimal design and operation of adsorbents in industrial and environmental contexts (Figure 4.9).

## 4.3.3. The Langmuir and Freundlich Isotherms

Langmuir and Freundlich isotherms are root models for studying adsorption processes, both providing unique insights into solutes' behavior on surfaces. The Langmuir isotherm assumes monolayer adsorption onto a surface with a finite number of equivalent sites and that following occupation of a site, no adsorption is possible at that site.

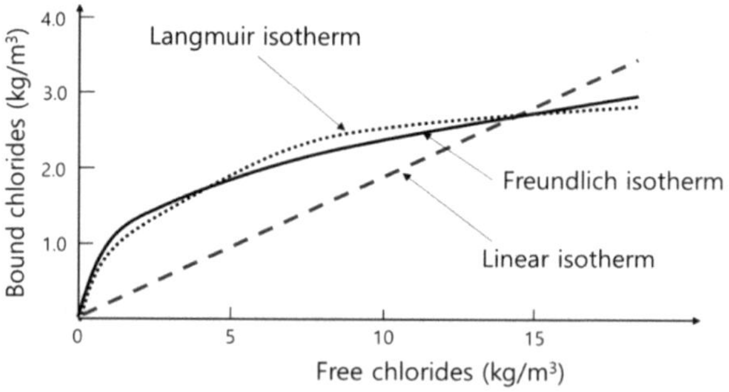


**Figure 4.10.** Comparison of linear, Langmuir and Freundlich isotherm models.

*Source:* https://external-content.duckduckgo.com/iu/?u=https%3A%2F%2Fwww.researchgate.net%2Fpublication%2F349972499%2Ffigure%2Ffig2%2FAS%3A1010733357731841%401617989019568%2FComparison-of-Linear-Langmuir-and-Freundlich-isotherm-models-72.png&f=1&nofb=1&ipt=fe181c859bbd73b2aa0a2bf264b971a6994e18f7e9cc6fe8b4490dc9a793e443.

This model is basic in the study of cases where surface coverage restricts adsorption and serves as a level of saturation. Langmuir model applicability has been demonstrated in numerous studies, such as the adsorption of bisphenol A on modified corn cob, where results were in good agreement with Langmuir isotherm prediction, providing the theoretical foundation for surface capacity and quantitative basis for its adsorption efficiency (Eze et al., 2025). The Freundlich isotherm, on the other hand, portrays heterogeneous surface energies, suggesting differential affinity surfaces multilayer adsorption. The model best fits non-ideal cases with such properties of differential surface or complex chemical interactions. The Freundlich model, as opposed to the Langmuir isotherm, suggests no finite number of equivalent sites but an exponential site energy distribution. Freundlich constants derived through Freundlich equation, especially Freundlich constant (Kf) and adsorption intensity (n), indicate the nature and mechanism of adsorption. In a study exploring the biosorption of lead ion by Zea mays biomass, Freundlich model was also effective, showing its capacity to describe the adsorption behavior under a certain condition (El-hoshoudy et al., 2024) (Figure 4.10).

Both isotherms are of fundamental significance in surface thermodynamics and kinetics, though their application is case-specific. While the Langmuir isotherm is especially suited to systems near saturation with homogenous adsorption sites, the Freundlich isotherm is appropriate for systems with heterogeneity in interaction energy across the adsorbing surface. Knowledge of the strengths and limitations of the models contributes towards our ability to design practical adsorbent materials for environmental remediation. In real-world cases, scientists must scrutinize which isotherm best represents their system to ensure successful optimization of adsorption conditions. By doing so, it further advances the science of surface thermodynamics into an understandable platform for improving material performance in adsorption technologies (Eze et al., 2025; El-hoshoudy et al., 2024).

## 4.4. KINETICS OF SURFACE REACTIONS

Understanding the kinetics of surface reactions is at the forefront of catalytic process development and improvement of system performance in emissions control and energy storage. Surface reactions are typically governed by the complex interaction among various thermodynamic parameters, such as concentration gradients and temperature. This interaction results in strong deviations from ideal behavior, for example, that of $Ce_{0.5}Zr_{0.5}O_{2-x}$ (CZO). The reaction rates at which reductants react with surface-adsorbed oxygen atoms (*) are significantly affected by the reactivity and stability of such species. As levels of O * drop, the free energy penalties incurred increase, highlighting the nonlinearity of reaction rates as a function of levels of O *. Such findings emphasize the necessity of developing theoretical models—like transition state theory—to simulate and predict reaction dynamics, particularly in nonideal systems where the thermodynamic theories are crucial (Hwang et al., 2024). The influence of surface charges on the reaction kinetics has been a subject of research, particularly in electrochemical applications. New advances in grand-canonical iteration algorithms, on the basis of density functional theory calculations, allow researchers to investigate this influence further. For example, when studying the hydrogen evolution reaction on S vacancies of

$MoS_2$, it has been seen that surface charges could modify the reaction kinetics significantly. Desorbed proton stability and the pathway of desorption may also be affected by the net electron motion toward or away from the catalyst surface, demonstrating that traditional kinetic models generated from the Butler-Volmer equation may be unable to adequately describe these advanced interactions. The mechanism shift in desorption processes has great influence on Tafel slope and overpotential, which implies that surface charge effects are universal phenomena in electrochemical reactions and must be investigated deeper (An et al., 2023). The combination of thermodynamic and kinetic models not only accounts for the mechanism of the surface reaction but also allows for the design of more effective catalytic materials. Realization of the fact that surface-bound intermediates' reactivity is sensitive to charge state and concentration gradient variations allows for strategic manipulation during catalysis. Through stringent scrutiny of the manner in which the thermodynamic activities of reactants vary, researchers are able to develop better models that predict transient stoichiometric reduction rates in systems characterized by extreme nonideality. With possible applications to carbon sequestration and environmental clean-up, data derived from surface reaction kinetic modeling can ultimately be applied to develop efficient technologies to combat pollution and make energy sustainable. Unraveling these complex interactions enhances our capability to invent in surface thermodynamics and kinetics, and push the boundaries of what is achievable in catalysis and electrochemical processes (Hwang et al., 2024; An et al., 2023).

### 4.4.1. Reaction Mechanisms on Surfaces

Understanding the mechanisms of surface reactions is at the heart of surface thermodynamics and kinetics since the surface reactions usually determine the efficiency of catalytic reactions. The surfaces play an important role in the adsorption and activation of the reactants, which can fundamentally alter the path of a chemical reaction. For instance, the surface composition, topography, and active site structure might determine the adsorption kinetics and energy that directs how a given reaction is conducted on a given substrate. Elucidating these steps of mechanisms is particularly important in catalysis applications, where maximizing conversion rates and selectivity is most desirable in industrial processes. The kinetic character of surface interaction and the ensuing changes that result from them highlight the subtle interplay between kinetic accessibility and thermodynamic stability of chemical reactions on surfaces. New experimental methods, e.g., in situ surface-enhanced infrared absorption spectroscopy (SEIRAS), have delivered detailed views into the mechanism of reaction on surfaces. This method allows direct visualization of the intermediate species present in catalytic reactions and provides valuable insights into the formation and stability of these species. These experiments have proven that pre-oxidation of copper surfaces can be made to enhance $CO_2$ reduction performance by more efficiently activating $CO_2$ and increasing surface coverage by reactive intermediates (Delmo et al., 2024). This indicates that pre-reaction surface state knowledge can have dramatic effects on optimizing catalytic efficiency. In addition, the capability of tracking and studying such changes in real-time enables better comprehension of how surface modification affects reaction pathways, ultimately resulting in improved catalyst design and application (Figure 4.11).

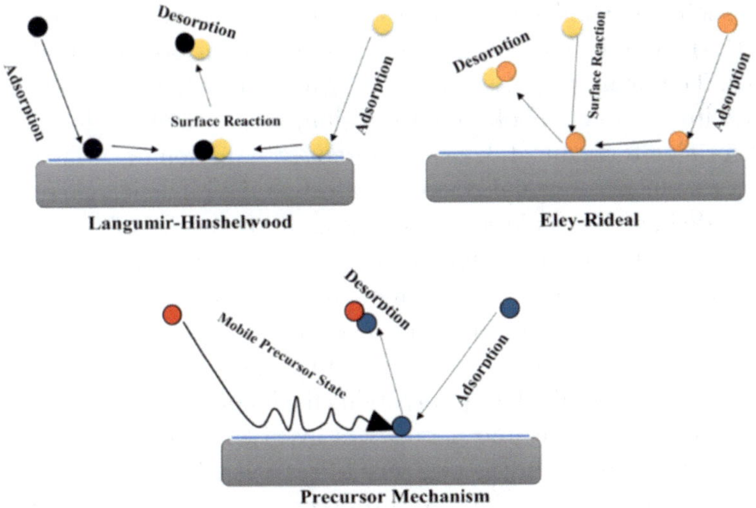


**Figure 4.11.** Surface reactions described by possible three mechanisms.

*Source:* https://www.researchgate.net/figure/Surface-reactions-described-by-possible-three-mechanisms_fig3_338564483.

Computational techniques, in particular computational simulations, have been instrumental in describing the energy landscape of surface chemistry. Techniques such as GaMD allow researchers to perform simulations of molecular interactions at faster velocities while capturing the thermodynamic and kinetic traits of complex biomolecular events with precision. These approaches enable examination of a variety of scenarios and their impact on the kinetics of reactions, as depicted by the development in modeling dimerization and ligand binding on nanoscale surfaces (Wang et al., 2021). Combination of such sophisticated computational approaches with experimental information further improves our understanding of surface reaction mechanisms and opens doors for innovative concepts in catalytic science. Collectively, the results here are significant for the formation of energy storage and conversion technologies and highlight interfacial effects as pivotal for establishing the future of catalysis.

### 4.4.2. Rate Laws and Their Applications

In the examination of surface thermodynamics and kinetics, the understanding of rate laws is paramount. Rate laws define the relationship between reactant concentration and the rate at which a reaction occurs, and provide useful information about mechanisms. In the majority of catalytic reactions, especially mixed conducting oxide-based reactions like cerium oxide ($CeO_2$–δ), kinetic constants such as the bulk oxygen diffusivity (DChem) and surface reaction rate constant (kS) are to be measured to ensure optimal performance under various conditions. The analysis provides a platform for researchers to associate the reactants with the rates of observed reaction, as required for the creation of catalysts for electrochemical devices. Moreover, variation in surface treatment or dopants can enhance catalytic activity, illustrated with ceria, where structural changes made huge boosts in reaction rates and efficiencies of thermochemical cycles (Gopal, 2015). Examination of redox processes illustrates the complexity of rate laws and their applications in numerous fields of science. One needs to know the redox potentials for different species to optimize chemical processes, particularly mineral surface reactions governing natural and industrial processes. Theoretical calculations involving the incorporation of computational techniques provide insight into parameters involved in redox processes, e.g., environmental conditions or the presence of species adsorbed on surfaces. The variation in the redox potential with different computational parameters like basis sets and DFT functionals is a realization of how difficult it is to properly model these occurrences. This renders the inclusion of reliable computational models make the predictability much better in the form of redox reaction kinetics, further solidifying the fundamental rate laws theories and their validity in the case of surface interactions (Arumugam et al., 2014). Finally, the technology of rate laws extends into environmental engineering, in which remediation strategy is guided by redox process kinetics. Advances in catalytic redox agents require knowledge of mechanisms in reduction and immobilization of harmful substances. Application of iron-sulfide mineral surfaces as catalysts, for instance, has been sparsely studied but holds tremendous potential for catalyzing natural redox reactions. Through comprehensive study of their kinetics, researchers are able to develop efficient pathways for the degradation and stabilization of toxic compounds. The ability to manipulate surface interactions for accelerated reaction rates not only promotes environmental sustainability but also improves industrial efficiency by optimizing reaction conditions (Arumugam et al., 2014).

Strong comprehension of rate laws is thus pivotal, bridging theoretical concepts to tangible advancements in catalysis and materials science.

### 4.4.3. Temperature Dependence of Reaction Rates

Temperature dependence of reaction rate is one of the fundamental concepts in the study of surface thermodynamics and kinetics since it significantly controls the character of chemical reactions. As Arrhenius equation postulates, reaction rates typically increase exponentially with a rise in temperature due to the greater kinetic energy of molecules. This increase of energy creates more frequent and effective collisions between reactants, leading to the greater capability of overcoming the activation energy barrier.

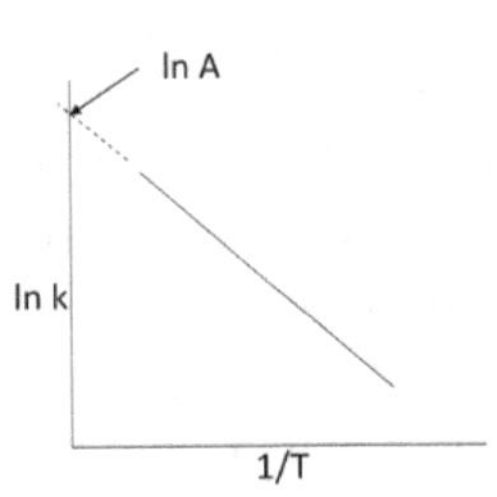


$$\ln k = -\left(\frac{Ea}{RT}\right) + \ln(A)$$

$$\text{slope} = -\left(\frac{Ea}{R}\right)$$

$$\text{y} - \text{intercept} = \ln(A)$$

Ea is the activation energy
R is the gas constant
T is temperature in degree Kelvin

**Figure 4.12.** Temperature dependence of reaction rates.

*Source:* https://external-content.duckduckgo.com/iu/?u=https%3A%2F%2Fimage1.slideserve.com%2F2939230%2Ftemperature-dependence-of-reaction-rates-l.jpg&f=1&nofb=1&ipt=ff44afe7950e9140161bfb49ef2942f5c00c48178e4f2baf5c6508c751d21672.

Furthermore, the temperature dependence is not only found in molecular collisions but also in structure and activity of surface atoms, and therefore catalytic reactions. Understanding this relationship between temperature, molecular motion, and reaction rates is relevant in making material performance and reaction efficiency judgments of a material in applications, particularly in industries where optimized conditions are required to increase product yield and lower energy costs. In addition, the interaction of multiple atomic-scale phenomena under varying temperatures can determine the characteristic properties of a material. Thermodynamically, with an increase in temperature, the entropy of a system gets higher, which can result in more varied reaction pathways, especially in advanced materials or catalysis systems. Studies confirm that most reactions include many different pathways, which might be enhanced or dampened under various temperatures. Therefore, the temperature not only regulates reaction kinetics for individual reaction but also changes the collective dynamics of these pathways. The resulting complexity demands advanced computational techniques and theoretical approaches able to capture the nuances of how temperature affects such micro-scale processes, demonstrating the need for modern methodology that unifies concepts from statistical mechanics and electronic structure theory to construct predictive models of material response (Gross et al., 2004) (Figure 4.12).

Temperature is critical in reactive systems, particularly in heterogeneous catalysis, to enhance self-organization and the emergence of dynamical order. Experiments conducted using field emission microscopy have established that nanoscale systems exhibit new dynamics under non-equilibrium conditions. Molecular structure on surfaces varies with varying temperatures, which maximizes self-organization, thereby optimizing catalytic activity. This is vital to the comprehension of how reactors may be designed for specific uses, e.g., improving conversion efficiency or selectivity in chemistries. The conditions under which these processes are conducted can be examined by scientists, who can then use the principles of nonequilibrium thermodynamics to design systems performing at maximum optimality under varying temperatures, as it gives insight into the very complex interplay between thermodynamic principles and catalysis (Gaspard et al., 2012).

## 4.5. CATALYSIS AND SURFACE REACTIONS

The intricate relationship between catalysis and surface reactions is supported by surface thermodynamics and kinetics principles in a fundamental manner. Catalysts, by definition, decrease the activation energy required for chemical reactions, thus making pathways potentially energetically unfavorable without them energetically accessible. The effectiveness of these reactions is largely a function of the surface properties of the catalyst, including surface area, morphology, and electronic structure. An in-depth understanding of these parameters allows one to tailor catalytic activity, a necessity in manufacturing processes. Recent studies employing operando techniques have clarified the dynamic behavior of catalysts under actual conditions, understanding changes in surface composition and catalytic performance that occur during reactions. This in situ observation highlights the importance of correlating experimental conditions with real-life operational conditions in order to achieve optimal utilization of catalysts in various industrial processes (Blomberg et al., 2021). Furthermore, kinetic modeling of catalytic reactions requires a high-level understanding of surface thermodynamics. For instance, correlation of reaction rates and surface temperatures can significantly influence catalytic behavior. Non-equilibrium thermodynamics offers the theoretical framework for the analysis of reaction heat and mass transfer, describing the effect of thermal gradients on reaction kinetics. When the oxidation of carbon monoxide over platinum is applied, surface temperatures from computation based on non-equilibrium thermodynamic models exhibit reaction rate-thermal driving force coupling, enhancing the knowledge of catalytic performance (Cruz et al., 2019). This kind of comprehension can be applied to classical reaction kinetics and can be utilized in the design of catalysts that will be highly effective over a range of temperatures and pressures and thereby maximize industrial production. In general, the study of catalysis and surface reaction is variegated and closely coupled with advances in surface thermodynamics and kinetics. Through the utilization of advanced techniques, such as operando spectroscopy, researchers are able to probe atomic-scale changes occurring on the surface of a catalyst during reaction and hence establish a direct link to catalytic activity. Advanced theory integrating surface temperature effects with reaction kinetics can, in principle, revolutionize our understanding of catalytic processes. This interaction between fundamental science and application not only stimulates the advancement of novel and improved industrial processes but also opens the door to the development of novel catalysts for targeted applications. As such, a good understanding of these interrelated concepts is crucial for catalytic science and engineering to make further progress in the future.

### 4.5.1. Role of Catalysts in Surface Kinetics

The role played by catalysts in surface kinetics is fundamentally based on their ability to lower activation energy barriers for chemical reactions. By providing an alternative reaction path, catalysts enhance the rate at which reactants become products. The effect is particularly significant at the surface of solid material, where most critical reactions, such as oxidation and reduction reactions, occur. Catalysts operate through the formation of transient complexes with reactants, stabilizing effectively the transition state and permitting conversion. The catalytic surfaces are thus intrinsic to their efficiency as the active site geometry and density dictate the surface reaction kinetics. An intricate understanding of the way these active sites influence different substrates must therefore be extracted in order to continue driving the progress of catalysis both in the laboratory and industrial settings. Recent research has highlighted the importance of non-active-site constituents in catalysis, suggesting that constituents can significantly influence surface kinetics. Constituents such as Mo, S, and N have been found to heavily influence not only catalytic site stability but also electronic states and reactivity. For instance, adsorption of thermodynamically stable oxyanions, which are formed from these elements, on the surface of active catalysts can control their behavior under working conditions. The consequence of this interaction is profound in its significance to critical reactions like the oxygen evolution reaction (OER), wherein the kinetics are highly responsive to changes in electronic environment. By considering these parameters, researchers are able to explore novel directions for catalyst design, with the intention of optimizing both efficiency and selectivity in catalytic reactions (Hausmann et al., 2022) (Figure 4.13).

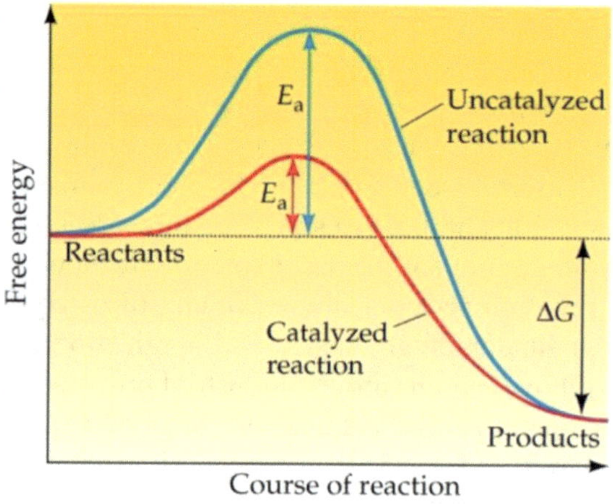


**Figure 4.13.** The role of catalysts in surface kinetics.

*Source:* https://external-content.duckduckgo.com/iu/?u=https %3A%2 F%2 Fslide-player.com%2Fslide%2F15136381%2F91%2Fimages%2F5%2FRole%2Bof%2Ba%2BCatalyst.jpg&f=1&nofb=1&ipt=0a4b565a2a30c5c371429d17e763fa82723b548d6f59271a1abe3f31bf8b0ef3.

The intricate play of interactions between catalysts emphasizes a deep transformation in catalytic science, especially in terms of surface electronic property manipulation. The interaction between active sites and anions has been found to totally transform reaction mechanisms such that catalysts are able to surmount traditional scaling relationships that limit performance. For example, the inclusion of (oxy)anions can aid in the adjustment of the electronic atmosphere of catalysts, thereby resulting in a more favorable adsorption of reaction intermediates. This process not only accelerates reaction rates but also maximizes the overall stability of catalytic materials under working conditions. With advances in the discipline, their translation to tangible applications is set to revolutionize catalysis, particularly in energy conversion technologies, thereby contributing significantly towards our decarbonization and sustainable energy goals (Hausmann et al., 2022).

### 4.5.2. Heterogeneous vs. Homogeneous Catalysis

The distinction between heterogeneous and homogeneous catalysis is intrinsic to the study of surface thermodynamics and kinetics, with significance in reaction pathway and efficiencies. Heterogeneous catalysis occurs at a phase boundary, usually solid catalysts with liquid or gaseous reactants. This type of catalysis will have unique properties of surfaces, such as high surface area and ability for surface property modification to enhance reactivity. For instance, recent research on single-atom catalysts has showcased the importance of metal-support interactions in modulating catalytic activity, particularly towards hydrogen evolution reactions. By modulating the oxidation states via these interactions, researchers have exhibited staggering performance enhancements, elucidating how structured surface environments can optimize reaction kinetics (Shi et al., 2021). Homogeneous catalysis, in contrast, involves catalysts that are in the same phase as the reactants, resulting in more homogeneously distributed and interacting catalyst-reactant molecules at the molecular levels. This process allows for greater control of reaction conditions such as concentration and temperature, and this usually leads to greater selectivity towards the desired products. Homogeneous catalysis mechanisms are easier to understand due to the absence of surface complexities inherent in heterogeneous systems. However, the challenge usually arises in recovery and separation of such catalysts after the reaction. Moreover, stability in the use of homogeneous catalysts in extreme conditions, such as those required for oxygen evolution reactions, poses significant challenges in devising useful catalytic protocols (An et al., 2021) (Figure 4.14).

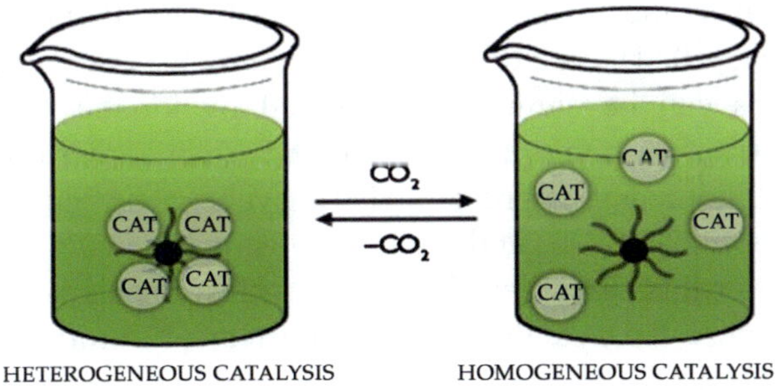


**Figure 4.14.** Differences between homogeneous catalysis and heterogeneous catalysis.

*Source:* https://external-content.duckduckgo.com/iu/?u=https%3 A%2F%2Fqsstudy.com%2Fwp-content%2Fuploads%2F2017%2F09%2FHomogeneous-catalysis.jpg&f=1&nofb=1&ipt=8ff74a6b1fbef221a5ec14bf5c122468dc1d1b6337d6aa05f6dfceb3d82c2221.

Lastly, the choice between homogeneous and heterogeneous catalysis is a function of specific reaction requirements and desired outcomes in catalytic activity. Understanding surface thermodynamics and kinetics of heterogenous systems can give one the most critical information to achieve optimum catalytic activity, whereas exploring characterizations of homogeneous systems allows further enhancement in selective reactivity. The dynamics of interaction between the two paradigms indicates that there is no closure to combining both paradigms' knowledge. As researchers tackle challenges in catalytic processes, their syntheses of heterogeneous and homogeneous catalysis knowledge will remain essential to future advancements in catalysis and applications towards sustainable energy (Shi et al., 2021; An et al., 2021).

### 4.5.3. Examples of Catalytic Reactions

In surface thermodynamics and kinetics, catalytic reactions are fundamental in describing several chemical processes. One of the most salient ones is the oxygen evolution reaction (OER), which has been widely emphasized due to its implementation in electrochemical energy conversion devices, such as water splitting. The OER can be analyzed by inspecting intermediates' activation energies, whose importance is a key factor determining the efficiency of heterogeneous catalysts. Advanced experimental methods, like time-resolved optical and vibrational spectroscopy, allow scientists to break down the discrete steps of the OER on electrode surfaces. For instance, an investigation of the n-doped $SrTiO_3$/aqueous interface has separated interaction dynamics of intermediates and uncovered details about the release of protons and electrons, crucial to the realization of efficient catalytic surfaces for energy production (Cuk, 2024). In addition, catalytic processes undergone during surface reactions can be significantly altered by kinetic models aimed at correlating the reaction rate with the energetics of the intermediates. Among them is the experimental approach where cyclic staircase voltammetry is simulated to examine surface catalytic mechanisms characterized by reversible chemical steps. In this direction, the complexities associated with reversible reactions are unraveled, yielding more informative information about the thermodynamics and kinetics of catalysis. These surface catalytic reactions are not only applicable for understanding energy conversion reactions like the OER but also have implications for other biochemical interactions, including drug-drug and drug-DNA interactions. Being a part of this assessment inquiry further enhances our knowledge of how structural and electronic factors come into play in catalytic efficiency (Petkovska et al., 2020). Besides the above examples, there is a need to examine catalytic reactions in environmental application, e.g., the oxidation of contaminants. Volatile organic compounds

(VOCs) catalytic oxidation on metal oxides are a prime example of the delicate interplay between thermodynamics and kinetics that occurs in surface reactions. The transition metals are exploited by catalytic converters in automotive use for the conversion of poisonous exhausts to less toxic substances through surface-bound pathways. Unexpectedly, the functioning of such catalysts is a function of the surface characteristics of the catalysts, e.g., the sites of active sites and electronic features. Enhanced activity normally arises from the surface properties of the catalysts having been specially optimized to achieve high adsorption and desorption processes, which in turn affect the overall kinetics of the oxidation reaction. Knowledge of these factors gives rise to more effective and sustainable catalysts for meeting environmental demands.

## SUMMARY

In surface thermodynamics and kinetics, the research highlights the tremendous advancement in the synthesis and application of new adsorbents. The adsorption capability of materials like sunflower stem activated carbon and graphene oxide proves the outstanding impact of surface properties tailored to particular functions on adsorption. As a case in point, the adsorption of crystal violet from water using graphene oxide as an adsorbent explains its satisfactory performance under optimized conditions with a monolayer adsorption capacity of 470.78 mg/g. These results highlight the pivotal position of surface interactions in the environmental clean-up and offer a possible way to minimize the impact of textile wastewater by efficient adsorption techniques. As such materials undergo intense characterization, implications extend beyond strictly theoretical development, to practical solutions for near-term environmental problems. Further information is from investigations of KOH-treated sunflower stem activated carbon, demonstrating extreme effectiveness for pollutant abatement. This process yielded a BET surface area of 452 $m^2/g$ and emphasized enhanced adsorption performance for bisphenol A and phenol, emphasizing the functionality of chemical activation in material performance. The spontaneity of the adsorption process along with the favorable thermodynamic properties makes KOH-SSAC a suitable alternative for scale-up operations in water treatment. This enables comparative assessment to sunflower stem biochar, illustrating that new approaches in surface treatment are able to lead to quite significant improvement in efficiency of adsorption—essential for the future of water cleaning technology. The overall conclusion is that optimization of such synthetic protocols holds important promise for widening the range available to effective environmental cleansing. Moreover, the implication of these results adds to greater understanding of surface thermodynamic principles and their kinetic directions. The use of models such as pseudosecond-order and Langmuir isotherm gives a platform for the prediction of adsorption phenomena, and therefore designing more efficient systems. As the investigation concludes, it is apparent that innovating synthesis and extensive exploration of adsorption materials, such as those highlighted within graphene oxide and sunflower stem activated carbon studies, provide a pathway for future research. Applying advanced characterization methodologies provides a deeper understanding of material interactions and functionalizations and facilitates achieving the sustainable water resource management. This research not only validates the role of surface thermodynamics and kinetics but also inspires ongoing innovation within the field towards solutions with the potential to meet the demands of the contemporary environment.

Surface thermodynamics and kinetics are at the center of explaining most physical and chemical phenomena occurring on material surfaces. At the center of explaining such phenomena is the definition of biosorbents, for instance, chicken bone biomass valorization for wastewater treatment. With the aid of advanced characterization techniques like Fourier-transform infrared spectroscopy (FTIR) and scanning electron microscopy (SEM), researchers have highlighted the influence of modification processes, for instance, calcination, on biomass's structural characteristics. The higher surface area from 247 $m^2/g$ for native biomass to 370 $m^2/g$ for calcined biomass optimizes adsorption capacities, as a vital correlation between surface characteristics and thermodynamic efficiencies for pollutant removal. Such research highlights the eco-friendliness of utilizing organic waste materials as effective biosorbents, thus improving environmental remediation and waste management protocols (Ouzani et al., 2025). In yet another area of surface thermodynamics and kinetics, material catalytic action, e.g., that of the undoped $Mo_2C$ MXene, has shown great potential in catalyzing the reverse water gas shift (RWGS) reaction. This theoretical study discusses various mechanistic pathways, i.e., redox and formate mechanisms, and how Gibbs free energy profiles control the thermodynamics of the reactions. Importantly, the relationship between reaction endergonicity and the operating parameters required predicts temperature sensitivity and reactant pressures in forcing the reaction (Dolz et al., 2024). The modeling process, through microkinetic simulations, signals the competitiveness of different mechanisms at differing conditions and hence the possibility of better realization of the surface interactions that govern overall catalytic efficiency. Both biosorption and catalysis studies show results that point out the significance of surface properties to kinetic and thermodynamic processes. For instance, higher adsorption activity in calcined chicken bone biomass suggests an in-built correlation between surface area and adsorption capacity, while reaction mechanisms in RWGS show the significance of optimal surface properties for catalytic performance. Also, the study of sorbent and catalyst behavior under different conditions not only adds to intellectual discourse in surface thermodynamics but also has real-life implications for environmental applications and industrial processes. This union of biosorption and catalysis best portrays the function of surface dynamics towards addressing problems of the current times, such as relieving pollution and energy conversion, offering windows to future research and technology developments (Ouzani et al., 2025; Dolz et al., 2024).

# REFERENCES


1. Akbashev, A. (2023). Electrochemical colloquium and beyond: Opportunities for online education in electrochemistry. *ECS Meeting Abstracts*. https://www.semanticscholar.org/paper/cc29418ca1ea0848c7056cc436ffe5c3a7beea26 (accessed on 3rd October 2025).
2. Alruwais, R. S. (2024). Investigating dye adsorption: The role of surface-modified montmorillonite nanoclay in kinetics, isotherms, and thermodynamics. Open Chemistry, 22. https://www.semanticscholar.org/paper/dd784ba0d7eb659a89ea0b26c0a65d631436f9f2 (accessed on 3rd October 2025).
3. An, L., Wei, C., Lu, M., Liu, H., Chen, Y., Scherer, G. G., Fisher, A. C., et al. (2021). Recent development of oxygen evolution electrocatalysts in acidic environment. Advanced Materials, 33. https://doi.org/10.1002/adma.202006328 (accessed on 3rd October 2025).
4. An, Y., Cao, W., Min, O., Chen, S., Wang, G., & Chen, X. (2023). Substantial impact of surface charges on electrochemical reaction kinetics on S vacancies of $MoS_2$ using grand-canonical iteration method. The Journal of Chemical Physics, 159(14). https://www.semanticscholar.org/paper/33753ca4a8edac76772892e2a4c7146ee62a7794 (accessed on 3rd October 2025).
5. Arumugam, A., & Becker, U. (2014). Computational redox potential predictions: Applications to inorganic and organic aqueous complexes, and complexes adsorbed to mineral surfaces. Minerals, 4, 345–387. https://www.semanticscholar.org/paper/ff64b8c1e355a4ab21795aa1f16b4deb2df6bb2d (accessed on 3rd October 2025).
6. Bazant, M. Z., & Fraggedakis, D. (2020). Tuning the stability of electrochemical interfaces by electron transfer reactions. http://arxiv.org/abs/2003.03057 (accessed on 3rd October 2025).
7. Blomberg, S., Hejral, U., Shipilin, M., Albertin, S., Karlsson, H., Hulteberg, C. P., Lömker, P., et al. (2021). Bridging the pressure gap in CO oxidation. ACS Catalysis, 11, 9128–9135. https://www.semanticscholar.org/paper/c0f2eb28f4388714c33bce65aa62b292a793459d (accessed on 3rd October 2025).
8. Bogaerts, A., Tu, X., Whitehead, J. C., Centi, G., Lefferts, L., Guaitella, O., Azzolina-Jury, F., et al. (2020). The 2020 plasma catalysis roadmap. *Journal of Physics D: Applied Physics, 53*, 443001–443001. https://doi.org/10.1088/1361-6463/ab9048 (accessed on 3rd October 2025).

9. Chen, G. (2024). Research progress in magnesium-based solid hydrogen storage. *Highlights in Science, Engineering and Technology*. https://www.semanticscholar.org/paper/e547726f251bba08b3a23e13c591de2ca6f4d214 (accessed on 3rd October 2025).

10. Chen, Z., Liu, H., Fan, S., Zhang, Q., Yuan, C., Peng, W., Li, Y., et al. (2024). Inhibition of vanadium cathode dissolution in zinc-ion batteries on thermodynamics and kinetics by guest pre-intercalation. Advanced Energy Materials, 14. https://www.semanticscholar.org/paper/aaa4ba97b04b395e7ff7b0158769c54c154f931b (accessed on 3rd October 2025).

11. Chesworth, W. (2008). Encyclopedia of Soil Science. https://www.semanticscholar.org/paper/2ec11307dddb3f1d9d265464d8b20582899073c5 (accessed on 3rd October 2025).

12. Cruz, A., Barragán, D., Magnanelli, E., Lervik, A., & Kjelstrup, S. (2019). Non-equilibrium thermodynamics as a tool to compute temperature at the catalyst surface. *Physical Chemistry Chemical Physics: PCCP*. https://www.semanticscholar.org/paper/def925d3c94a69c8ea6099f29423d6e37400f339 (accessed on 3rd October 2025).

13. Cuk, T. (2024). (Invited) Resolving catalytic mechanism at an electrode surface: Thermodynamics & kinetics of reaction steps. ECS Meeting Abstracts. https://www.semanticscholar.org/paper/7861896d21bf18784a2768aeef18efb1b922e9bd (accessed on 3rd October 2025).

14. Delmo, E. P., Wang, Y., Song, Y., Zhu, S., Zhang, H., Xu, H., Li, T., et al. (2024). *In situ* infrared spectroscopic evidence of enhanced electrochemical $CO_2$ reduction and C–C coupling on oxide-derived copper. *Journal of the American Chemical Society, 146*, 1935–1945. https://doi.org/10.1021/jacs.3c08927 (accessed on 3rd October 2025).

15. Dolz, D., De Armas, R., Lozano-Reis, P., Morales-García, Á., Viñes, F., Sayós, R., & Illas, F. (2024). Understanding the reverse water gas shift reaction over $Mo_2C$ MXene catalyst: A holistic computational analysis. *ChemCatChem, 16*. https://www.semanticscholar.org/paper/15e3fe9b3cfbc2a4867ccb439a6d96dd0857e20d (accessed on 3rd October 2025).

16. Dziejarski, B., Serafin, J., Andersson, K., & Krzyżyńska, R. (2023). $CO_2$ capture materials: A review of current trends and future challenges. *Materials Today Sustainability, 24*, 100483–100483. https://doi.org/10.1016/j.mtsust.2023.100483 (accessed on 3rd October 2025).

17. El Bourachdi, S., El Ouadrhiri, F., Moussaoui, F., Saleh, E. M., El Amri, A., Althomali, R. H., Kassem, A. F., et al. (2024). Enhancing graphene oxide production and its efficacy in adsorbing crystal violet: An in-depth study of thermodynamics, kinetics, and DFT analysis. International Journal of Chemical Engineering. https://www.semanticscholar.org/paper/dc5141288f1aa218a507a580694f136fe363f6b3 (accessed on 3rd October 2025).

18. El-Hoshoudy, A. N., Matallah, M., Gouzi, H., Saidat, B., Khane, Y., Chabani, M., & Nouacer, I. (2024). Bio removal of lead ion from the aquatic environment using lignocellulosic (*Zea mays*), thermodynamics modeling, and MC simulation. *International Journal of Environmental Science and Technology*. https://www.semanticscholar.org/paper/2743ba161861a8cd9a50fd9ca40411a8ea8dadb4 (accessed on 3rd October 2025).

19. Eze, N. O., Ogbonna, O. N., & Samuel, N. (2025). Isotherm, kinetics and thermodynamics of adsorption of bisphenol A onto modified and unmodified corn cob: A comparative study. World Journal of Advanced Research and Reviews. https://www.semanticscholar.org/paper/b461e21c915c6cbbab1600da22732562f5526648 (accessed on 3rd October 2025).

20. Gaspard, P. (2012). Self-organization at the nanoscale scale in far-from-equilibrium surface reactions and copolymerization's. http://arxiv.org/abs/1203.0972 (accessed on 3rd October 2025).

21. Gill, I. S., Mukhopadhyay, S. S., & Parshad, V. R. (2009). Nanoscience and nano-technology: Cracking prodigal farming. https://core.ac.uk/download/pdf/288475.pdf (accessed on 3rd October 2025).

22. Gong, E., Ali, S., Hiragond, C. B., Kim, H. S., Powar, N. S., Kim, D., Kim, H., et al. (2021). Solar fuels: Research and development strategies to accelerate photocatalytic $CO_2$ conversion into hydrocarbon fuels. *Energy & Environmental Science, 15*, 880–937. https://doi.org/10.1039/d1ee02714j (accessed on 3rd October 2025).

23. Gopal, A. (2015). Ab-initio and experimental techniques for studying non-stoichiometry and oxygen transport in mixed conducting oxides. https://www.semanticscholar.org/paper/d94724c7c60a9b52f69d8fe768d024db07daef20 (accessed on 3rd October 2025).

24. Gorban, A. N., Barenblatt, G., Barenblatt, G., Bertini, L., Blesgen, T., Boudart, M., Briggs, T., et al. (2011). Quasichemical models of multicomponent nonlinear diffusion. http://arxiv.org/abs/1012.2908 (accessed on 3rd October 2025).

25. Gross, A., Michaelides, A., Van de Walle, A., Zangwill, A., Zunger, A., Bortz, A. B., Voter, A. F., et al. (2004). Ab initio atomistic thermodynamics and statistical mechanics of surface properties and functions. http://arxiv.org/abs/cond-mat/0404510 (accessed on 3rd October 2025).
26. Guo, K. (2023). Applications of metal-organic frameworks in methane storage. Highlights in Science, Engineering and Technology. https://www.semanticscholar.org/paper/addc8cc43aeaccd94a9f6bc856e3b0e5be37225a (accessed on 3rd October 2025).
27. Hähnel, T., Frenzel, S., Möllmer, J., Reichenbach, C., & Kalies, G. (2017). Immersion enthalpies and adsorption isotherms of liquids on carbon molecular sieves. Diffusion Fundamentals. https://www.semanticscholar.org/paper/5fecfb6898f42ed347ce73fa82a117d04f6012a7 (accessed on 3rd October 2025).
28. Hausmann, J. N., & Menezes, P. W. (2022). Effect of surface-adsorbed and intercalated (oxy)anions on the oxygen evolution reaction. *Angewandte Chemie International Edition, 61*. https://doi.org/10.1002/anie.202207279 (accessed on 3rd October 2025).
29. He, J., Zhang, H., Yue, T., Sun, W., Hu, Y., & Zhang, C. (2021). Effects of hydration on the adsorption of benzohydroxamic acid on the lead-ion-activated cassiterite surface: A DFT study. *Langmuir, 37*, 2205–2212. https://doi.org/10.1021/acs.langmuir.0c03575 (accessed on 3rd October 2025).
30. Hwang, A., Getsoian, A., & Iglesia, E. (2024). Kinetics, mechanism, and thermodynamics of ceria-zirconia reduction. *ACS Catalysis*. https://www.semanticscholar.org/paper/3d7d9dee019c141a3c12e54bcbe1e80587f82501 (accessed on 3rd October 2025).
31. Jabłońska, B., Jabłoński, P., & Gęga, J. (2025). Kinetics and thermodynamics of Pb (II), Zn (II), and Cd (II) adsorption from aqueous solutions onto activated biochar obtained from tobacco waste. *Materials, 18*. https://www.semanticscholar.org/paper/b0216a3400d95cd1b747211d311183871a3ac584 (accessed on 3rd October 2025).
32. Joudeh, N., & Linke, D. (2022). Nanoparticle classification, physicochemical properties, characterization, and applications: A comprehensive review for biologists. Journal of Nanobiotechnology, 20. https://doi.org/10.1186/s12951-022-01477-8 (accessed on 3rd October 2025).

33. Kumar, S. S., & Lim, H. (2022). An overview of water electrolysis technologies for green hydrogen production. Energy Reports, 8, 13793–13813. https://doi.org/10.1016/j.egyr.2022.10.127 (accessed on 3rd October 2025).

34. Libbrecht, K. (2019). Snow crystals. arXiv: Materials Science. https://www.semanticscholar.org/paper/e28dda715134972b20f2a204003944b0c539fbe6 (accessed on 3rd October 2025).

35. Lingamdinne, A. P., Angaru, G., Kumar Pal, C. A., Koduru, J. R., Karri, R., Mubarak, N. M., & Chang, Y. Y. (2024). Insights into kinetics, thermodynamics, and mechanisms of chemically activated sunflower stem biochar for removal of phenol and bisphenol-A from wastewater. Scientific Reports, 14. https://www.semanticscholar.org/paper/dad88a3808e28da994c66bcdd71caeaa42be1025 (accessed on 3rd October 2025).

36. Liu, J., Xu, N., Shi, D. D., Cao, Y., Zhang, T. Y., Zhong, K. P., Zhao, Q., et al. (2024). Preparation, characterization, and adsorption performance of $H_2TiO_3$ lithium-ion sieves with homogeneous nanoparticles: Kinetics, thermodynamics, and mechanism analysis. Chemistry Select. https://www.semanticscholar.org/paper/016a6e396db7f053953e47c55567e804ebf316f0 (accessed on 3rd October 2025).

37. Masoudi Soltani, S., Lahiri, A., Bahzad, H., Clough, P. T., Gorbounov, M., & Yan, Y. (2021). Sorption-enhanced steam methane reforming for combined $CO_2$ capture and hydrogen production: A state-of-the-art review. Carbon Capture Science & Technology, 1, 100003–100003. https://doi.org/10.1016/j.ccst.2021.100003 (accessed on 3rd October 2025).

38. Mejuto, A. (2020). Introducing *Compounds*: An interdisciplinary open access journal. *Compounds, 1*, 1–2. https://www.semanticscholar.org/paper/523162333a366a6a8c086bf7978554b691860673 (accessed on 3rd October 2025).

39. Noé, F., Tkatchenko, A., Müller, K. R., & Clementi, C. (2020). Machine learning for molecular simulation. *Annual Review of Physical Chemistry, 71*, 361–390. https://doi.org/10.1146/annurev-physchem-042018-052331 (accessed on 3rd October 2025).

40. Ouzani, A., Zouambia, Y., Maachou, H., Krea, M., Assadi, A. A., Khezami, L., Benguerba, Y., et al. (2025). Sustainable removal of basic fuchsine and methylene blue dyes using chicken bone biomass: Thermodynamics, kinetics, and insights from experimental studies and decision tree with least squares boosting predictive modeling. *Water*. https://www.

semanticscholar.org/paper/6c008f8f3601a0e6839dcb1280f7166d56dc6846 (accessed on 3rd October 2025).

41. Petkovska, S., & Gulaboski, R. (2020). Theoretical analysis of a surface catalytic mechanism associated with reversible chemical reaction under conditions of cyclic staircase voltammetry. Electroanalysis. https://www.semanticscholar.org/paper/606b5be5bf69aa1fd9555e5e600a36508d3178c8 (accessed on 3rd October 2025).
42. Qi, Y., & Zhang, B. (2021). Chromatin network retards nucleoli coalescence. Nature Communications, 12. https://www.semanticscholar.org/paper/855f1b3e8f07d1389b15ec224880980c9c369396 (accessed on 3rd October 2025).
43. Qiu, R. (2019). Crystal growth for beginners. Acta Crystallographic Section B: Structural Science, Crystal Engineering and Materials, 75(Pt 6), 1226–1227. https://www.semanticscholar.org/paper/95e6d71c7fd885e70d806182f1d51e0bd805ba2b (accessed on 3rd October 2025).
44. Roque-Malherbe, R. (2009). The Physical Chemistry of Materials: Energy and Environmental Applications. https://www.semanticscholar.org/paper/6cec86d5b2c30e2b9ae38c39b50249bd3e4ee312 (accessed on 3rd October 2025).
45. Samanta, S. K., Adak, S., & Tripathy, T. (2024). Adsorption studies of toxic organic pesticides from aqueous medium by sodium alginate-cl-poly (allyl amine-co-acrylic acid)/copper oxide nanocomposite hydrogel: Isotherms, kinetics, and thermodynamics. Polymer Engineering & Science. https://www.semanticscholar.org/paper/3b32cb7829a15a1f2b7e9358cb479f0142ee1702 (accessed on 3rd October 2025).
46. Shi, Y., & Beck, T. (2020). Absolute ion hydration free energy scale and the surface potential of water via quantum simulation. Proceedings of the National Academy of Sciences, 117, 30151–30158. https://www.semanticscholar.org/paper/5828cfd101d69fb16de7942f83e344960b3c01ca (accessed on 3rd October 2025).
47. Shi, Y., Ma, Z., Xiao, Y., Yin, Y., Huang, W., Huang, Z., Zheng, Y., et al. (2021). Electronic metal–support interaction modulates single-atom platinum catalysis for hydrogen evolution reaction. Nature Communications, 12. https://doi.org/10.1038/s41467-021-23306-6 (accessed on 3rd October 2025).
48. Tan, M. J. (2008). Effect of alumina-formers addition on the isothermal oxidation of Ti-Al based intermetallic. https://core.ac.uk/download/11785171.pdf (accessed on 3rd October 2025).

49. Wang, A., Kadam, S., Li, H., Shi, S., & Qi, Y. (2018). Review on modeling of the anode solid electrolyte interphase (SEI) for lithium-ion batteries. *npj Computational Materials, 4.* https://doi.org/10.1038/s41524-018-0064-0 (accessed on 3rd October 2025).

50. Wang, J., Arantes, P., Bhattarai, A., Hsu, R. V., Pawnikar, S., Huang, Y. M. M., Palermo, G., et al. (2021). Gaussian accelerated molecular dynamics: Principles and applications. Wiley Interdisciplinary Reviews: Computational Molecular Science, 11. https://doi.org/10.1002/wcms.1521 (accessed on 3rd October 2025).

51. Zein, R., Mashelina, R., Emriadi, Fauzia, S., & Ramadhani, P. (2025). Lignocellulosic-based materials from kapok husk (Ceiba pentandra L. Gaertn) wastes for indigo carmine dye removal: Isotherms, kinetics, thermodynamics, and reusability studies. Environmental Progress & Sustainable Energy. https://www.semanticscholar.org/paper/79e42a65a57202377c552378171498af9b57fa97 (accessed on 3rd October 2025).

52. Zhang, J., Chen, H., Duan, X., Sun, H., & Wang, S. (2023). Photothermal catalysis: From fundamentals to practical applications. Materials Today, 68, 234–253. https://doi.org/10.1016/j.mattod.2023.06.017 (accessed on 3rd October 2025).

CHAPTER 5

# CATALYSIS AT SURFACES

## Contents



## INTRODUCTION

In surface catalysis, an understanding of the fundamental principles and recent research is essential to scientific advance as well as day-to-day uses. Catalysis, the process of making a chemical reaction proceed at a faster rate through the introduction of a substance that is consumed neither in the reactants nor in the products, is a vital process in many industrial processes, such as petrochemical refining and pharmaceuticals. Surfaces are the focus due to extensive interactions at solid-liquid or solid-gas interfaces, where reactants are transformed into products. As there is more demand for more efficient and selective catalytic systems, researchers are exploring new directions that utilize surface properties to enhance catalytic activity. This preface gives the background for comprehensive treatment of the materials and processes that are the cause of catalysis at surfaces, focusing on the necessity for innovation within this fast-moving field. The synthesis and characterization of new catalytic materials are intrinsic aspects of the exploration of catalysis at surfaces. Advances within the literature have highlighted the necessity for increased nuclearity clusters, particularly as pertaining to ruthenium-containing metal complexes. For example, the studies done in Guo et al. (1999) investigate the synthesis of arene-ruthenium-sulfur clusters and their corresponding nitrogen analogs. Such studies not only give insight into the structural complexity of the clusters through techniques like X-ray crystallography but also offer information about their electronic properties and catalytic performances. Through the investigation of the special features of such complexes, researchers gain additional information on how the electronic structure and the environment influence their catalytic activity. Such a combination of experimental data and theoretical models serves as the foundation for the design of more efficient catalytic systems. Despite the advances in homogeneous catalysis, it remains a difficult problem to model heterogeneous catalysts due to the complexity of reaction pathways and intermediate species involved. The necessity of a wider theoretical foundation for heterogeneous catalysis is brought out in (Voecks, 2025), which discusses the difficulties in industrial processes that demand high-performance catalysts. These emanate from the need to combine empirical experimental information and computational modeling in being able to correctly predict the dynamics of surface reaction. While scientists strive towards optimizing catalysts for higher selectivity and stability, concerted efforts towards knowledge translation from homogeneous catalysis can make important guidance. The future of surface catalysis lies in the interaction of experimental validation and advanced computational methods, resulting in innovative solutions that suit the demands of today's chemical industries.

## *Definition of Catalysis*

Catalysis, a fundamental concept of chemistry, is the process by which the speed of a chemical reaction is sped up by the introduction of a substance known as a catalyst. Unlike reactants, catalysts participate in reactions but are not consumed, which allows them to catalyze repeated cycles of reaction. This kind of characteristic makes catalysis important to most industrial processes and natural phenomena, particularly at surfaces where most reactions occur. In heterogeneous catalysis, which is at the interface of a solid catalyst and liquid or gaseous reagents, the surface character of the catalyst dictates its performance. The interplay between surface area, morphology, and electronic properties decides the catalytic activity and selectivity required for productive chemical conversions. Catalysis significance comes to light when one looks at its applications in industries, from petrochemicals to pharmaceuticals. Apart from enhancing reaction rates, catalysts allow for the process of reactions to take place under less severe conditions, hence saving energy and fostering higher environmental sustainability. Zeolites are a case in point that has evolved to be central as catalysis materials, with the platform used to purify processes and catalytic synthesis of complicated molecules.

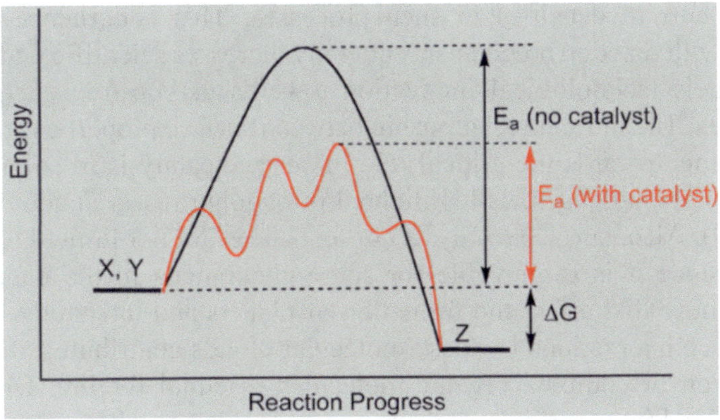


**Figure 5.1.** *A catalyst permits a different energy pathway for a chemical reaction which has a lower activation energy. The catalyst is not consumed in the chemical reaction.*

*Source: https://www.thoughtco.com/definition-of-catalyst-604402*

## Importance of Surface Catalysis in Chemical Reactions

Surface catalysis is of significant importance in making chemical reactions more efficient, especially in industrial applications where reaction rate and selectivity are of utmost importance. Through the creation of reactive sites, catalysts greatly reduce activation energy barriers, enhancing reaction rates. This is predominantly important in heterogeneous catalysis, where catalysts are in a phase different from reactants. For instance, metal nanoparticle-supported catalysts are commonly applied to reactions such as hydrogenation or oxidation. Because of their unique surface features, they facilitate maximum adsorption of reactants and hence make the breaking and formation of chemical bonds simple. Thus, the production of catalysts with designed surface properties is key to attaining maximum efficacy and intended reaction product, making surface catalysis a vitally important field in modern chemical engineering and industrial chemistry. Additionally, the quest for very selective catalysts has become the central issue in research aiming to minimize byproduct formation and maximize process sustainability. A pertinent area is the electrochemical reduction of carbon dioxide employing surface catalysis to convert wasteful emissions into useful carbon products. Here, determining the overriding mechanistic pathways on the catalysts surface is the key for surface optimization and maximization of selectivity and efficiency. For example, recent developments in catalytic design have focused on optimizing surface structures for reducing overpotentials but increasing current densities in such processes. This congruence of surface properties with reaction mechanisms not only increases scientific understanding but also fuels technological innovation towards carbon-free energy storage technologies. The intricate relationship between surface properties and catalytic activity points towards the critical role of surface catalysis in solving today's environmental challenges, as highlighted in ongoing research activities (Luna et al., 2019). Also, surface catalysis consequences are not limited to industrial processes since it is responsible for some phenomena in the natural world, including those that affect the formation of planets and interstellar chemistry. Grain surface interactions in dense molecular clouds contribute extensively to the formation of complex organic molecules essential for life. Observations have explained how surface catalysis can maintain gas phase element ratios, influencing the chemical development of such clouds as they progress into denser regimes where star and planet formation occurs. These results are of utmost relevance for understanding both astrochemical processes and the principles governing surface reaction. As research continues to untangle the complexities of surface interactions, it more clearly emerges that the importance of surface catalysis transcends traditional chemical limits to influence everything from materials science to planetary formation theories (Blake et al., 1988).

## 5.1. FUNDAMENTALS OF CATALYSIS

Surface catalysis is at the forefront of advancing our understanding of a broad variety of chemical reactions, particularly in industry. A number of reactants and catalysts most commonly interact at solid surfaces, where the catalytic character enhances reaction rate over non-catalytic conditions. Surface-based catalysis allows for more efficient utilization of energy as well as product formation. For instance, the employment of mesoporous silica as chiral ketone catalyst support demonstrates a significant achievement in asymmetric synthesis, when such heterogeneous catalysts have exhibited satisfactory enantioselectivities, up to 90% e.e, in the epoxidation of olefins with Oxone (Adam et al., 2013). These findings emphasize the top priority of surface properties in deciding the efficiency of catalytic reactions and show how the modification of structure can lead to improved catalytic performance. Having an idea about catalysis basics also entails an awareness of the structural and electronic features of catalysts, particularly in emerging materials like graphene. Graphene-like materials have gained widespread attention in electrocatalysis because they possess remarkable electrochemical properties. Such materials promote efficient oxygen electroreduction, which is critical for fuel cell technology. Research into these revolutionary catalysts determines the importance of creating catalytically active sites by adding non-metallic heteroatoms, which can potentially optimize catalytic performance and stability (Alammari et al., 2018). Scientists can also enhance not only the efficiency but also the longevity of catalysis at surfaces by designing hierarchical nanostructures and controlling active site distribution. This advancement expands the range of catalysis and sheds light on the intricacy of surface reactions. Furthermore, the study of catalytic reactions is inherently connected with a detailed exploration of adsorption processes, which governs the interaction among reactants and active sites on catalyst surfaces. This interaction is significant because it dictates reaction selectivity and rate and thus adsorption is also an inherent component of catalysis science. Advanced surface characterization methods, such as microscopy and spectroscopy, allow scientists to probe the interactions at the atomic level, revealing how the direction of molecules and the surface topology affect catalytic activity. Moreover, the inclusion of theoretical approaches, such as density functional theory, can be utilized to precalculate and rationalize the performance of catalysts in various reaction conditions. Jointly, they do not only further our understanding of surface catalysis but also pave the way for developing tomorrow's catalysts specifically for specific industrial purposes.

## 51.1. Mechanism of Catalytic Processes

The mechanism of catalytic process is based on the contact of the reactants with the active sites of the catalyst. Catalysts, being primarily metals or metal oxides, operate by providing an alternate reaction pathway with lower activation energy. This is particularly emphasized in heterogeneous catalysis, where reaction occurs at the interface between a solid catalyst and liquid or gaseous reactants. Surface structure and composition of the catalysts are significant in the determination of their efficiency. For instance, in alloy systems such as Ag-Cu, segregation of copper at the surface can enhance selectivity in reactions such as ethylene epoxidation, showing the advantage of tailoring surfaces to deliver the best catalytic performance (Bocquet et al., 2008). Hence, identification of these mechanisms at the molecular level is important for the rational design of more efficient catalysts. The examination of single reactions can unveil the richness of catalytic mechanisms. The distinctive catalytic behavior of different crystal faces is reported, for example, by oxidation of carbon monoxide (CO) on $RuO_2$ surfaces. Another study based on density-functional theory and kinetic Monte Carlo simulations underlined that the (111) face of $RuO_2$ is characterized by varying binding energetics and elementary processes in contrast to the (110) face. Despite these microscopic differences, both faces displayed comparable catalytic activity under certain reaction conditions, presenting a challenge to conventional views of structure sensitivity in catalysis (Reuter et al., 2015). This finding emphasizes the relevance of considering not only the geometric factors of the catalyst but also the thermodynamic and kinetic factors governing the reaction regime. The connection between surface structure and catalytic activity is supplemented by other factors such as temperature, pressure, and composition of the reactants. Through the knowledge of these interactions, it is attainable to foretell and design catalytic properties to develop more efficient catalysts. For the Ag-Cu alloy system, the surface phase diagrams constructed under different conditions can identify which surface structures are most stable and suitable for specific reactions. Ongoing progress in computational techniques, including first-principles calculations, enhances our predictive capabilities toward surface catalytic activity. Such information is instrumental to the advancement of catalysis research, where catalyst synthesis with desired functionalities can have significant economic and environmental consequences on industrial processes, thus resulting in cleaner chemical practice (Figure 5.2).

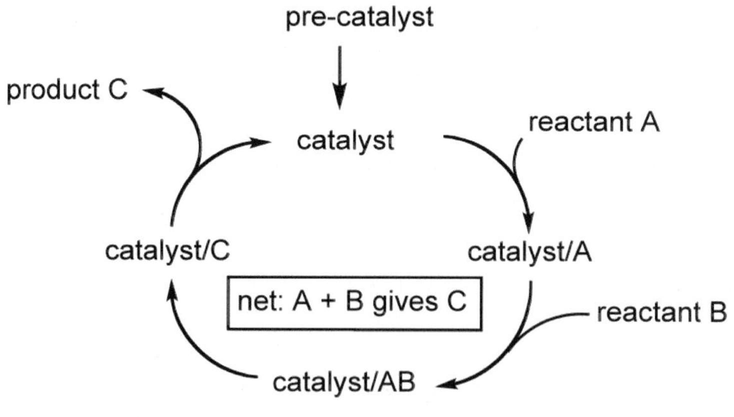


**Figure 5.2.** Catalytic cycle for conversion of A and B into C.

Source: https://upload.wikimedia.org/wikipedia/commons/1/10/Catcycle.png.

## 5.1.2. Types of Catalysts: Homogeneous vs. Heterogeneous

In catalysis, catalysts are typically divided into two types: homogeneous and heterogeneous. Homogeneous catalysts are those which are in the same phase as the reacting species, typically dissolved in a solvent. The homogeneity facilitates identical interaction between the catalyst and reactants and leads to steady kinetics and enhanced selectivity for most reactions. However, the primary drawback is presented by their difficulty of separation from the reaction mixture after the process, which may limit their applicability. Besides, homogeneous catalysts are prone to sensitivity to alteration of their reaction environment, and this can lead to loss of selectivity or activity. This sensibility is different from heterogeneous catalysts, which are in a different phase, usually solid, and interact with gas or liquid reactants on their surface, thus easily separable and recyclable after reaction. Heterogeneous catalysts have gained more significance in most industrial uses due to the fact that they are sturdy and do not demand hard handling. As opposed to homogeneous catalysts, heterogeneous catalysts typically consist of active sites embedded within a solid matrix so that reactions occur on exposed surfaces. Not only does this design improve the stability of the catalysts, but it can lead to increased reaction speeds due to the larger surface area in contact. For instance, metal-organic frameworks (MOFs) are new heterogeneous catalysts with tailor-made properties, mimicking enzymatic activity but with higher stability and recyclability, as in Wang et al. (2023).

The confinement of MOFs is used to enable new catalytic processes and increase selectivity, which renders them highly useful in complex reactions where precise control over product distribution is more than necessary. The differences between homogeneous and heterogeneous catalysts also encompass the mechanism underlying them, as well as the type of reaction that they catalyze. Copper complexes are effective electrocatalysts for the reduction of carbon dioxide to desirable chemicals when employed in electrochemical operations, varying in selectivity based on whether they exist under homogeneous or heterogeneous conditions. Small mononuclear complexes favor certain products, while dinuclear complexes favor more complex hydrocarbons on immobilized surfaces for heterogeneous catalysis, as demonstrated in Kim et al. (2022). This flexibility highlights the necessity of catalyst design, where the choice between homogeneous and heterogeneous systems hinges on factors like efficiency, selectivity, and catalyst recovery ease. Lastly, catalyst type identification is a major consideration in determining the efficiency and sustainability of chemical reactions, emphasizing the need to understand their respective strengths and weaknesses (Figure 5.3).

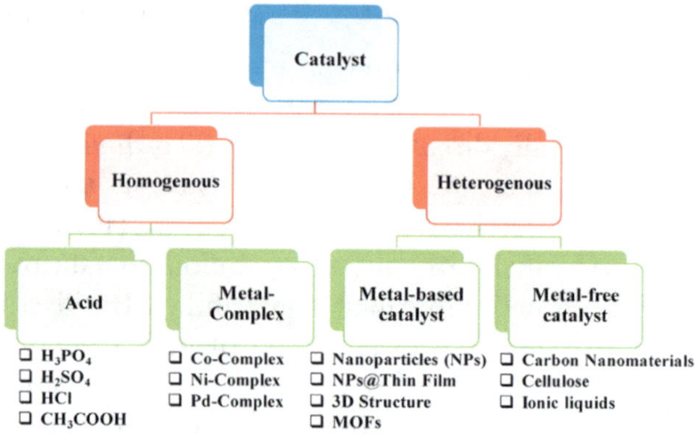


**Figure 5.3.** Types of catalysts.

*Source:* https://www.researchgate.net/profile/Hani-Abdelhamid/publication/346884387/figure/fig3/AS:972957757231105@1608982614983/e-Types-of-catalysts-used-for-the-hydrogen-generation-using-the-hydrolysis-of-NaBH-4.jpg.

## 5.1.3. Role of Activation Energy in Catalysis

Understanding the role of activation energy in catalysis is crucial to describe how chemical reactions can be accelerated. Activation energy is a threshold amount of energy required to initiate a chemical reaction, and it is a barrier that reactants must overcome for an alteration to occur. In surface catalysis, this concept takes center stage, as catalysts deliver active sites with the potential to lower activation energy significantly, facilitating reactions at far lower temperatures and pressures than would otherwise be the case. By transforming the landscape of energies, catalysts can better direct the creation of transition states than reactions that are uncatalyzed, and hence optimize the reaction rates and selectivity. This catalytic efficiency is essential in most industrial processes in which energy consumption and rates of reactions must be restricted. Processes by which catalysts lower activation energy involve various interactions involving the reactant molecules and the catalyst at the surface. When a substrate binds to a catalyst, it has a tendency to form intermediates which possess lower activation energy barriers compared to the uncatalyzed process. This can typically be observed in electrochemical reactions, with catalytic surfaces enabling carbon dioxide to be utilized in the form of valuable chemical products. In such cases, the organization of catalyst materials needs to be conceived with a deep understanding of the mechanism underlying reduction in activation energy. For instance, scientists are able to engineer surface features in a strategic manner to further improve substrate adsorption properties to convert them more easily. The rivalry of strength and geometry in these catalytic sites can immediately influence the overall efficiency of the sought-after reactions, and thus the supreme importance of nuanced mechanistic insight. Emerging advances mark the intersection of theoretical and computational methods to probe activation energy dynamics in catalysis. With artificial intelligence and high-throughput techniques, researchers are distinguishing optimum surface architectures that maximize catalytic functionality. For example, the process of carbon dioxide reduction has been shown to be evidence for weakening the C-O bonds through binding molecular oxygen atoms to surface cations, a prelude to effective activation. This finer understanding gives rise to developing new catalysts with improved activity and selectivity through an attentive approach. By focusing on attributes such as bond length and stabilizing transition states, scientists can construct catalysts for specific reactions strategically, thereby broadening the science of catalysis at the surface and facilitating chemical transformations to be more sustainable (Luna et al., 2019; Ghiringhelli et al., 2020) (Figure 5.4).

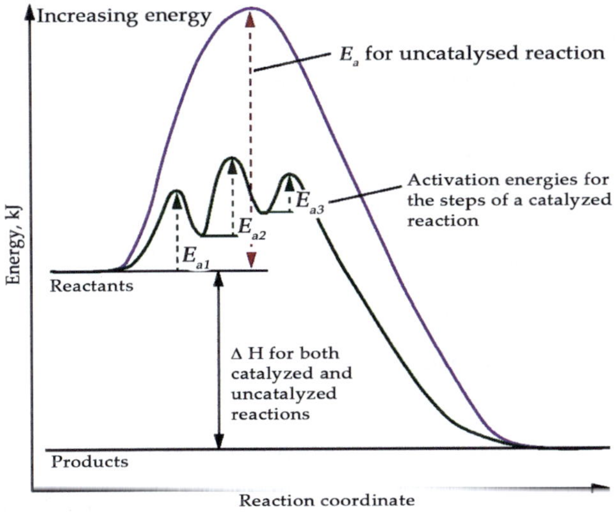


**Figure 5.4.** The role of activation energy in catalysis.

*Source:* https://www.chemengonline.com/catalysis-fundamentals/.

## 5.2. SURFACE PROPERTIES AND THEIR INFLUENCE

The role of surface properties in catalysis is barely capable of exaggeration, as the properties tend to govern the efficiency and effectiveness of catalytic processes. Catalysts operate predominantly at surfaces, where reactants interact and products are formed. Surface roughness, morphology, and electronic structure can significantly alter catalytic performance. A high surface area catalyst, for instance, can offer greater exposure to reactants, possibly resulting in increased reaction rates. Moreover, surface bonding characteristics and active sites dictate the turnover frequency and selectivity of the catalysts and hence surface engineering presents itself as a significant aspect in catalytic design for efficient applications. As evidenced in the literature, ideal surface characteristics can lead to significant breakthroughs in catalytic efficiency, thus impacting the overall performance of catalytic systems applied to industrial processes. Surface design of properties for specific catalytic reactions typically needs a general insight into the mechanism involved.

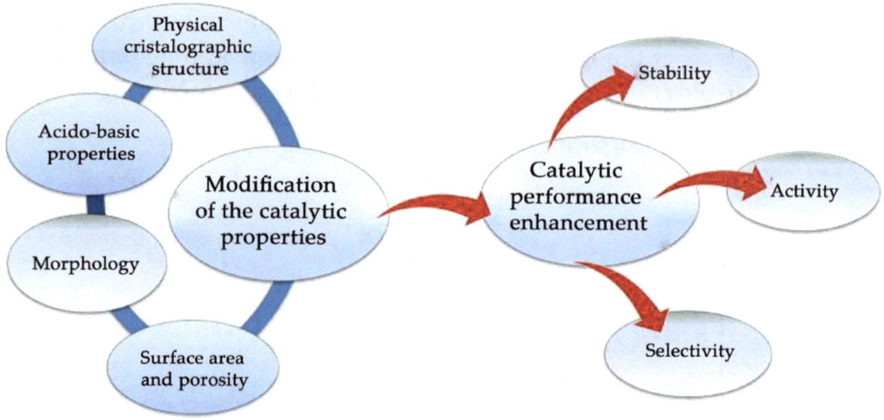


**Figure 5.5.** The impact of catalytic properties modification on the catalytic performances.

*Source:* https://www.mdpi.com/2073-4344/12/12/1637.

The researchers have come to stress the integration of catalytic design with mechanistic information in order to enhance performance measures such as selectivity and overpotential. The synthesis of catalysts able to efficiently reduce carbon dioxide to valuable hydrocarbons is an example of how certain surface features can lead to enhancement in green chemistry. According to Luna et al. (2019), the design of new materials that maximize surface characteristics can result in catalysts operating at lower overpotentials with high current densities. Such a blending of information and design is crucial to addressing problems in the field and creating technologies towards greener processes. Surface analysis techniques are also key to the description of reactant-catalyst interactions. Advanced methods like scanning vibrating electrode technique (SVET) allow real-time interrogation of catalytic processes at the micro-scale and yield valuable insights into how reaction dynamics are influenced by surface topography. For example, using SVET in experiments, it has been demonstrated that the presence of enzymes, like hemin, on electrode surfaces significantly enhances cathodic current during oxygen reduction reactions, demonstrating a unique catalytic effect (Basséguy et al., 2008). Comprehension of these local interactions is critical for surface condition optimization and prevention of harmful effects like local corrosion. Therefore, combining surface property evaluation with catalytic performance measurement is a critical field of research that holds the potential to further improve catalyst design and usage in various chemical conversions (Figure 5.5).

## 5.2.1. Surface Area and Porosity

Porosity and surface area are natural characteristics that play a powerful influence on the performance and efficiency of catalysts in surface reactions. High surface area allows for the presence of more active points, facilitating the reactant-catalyst interaction to occur. This is particularly crucial in heterogeneous catalysis, where reaction occurs at an interface between two phases. Moreover, porosity allows such active sites to be accessed easily, offering a structural framework that allows reactants to penetrate and approach the catalyst material effectively. The interconnection between surface area, porosity, and catalytic activity calls for the need for precise synthesis methods that are able to fine-tune these characteristics to optimize them in the quest for next-generation catalytic materials. Characterization techniques for the purposes of measuring surface area and porosity are important factors in the determination of catalyst performance. Techniques such as Brunauer-Emmett-Teller (BET) measurement provide quantitative surface area, while techniques such as scanning electron microscopy (SEM) and X-ray diffraction (XRD) can give porosity and structural integrity information. Such thorough characterization is essential in the study of new catalysts for specific reactions, such as electrochemical $CO_2$ reduction, where design not only seeks to optimize surface area but also enhance catalyst stability and selectivity. Like in recent studies, the integration of catalytic design and mechanistic understanding provides improved performance factors, enabling more useful pathways toward industrial application in renewable energy (Luna et al., 2019). There is evidence that the morphology of catalyst materials can be tailored for optimal surface area and porosity to enhance catalytic activity. One such demonstration of the effect of metal content variation on catalyst properties and performance is the use of metal-based catalysts, such as those with iron (Fe) content in selective catalytic reduction. Extension of the surface area through the formation of porous structures has been found to improve the efficiency of NOx conversion processes by a vast amount. However, it is also observed that metal overloading can adversely affect performance, which is a commentary on the delicate balance between composition, surface area, and porosity (Xiao et al., 2009). Therefore, ongoing research on the interplay between these factors is relevant for the design of catalysts that can meet the requirements of current catalytic applications and promote environmental sustainability.

### 5.2.2. Chemical Composition and Reactivity

The intricate relationship between reactivity and chemical composition is of prime relevance in the field of surface catalysis, namely in how catalysts interact with reactants. A catalyst's structure may greatly influence its effectiveness and selectivity to enable chemical reactions. For instance, metal catalysts have distinct electronic and geometric features depending on the atomic arrangement. These properties dictate the extent to which the reactants adsorb on the surface and how they subsequently undergo subsequent reaction channels. Understanding the inherent nature of surface chemistry, including electron transfer and interaction of bonding, is essential in maximizing the efficiency of catalysts. Hence, scientists are forced to investigate the molecular-level interactions that dictate the entire catalytic process, which are important in enhancing catalyst design (Luna et al., 2019). Technological progress in molecular characterization techniques also assists this research, allowing one to identify important structural features to be matched to reactivity. As catalysis at surfaces advances, it becomes increasingly important to bring in an understanding from other fields of science. Chemical composition not just implies the active catalytic phases but is also a function of the substrate and reaction environment influence. Impregnation with some functional groups or contaminants can significantly alter reaction kinetics by influencing the electronic surrounding of the catalyst. For instance, it has been demonstrated in experiments that doping some catalysts with heteroatoms enhances reactivity via the creation of new active sites or optimization of desirable channels and thus overall reaction efficiency (Blake et al., 1988). Further, understanding of surface and gas-phase species roles in catalysis leads to improved catalytic performance predictions. Studies of surface reactivity should therefore consider the dynamic nature of chemical interactions and their implications in the construction of new catalytic systems. Additionally, catalytic material aging over extended operation periods poses central challenges in matters relating to stability and reactivity of the chemical composition. The degradation of catalysts typically results in changes to their compositional composition and structural integrity over time, leading to reduced catalytic activity or selectivity with age. Studding these modifications is key to achieving long-term reliability in catalytic systems. New methodologies, including in situ catalytic characterization techniques, are being employed to monitor the chemical state and structural evolution of catalysts under operating conditions. These methods provide improved understanding of how the composition's modifications correspond with reactivity variations under actual conditions, paving the way for the development of more efficient and robust catalysts. Ultimately, an integrative knowledge of chemical composition and reactivity is a key driver in advancing the field of catalysis to sustainable and efficient energy alternatives (Luna et al., 2019).

### 5.2.3. Surface Defects and Their Catalytic Effects

Surface defects in catalytic materials are central to altering reaction pathways and enhancing catalytic activity. Surface defects like steps, kinks, and vacancies introduce new electronic environments with a dramatic impact on the adsorption energies of reactants and intermediates. For instance, undercoordinated surface atoms on defects can facilitate bond breaking and charge-carrier trapping, which are significant processes in catalysis. The interaction of these defects with the surrounding species, i.e., oxygen, could result in changes in material properties ultimately affecting its catalytic activity. As has been demonstrated in experiments on p-doped MgO, the controlled tuning of coordination in the surface site vicinity may enable researchers to control the stability of reduced vs. oxidized state, further emphasizing the key role of defects in catalysis (Berger et al., 2017). Perovskite oxynitrides' catalytic activity is also an exhibition of the influence of surface defects on regulating reactivity. The materials exhibit fluctuations in catalytic properties in response to variations in surface stoichiometry and concentrations of defects with time. With the self-oxidation of nitrogen ions under working conditions, their surface vacancies form, which affect adsorption and desorption kinetics critical for catalytic processes. Density functional theory calculations determine that surface anion vacancies are restored preferentially by reactive surface species such as *O and *OH under oxygen evolution reaction (OER) conditions. This dynamic surface defect reorganization not only alters electronic structure but is also accompanied with decreased catalytic activity, underlining the complex relationship between surface stability and function as a catalyst (Aschauer et al., 2019). Besides, the influence of surface defects extends to beyond localized change in reactivity to impact long-term stability and material efficiency. Dislocation strain due to defect generation can create localized charge buildup, directly affecting binding energies of surface reaction intermediates. The modification can increase overpotential with preferred reactions, and more work would be required to clarify the cause of the effects. Identification of intrinsic correlation between surface defects and catalytic activity provides an opportunity to design catalytic systems that operate at maximum efficiency and lifetime. Therefore, thorough investigations on the behavior of surface defects have the potential to enhance catalysis at surfaces and innovate processes in industries from energy conversion to environmental clean-up (Berger et al., 2017; Aschauer et al., 2019).

## 5.3. TYPES OF CATALYTIC REACTIONS ON SURFACES

The diversity of the catalytic reactions occurring on the surface accounts for the effectiveness of various catalytic processes. The most evident among them includes oxidation reactions, where adsorbed reactants on the surface transfer electrons to form oxides. The reaction is significant in industrial applications, such as in the oxidation of carbon monoxide to carbon dioxide. New computational methods have now been applied in research, which has revealed that thermodynamics and kinetics of such reactions can differ significantly depending on the exposed crystalline facets. For instance, for $RuO_2$, although (111) and (110) surfaces exhibited qualitative variation in binding energetics, both were equivalent in catalytic activity under some conditions, challenging traditional understanding of structure sensitivity in catalysis (Reuter et al., 2015). Therefore, a knowledge of the nature of such surface interactions is essential for creating more efficient catalytic systems. Hydrogenation reactions constitute a second significant category and are the heart of fuel and chemical feedstock production. The process usually involves adsorption of hydrogen and other reactants on active sites and reaction at the surface to yield products. These efficiencies of reaction are highly dependent on the electronic, morphological, and surface composition of the catalyst. Certain transition metal surfaces have been found to uniquely increase hydrogen activation by favorable binding interactions. In addition, advances in artificial intelligence and computational methodology have made it possible to explain optimum reaction pathways by determining the geometrical and electronic structural factors responsible for catalytic activity. Most significantly, oxide-based catalysts' synthesis proves that a shift from metal to oxide surfaces can bring new and effective hydrogenation processes, in addition to the sustainability of chemical manufacturing (Ghiringhelli et al., 2020). Finally, catalytic reactions can be divided based on their specificity to chemical groups or surface intermediates. For example, certain catalysts may be engineered to selectively activate certain bond types, for example, C–O or C–C bonds, which is significant in application fields like polymer synthesis or biomass conversion. Catalyst molecular design is what makes increased selectivity and control over yield possible. It has been introduced in a recent study that the activation of carbon dioxide, which plays a vital role in the fight against climate change, includes certain surface interactions that stabilize the carbon-oxygen bond, facilitating future reactions. This suggests molecular-level understanding in the design of catalytic activity.

With continued study, the combination of computational tools will continue to enhance surface catalyst design, with creative solutions for complex chemical transformations on the horizon (Ghiringhelli et al., 2020).

### 5.3.1. Oxidation Reactions

The fundamental processes involved in oxidation reactions play a pivotal role in surface catalysis, and particularly heterogenous catalysis. These reactions typically form electron transfer between an oxidizable compound to an oxidizing reagent, often with catalyst surfaces facilitating specific active sites for the process to take place. In heterogeneous catalysis, catalytic activity also usually depends upon the nature of surface coverage, as modifications in the distribution of the reactants can result in significant impacts on reaction rates and selectivity. As a case in point, when oxidation reactions are examined, the presence of oxygen species at the surface can significantly influence the overall reaction dynamics, demonstrating the intricate interdependence of surface chemistry and catalytic activity. It is important to understand such interactions in designing catalysts that are specially designed for specific oxidation reactions, and such helps in improved efficiency of chemical conversion and energy utilization. Surface coverage and reaction kinetics synergy is an area of special interest while understanding oxidation reactions. One of the fascinating phenomena observed here is the compensation effect, in which surface coverage changes result in a linear relationship between pre-exponential factors and activation energy. This indicates the direction that the contribution of the surface coverage is significant in determining the energy landscape of oxidation reactions. Compensation effects research illustrates that changed surface interactions, such as those seen in the oxidation of HCl to $Cl_2$ over $RuO_2$ catalysts, can have a dramatic influence on the observed kinetics that are implicated in such processes. Using advanced analytical techniques, scientists can make inferences regarding the relationship between these compensation effects and the underlying mechanisms for oxidation processes and, in turn, gain a better understanding of surface catalysis (Farra et al., 2012). With growing interest in renewable energy and sustainable energy, the employment of oxidation reactions is not restricted to traditional catalysis anymore but encompasses electrochemical applications, such as carbon dioxide recycling. Effective oxidation reactions are required for recycling to convert $CO_2$ to valuable fuels and feedstocks. Studies currently recognize important performance parameters, including reaction selectivity and current densities, and emphasize the requirement of catalysts that can operate under mild conditions with minimal energy input. The development of these catalysts needs an in-depth understanding of the oxidation process mechanisms, thereby integrating theoretical and experimental approaches to catalysis. Advances in this field promise not only understanding the complexity of oxidation reactions but also the potential for designing innovative solutions to the problem of climate change and the provision of carbon-neutral energy systems (Luna et al., 2019) (Figure 5.6).

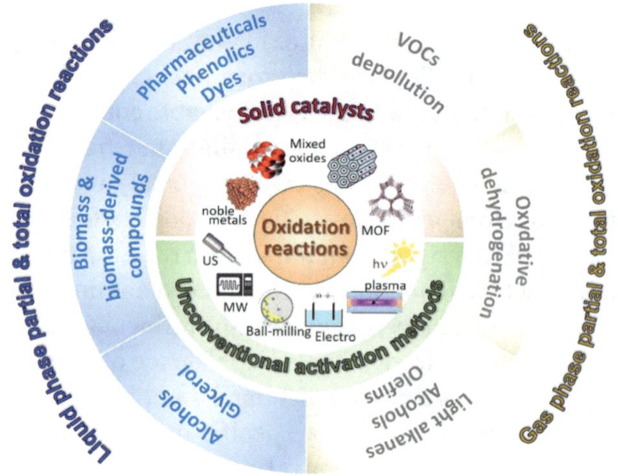


**Figure 5.6.** Oxidation reactions in heterogeneous catalysis.

*Source:* https://www.mdpi.com/2073-4344/8/10/483#.

## 5.3.2. Hydrogenation Reactions

The significance of catalytic hydrogenation reactions comes about through their extensive application in organic synthesis and industrial processes. Hydrogenation is the addition of hydrogen to substrates with unsaturation such as alkenes or aromatics to form saturated hydrocarbons. It is conducted using metal catalysts with adsorption sites for both the substrate and hydrogen. The efficiency of the reaction largely depends on the ability of the metal to activate hydrogen molecules as well as stabilize the reaction intermediates. For instance, catalytic properties of platinum, palladium, and nickel metals are an example of their outstanding electronic characteristics, which create conditions for effective cleavage and bond formation during the process. Thus, understanding the nuances of surface interaction and stabilization of the intermediates is a prerequisite to make hydrogenation reactions effective in industrial processes. In addition to the choice of metal catalyst, structural surface properties play an important role in the efficiency and selectivity of hydrogenation reactions. Advances in catalyst synthesis, such as alloy nanoparticles or supported catalysts, have been promising for raising catalytic activity. The changed surfaces can be engineered to favor particular pathways of reaction, thus influencing product distribution. In addition, the morphology of the catalyst and surface defects have also been found to influence adsorption energies and consequently the hydrogenation kinetics. This has been illustrated using studies, where it was shown that the structure and the saturation of the active sites can significantly alter the catalytic activity, further emphasizing the importance of advanced characterization techniques. An understanding of the surface phenomena can result in improved catalyst designs, which meet the green chemistry and sustainable production requirements (Guo et al., 1999). Hydrogenation reactions not only represent the foundation for production of valuable chemicals but also have environmental benefits through the capacity for converting hazardous intermediates to less harmful products. For instance, hydrogenation of carbon dioxide to hydrocarbons is a promising route for the mitigation of greenhouse gases and the production of renewable chemicals and fuels. The complexity of such reactions was recently investigated in $CO_2$ electrochemical reduction studies where the reaction pathways mostly depend on the surface properties of metals such as metal-carbon affinities and metal-oxygen affinities (Andreussi et al., 2019). Conclusions drawn through such research help build a comprehensive understanding of reaction processes, enabling rationale-based design for more effective and selective catalysts. By incorporating such information, researchers can improve hydrogenation reactions to enable the development of a greener chemical sector and move toward carbon neutrality (Figure 5.7).

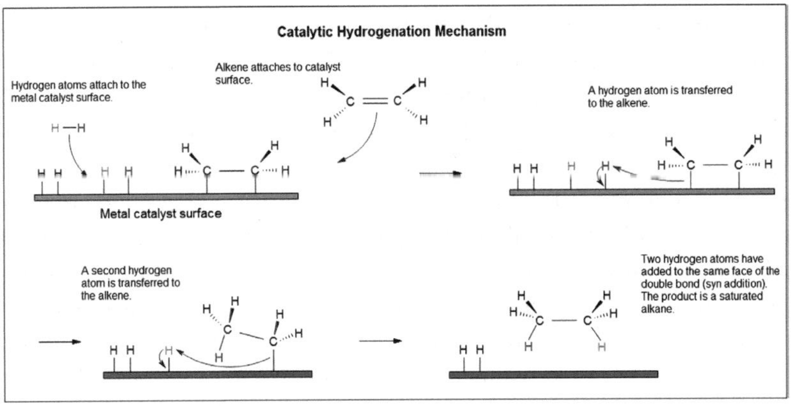


**Figure 5.7.** Catalytic hydrogenation mechanism.

*Source:* https://chem.libretexts.org/Bookshelves/Organic_Chemistry/Supplemental_Modules_%28Organic_Chemistry%29/Alkenes/Reactivity_of_Alkenes/Catalytic_Hydrogenation.

### 5.3.3. Decomposition Reactions

Decomposition reactions are a significant category of chemical processes in which complex molecules are broken down into less complex molecules, typically with the assistance of different types of surface catalysts. Decomposition reactions are significant in industrial processes, where they enable effective production of useful compounds through the formation of intermediate forms that can hinder or facilitate subsequent reaction. The molecular behavior of decomposition reactions may also be influenced greatly by the nature and condition of the catalytic surface. For instance, the electronic character and morphology of a catalyst, such as metal nanoparticles, may create specific reaction sites that determine the rate and orientation of decomposition processes. Clarification of such interactions is crucial in order to optimize catalyst design and optimize overall reaction efficiency in heterogeneous catalysis, especially considering the impact of environmental conditions like solvent and temperature. Moreover, interfacial dynamics between the catalyst surface and environment, e.g., water or organic solvents, can have a significant impact on decomposition mechanisms. Emerging research indicates water creating complex liquid-solid interfaces with catalytically active surfaces and, consequently, shaping reaction mechanisms. For instance, in the case of a titania-supported gold nanocatalyst, it was illustrated that water stabilizes non-gas phase charge states that differ from gas phase charge states. Such charge reorganization mediated by the solvent allows for the creation of new active sites and accordingly enhances catalytic performance in decomposition reactions. This solvent engineering holds intriguing potential for the optimization of reaction conditions and the improvement of yields via engineered interfaces between liquid solvents and catalyst surfaces, ultimately to more sustainable catalytic processes in liquid-phase heterogeneous catalysis (Camellone et al., 2013). In investigating oxidative stability in terms of decomposition reactions, individual interaction of chemical structure and catalytic activity cannot be excluded. Research on certain perfluoropolyalkyl ethers (PFPAE) reveals that there is an intrinsic connection between molecular structure and oxidative stability, and between reactive metal and oxidative stability. The findings reveal that the structure of difluoroformal groups determines the stability directly, and trifluoromethyl group addition improves resistance to decomposition even more. Furthermore, findings emphasize that active metal surfaces, particularly in conjunction with the addition of an oxidative agent like air, are necessary for promoting oxidative decomposition. These observations mean that passivation of reactive surfaces and optimization of the metal support can inhibit undesirable decomposition and therefore enhance the effectiveness and longevity of catalytic systems (Helmick et al., 1992). This information is essential in the development of future means to specifically control decomposition reactions for various chemical applications and provide sustainable and effective processes in industrial catalysis.

## 5.4. INDUSTRIAL APPLICATIONS OF SURFACE CATALYSIS

The contribution of surface catalysis to industrial processes is apparent throughout several industries, especially in energy conversion and environmental cleanup. Catalysts facilitate the ability to accelerate reaction rates without consuming excessive energy or producing unwanted byproducts. An example is the advance in electrocatalysts for fuel cells, which illustrates the ability of surface characteristics to make a great performance difference. High-performance materials, such as graphene-type structures, are being explored for their ability to create active catalytic sites, which play a central role in oxygen reduction reactions. Scientists are achieving improved efficiencies for fuel cells by introducing non-metallic heteroatoms into graphene's structure. This highlights the need for surface property optimization to drive industrial applications, such that the frameworks not just function efficiently but also sustainably respond to future energy demands (Alammari et al., 2018). One of the major fields in surface catalysis research is in catalytic materials interacting with solvents and contributing significantly towards liquid-phase reactions. The behavior of solvents, particularly water, can determine metal catalyst reactivity and charge distribution, as demonstrated by experiments on gold nanocatalysts supported on titania. Such liquid-solid interface interaction can potentially generate new active sites, which increase overall catalytic activity in relevant industrial processes.

By understanding how solvents, such as water, affect catalytic activity, researchers can create solvent systems to tailor catalysts for specific applications. This subtle nicety of solvent engineering provides more effective operation in diverse reactions, particularly in situations such as organic synthesis and waste treatment (Camellone et al., 2013). Furthermore, the technological repercussions of surface catalysis extend into the field of materials science, where the development of hierarchical nanostructured catalysts is a significant achievement. These structures not only increase the catalytic efficiency but also stand to be useful on the stability and recyclability fronts, which are significant parameters for industrial applications. The constraint here still remains the scaling-up of these novel materials from the lab to real-world applications. Interaction with novel fabrication techniques and optimization of the synthesis procedures can bridge this gap so that highly efficient catalytic systems can be adopted by industries. As industries become more sustainability-focused, the role of surface catalysis in reducing waste and energy is all the more significant, keeping in view its role in formulating future industrial practices and technology (Alammari et al., 2018).

### 5.4.1. Catalytic Converters in Automobiles

Catalytic converters are integral components of modern automobiles and serve the function of minimizing the harmful emissions released by internal combustion engines. Catalytic converters work by employing a broad range of catalysts, typically constituted by precious metals such as platinum, palladium, and rhodium, to catalyze chemical reactions that convert lethal gases like carbon monoxide, hydrocarbons, and nitrogen oxides into less harmful chemicals. The effectiveness of these catalytic reactions is based on the surface area and position of the catalytic material, which offers optimal contact with the exhaust gases. Technological innovation in catalytic converter technology is required constantly due to increasing regulatory pressure to reduce automobile emissions. In order to achieve higher efficiency and improved compliance with environmental standards, advanced catalytic materials and improved modeling techniques for exhaust aftertreatment systems are called for (Walander et al., 2020). Further research into catalytic processes in the converters proved to be crucial, including modeling and experimental validation. Describing the performance of these systems by experimental techniques such as gravimetric analysis, temperature-programmed desorption, and scanning electron microscopy provides evidence regarding the catalytic activity and life in conditions that exist in real service. With this, scientists are increasingly paying attention to examining the influence of agents such as poisoning agents which may influence the effectiveness of the catalyst. For example, poisoning by phosphorus has been identified as a significant issue, as it not only deactivates catalytic activity but also surface properties, and there is work required to design catalysts that are resistant to this type of degradation. Identification of the surface chemistry not only leads to better emission control technologies but also to the complicated balance between catalyst formulation and environmental performance for the automotive sector (Kröger et al., 2007). The ongoing evolution of catalytic converters is a reflection of the broader catalytic science picture, particularly surface catalysis. Researchers employ advanced modeling techniques to create more realistic simulations of how the catalysts will operate, and this will eventually lead to more efficient converters. They account for aspects such as mass transport, heat distribution, and catalytic activity, which all have a composite impact on the performance of the converters. Continuous updating of such models is inevitable, in sync with the automotive sector's vision for building emission-free transport. As regulations continue to get stricter, it is a must that catalytic converter technology keeps up with developments. Therefore, further studies must address not only the short-term efficiency of catalytic converters but also their long-term stability and durability under various operating conditions so that they can remain functional in reducing car emissions over the long term (Walander et al., 2020) (Figure 5.8).

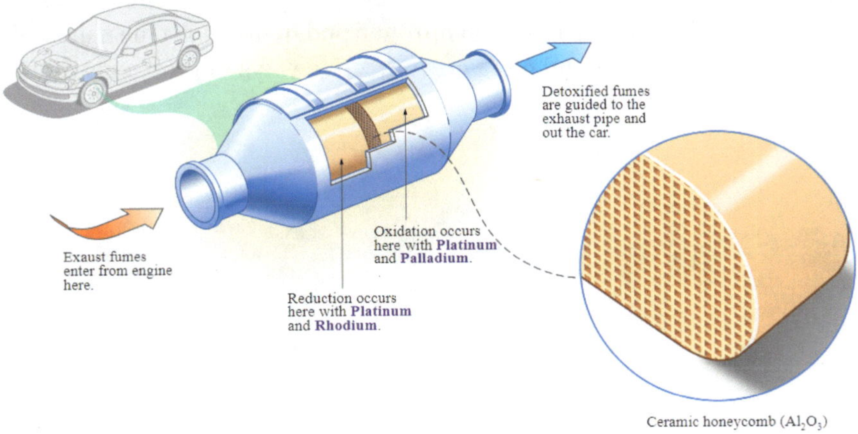


**Figure 5.8.** Catalytic converters in automobiles.

*Source:* https://chem.libretexts.org/@api/deki/files/456422/003.svg?revision=1&size=bestfit&width=1000&height=485.

## 5.4.2. Haber-Bosch Process for Ammonia Synthesis

The Haber-Bosch process is an industrial leap forward in the sense that it provides a methodical method of manufacturing ammonia from nitrogen in the air and hydrogen gas. The high-temperature (around 400–500°C) and high-pressure (150–300 atmospheres) process employs iron catalysts to accelerate the reaction rate. While very effective, the Haber-Bosch process is vastly energy expensive and a significant contributor to industrial carbon dioxide emissions. Consequently, its impact on the environment generates a boom of interest in searching for alternative synthesis routes that would cut down on energy consumption without compromising efficiency. Such alternatives can be more clearly contemplated with a solid base in surface catalysis, as enhancing this area could revolutionize the manufacture of ammonia and enable a cleaner way of satisfying global agricultural demand and energy needs. Recent studies have witnessed a paradigm shift to heterogeneous catalysis being the solution to the limitations of the Haber-Bosch process. The paradigm shift aims at leveraging solid catalysts that operate under less severe conditions, which might reduce the energy consumption in synthesizing ammonia. For instance, novel catalyst designs with transition metals hold great promise for nitrogen reduction reaction (NRR) activities, offering avenues that promise to be superior to conventional methods. Recent developments in molecular catalysis have also brought to the fore the importance of certain catalyst surface features in optimizing ammonia synthesis efficiency. Understanding these new catalyst architectures, especially in the context of their interaction with nitrogen and hydrogen at the nanoscale, is critical in creating cleaner and more cost-effective ways of producing ammonia (Duda et al., 2018).

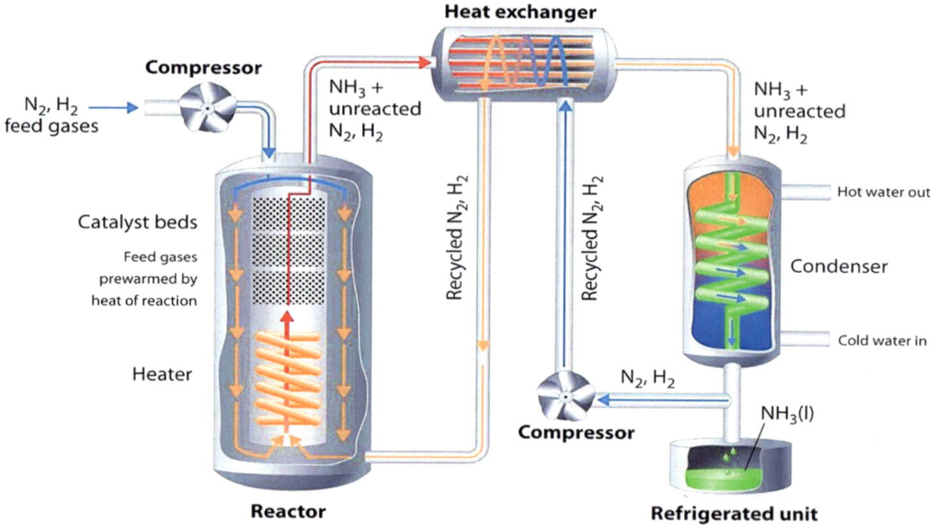


**Figure 5.9.** Ammonia synthesis (Haber-Bosch process).

*Source:* https://www.researchgate.net/figure/Ammonia-synthesis-Haber-Bosch-process_fig1_380124128.

Clarification of the mechanistic intricacies involved in surface catalysis will be the source of unraveling efficiency as well as selectivity in NRR processes. Furthermore, since the agriculture sector becomes increasingly reliant on ammonia-based fertilizers, there has been a greater need for sustainable synthesis processes. Dependence on the Haber-Bosch process is prevalent, yet necessity is highlighted by rising energy costs and concerns regarding global warming. New technologies such as electrochemical nitrogen reduction demonstrate the potential of harvesting renewable energy sources to synthesize ammonia while significantly reducing the carbon footprints of conventional production (Ford et al., 2019). Recent studies on functionalization of organic ligands over catalysts to enhance selectivity of NRR reveal a productive research avenue. Such advances would be readily compatible with existing agriculture and ensure the steady supply of ammonia and be consistent with sustainability requirements for the environment. Additional research in catalytic methods and surface interactions is still critical to the achievement of twin objectives of reducing energy usage and maintaining agriculture output (Figure 5.9).

### 5.4.3. Catalysis in Petrochemical Industries

The petrochemical sector relies significantly on catalytic reaction to enable the conversion of crude oil and natural gas into other valuable chemicals and fuels. Surface catalysis plays a critical role in enhancing the rate and selectivity of the reaction, making the process efficient and energy-conserving. Heterogeneous catalysis is particularly dominant within this sector, given that the reaction occurs between solid catalysts and gas or liquid reactants at the surface. These interactions facilitate advanced mechanisms, such as cracking and reforming, to produce hydrocarbons from heavier molecules, optimize the yield of desired products, and reduce byproducts. With the world's demand for energy and petrochemical feedstock growing every day, advancement in catalytic technologies remains critical to optimize processes and reduce environmental impacts associated with petrochemical production. In recent years, focus has been placed on increasing catalytic efficiency through the development of new catalysts with improved performance at reduced operating pressures and temperatures. Such technological advancements can be applied to make petrochemical processes more sustainable overall, furthering initiatives towards reducing carbon footprints worldwide. Design advancements like the employment of nanoscale materials and mixed metal oxides have shown potential in improving catalyst activity and stability for industrial environments. Knowledge of the surface effects, including adsorption and accessibility of catalytic sites, has a direct relation with the design of long-lasting catalysts that are resistant to petrochemical environments, which in turn is essential towards the integration of high-throughput and cost-effective production processes within the industry. Moreover, catalysis plays a pivotal role in attempts to mitigate pollution problems surrounding petrochemical processes. The photo-degradation of phenolic substances found in petrochemical production wastewater, for instance, has been studied using titanium dioxide photocatalysts. All these examples remind us of the duality of catalysis—not just is it essential for optimal output production, but also takes on a significant role in addressing environmental hazards accompanying industrial processes. By creating catalytic methods for pollutant removal, companies can achieve environmental regulation at the expense of implementing more eco-friendly procedures. Lastly, the evolution of catalysis in the petrochemical industry is still an active area of research, showcasing its profound potential for driving innovation and solving important environmental challenges (Figure 5.10).

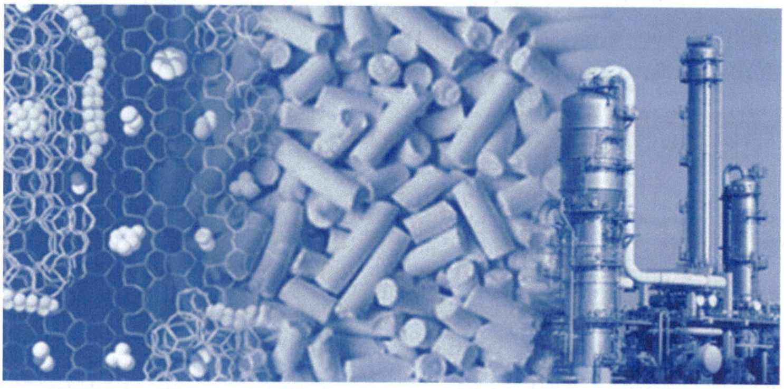

**Figure 5.10.** Zeolites are among the most widely used catalysts in the petrochemical industry. Pictured left is a representation of the microscopic structure and center the zeolite catalysts in pelleted.

*Source:* https://ethz.ch/en/news-and-events/eth-news/news/2014/05/better-catalysts-for-the-petrochemical-industry.html.

## 5.5. CHALLENGES IN SURFACE CATALYSIS

The intricate world of surface catalysis presents unique challenges that limit the creation of efficient catalysts and the broad applicability of this technology. One of the principal hurdles is the need for catalysts to be highly selective while maintaining low overpotentials during reactions. A case in point is the electrochemical reduction of carbon dioxide, where the challenge is presented by catalysts needing to be developed to produce valuable products from carbon efficiently without energy loss. The sacrifice of such performance measures with the failure to make headway in disentangling underlying mechanisms that lie in these processes is a critical area of study. The achievement of more insight into these mechanisms can potentially contribute to maximally optimized catalysts, eventually making it possible to make efficient conversion of renewable energy into sustainable forms of fuel capable of addressing world energy requirements (Luna et al., 2019). Another element of complexity in surface catalysis arises from the vast number of reaction intermediates and elementary reactions available on a catalyst surface. In order to be able to comprehend reaction networks fully and enhance catalytic efficiencies, it is vital to discover all the significant intermediates. Current algorithmic methodologies have several ways of probing these pathways of reaction;

however, the efficiency of these methods is highly variable. Challenges in such computational studies usually stem from the requirement of a deep analysis with minimal reliance on heuristics that could lead to incomplete or biased results. Such reliance on human input to establish parameters also causes variations, limiting reproducibility and range. A more advanced approach that minimizes human bias yet capitalizes on computational methods could enable us to gain a deeper insight into catalyst behavior on surfaces (Reiher et al., 2018). The relationship between surface physical characteristics and catalytic activity is another problem of concern. Surface roughness, morphology, and electronic properties are some of the properties that affect catalytic efficiency and product selectivity by a large margin. These problems require a synergistic solution combining developments in material science with surface chemistry to produce innovative catalytic systems. Advances with the addition of nanoscale materials are likely to improve performance, but issues related to stability and longevity under real-world conditions have a tendency to undercut such efforts. As scientists work to overcome these issues, the ultimate goal continues clear: the development of stable, efficient catalysts for sustainable chemical reactions, particularly those that facilitate carbon neutrality and the transition to renewable sources of energy (Luna et al., 2019).

### 5.5.1. Catalyst Deactivation and Regeneration

Catalyst deactivation is a significant issue in heterogeneous catalysis, particularly with regard to its effect on efficiency and cost-effectiveness in the industries. The deactivation can occur through various mechanisms, including fouling, sintering, and chemical poisoning. Fouling is deposition of unwanted material onto the surface of the catalyst, preventing reactant access. Similarly, sintering forms agglomeration of active sites, and this reduces the available surface area. Chemical poisoning is observed when some reactants or products irreversibly adsorb on active sites and inactivate them. Understanding such phenomena is critical to develop ways to avoid deactivation and increase catalyst life, maintaining catalytic reactions with high turnover and low energy consumption in the long term in real-world applications. Catalyst regeneration is a significant process to restore catalytic activity after deactivation has occurred. Regeneration methods may also vary significantly depending on the deactivation mechanism. Heat regeneration, for instance, which is normally employed for the desorption of carbonaceous deposits from the catalyst surface, utilizes elevated temperatures to incinerate organic foulants. Chemical regeneration processes, however, can incorporate solvents or reagents that will dissolve poison molecules. Current research has shown that regeneration not only enhances the lifetime of catalysts but also leads to enhanced performance characteristics because catalyst surfaces can be positively altered during regeneration. With these methods, researchers are able to develop protocols that are cost-effective and economical in nature, thus opening up doors to valuable knowledge on sustainable industrial catalysis. New catalyst deactivation and regeneration solutions have emerged, particularly with the advent of photocatalytic reactions. Photocontrolled processes have been under study, as referenced in literature, to prove the worth of using light as a means to mediate catalysis reactions. More so, photocatalysts have evolved well in regenerating deactivated catalyst sites in gentle conditions with less energy input and improved efficiency (Chen et al., 2015). For example, ethylene epoxidation studies revealed the capacity of plasmonic catalysts to be catalytically efficient under near-atmospheric conditions, providing new opportunities for regeneration through photogenerated intermediates such as graphene (Heo et al., 2018). These breakthroughs emphasize the need to introduce novel materials and methodologies to enhance the stability of catalysts while opening new directions to addressing the issue of deactivation. With more research on these new solutions, the catalysis field can move in a positive direction towards efficient and sustainable chemical processing (Figure 5.11).

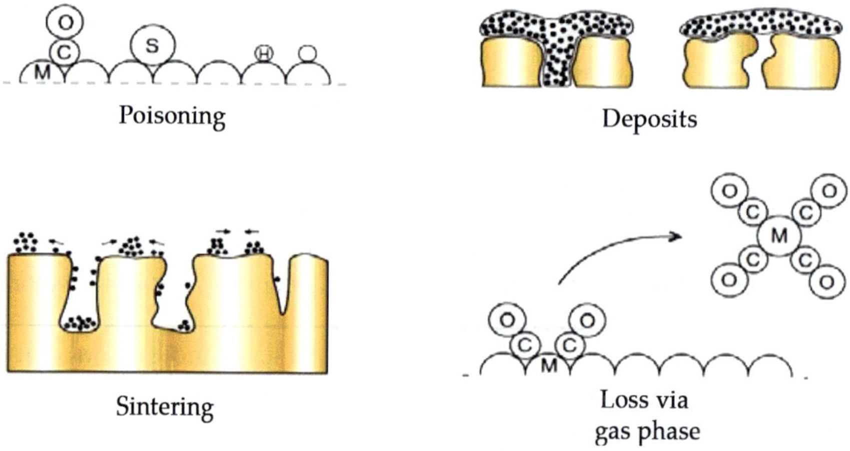


**Figure 5.11.** Mechanism of catalyst deactivation.

*Source:* https://ammoniaknowhow.com/catalyst-deactivation-common-causes/.

## 5.5.2. Selectivity and Product Distribution

Selectivity and product distribution are crucial factors regulating the efficiency of catalytic processes, particularly in surface catalysis. The intrinsic property of catalysts regulates the pathways along which the reactants are converted into desired products to a significant extent. In the case of electrochemical processes such as carbon dioxide reduction, selectivity is extremely crucial due to the variety of possible products, which can be developed from a single reactant. The ability of a catalyst for preferential reaction pathway with respect to any other pathway is not merely founded on its surface characteristics but also on the interaction energies between the reactants, intermediates, and products. High selectivity can lead to effective consumption of reactants and minimized byproducts, thus maximizing the efficiency of catalytic systems and their practical applications in industry. Innovative materials' designs are now central means of improving selectivity in catalyzed reactions. CORR, for instance, is currently in the spotlight of research where polycrystalline copper surfaces have displayed remarkable catalytic activity. Such surfaces hold immense value in the conversion of $CO_2$ to alcohols and higher-order hydrocarbons. However, achieving a desired product distribution is still an uphill task. Researchers have also been interested in the intermediates' function, i.e., the $OCH_3$ and C-bound CO, as selectivity determinants. Application of more advanced techniques like surface enhanced infrared absorption spectroscopy (SEIRAS) and ambient-pressure X-ray photoelectron spectroscopy (APXPS) has shed light on these processes' mechanisms, allowing rational catalyst design to produce products with specificity, thus addressing the major issue of selectivity (Andreussi et al., 2019). Besides, enhancement of metal-catalyst interactions has been identified as a crucial step in the design of improved product distributions in surface-catalyzed reactions. Metal surface reactivity with different geometries of bound reactants plays a significant role in product formation. As presented in the literature, it has been addressed that harmony between metal-carbon (M-C) and metal-oxygen (M-O) interactions can lead to improved performance parameters for catalysts. For example, C1 hydrocarbons arise from O-bound intermediates, while C2 hydrocarbons are suggested to arise from C-C coupling of C-bound species. Such types of detailed mechanistic insights underscore the importance of catalyst design around specific surface properties conducive to selectivity and optimal product distributions. The interface between chemical engineering and fundamental surface science thereby becomes essential in moving towards efficient catalytic systems for a variety of industrial applications (Luna et al., 2019).

### 5.5.3. Environmental and Economic Considerations

The relationship between economic viability and environmental sustainability in surface catalysis is more critical in current studies and commercial operations. As demands for cleaner energy technologies increase, catalytic processes that conserve energy and waste are the most significant. Microbial fuel cells (MFCs) are one such example, which is very promising in producing electricity from the catalysis of organic substrates by microorganisms at the electrode surface. This technology not only generates energy but also treats waste, integrating ecological benefit with economic gain. The disadvantage of slow kinetics of electrochemical ORR in MFCs hinders efficiency, which warrants more research into microbial catalysis processes that can overcome such shortcomings. Overcoming such limitations can enhance the economic feasibility of MFCs by lowering operation costs in the form of energy input and material requirement (Bergel et al., 2012). In understanding the economic dimension of catalytic processes, there is a necessity to consider the role of material selection and structural sophistication in maximizing performance. For instance, certain enhancements on the design and synthesis of microcrystalline catalysts, such as silver and palladium, have exhibited better catalytic activity that can lead to more effective reactions. The formation of more finely cut, smaller crystals increases active surface area, thereby allowing for higher reaction rates at lower energy costs. Further, novel production techniques in low-gravity environments remove convection in electrodeposition, offering a novel approach to the synthesis of catalyst material with enhanced properties (Grodzka et al., 2025). These technological innovations, in addition to enhancing the process efficiencies of catalysis, also signify possible economic advantages in terms of reduced resource utilization and enhanced product efficiencies, meeting ecological and economic goals simultaneously. Innovation of catalytic technology and environmental policies must be adopted to drive increased sustainable development. Policymakers should motivate researchers to investigate catalytic methodology with emphasis on reducing carbon footprints and optimizing energy efficiencies. Clean catalytic processes commercialization has the potential to bring significant economic growth by promoting new markets and job creation in green technology sectors. Additionally, as sustainability consciousness in the population and the regulatory landscape increasingly converge with sustainability, catalysis research companies can gain a competitive edge by ensuring their business is aligned with such principles. Here, surface catalysis presents itself as a focal point of emphasis toward achieving environmental sustainability and economic growth, justifying the imperative for collective effort on the part of researchers, business stakeholders, and policymakers to stimulate progress with greater social value (Bergel et al., 2012; Grodzka et al., 2025).

## SUMMARY

In summarizing a study of catalysis at surfaces, one needs to thank the nuances and subtleties inherent to this significant area of research. The significance of the interactions at the surface should not be overestimated, given that these play a crucial role in determining the efficiency and selectivity of catalytic processes. A knowledge of the underlying principles that govern surface reactions provides access to designing more effective catalysts. Current studies, particularly through the use of state-of-the-art computational methods, have emphasized the importance of phase diagrams in portraying the properties of materials under varying states of temperature and pressure. Such information plays a crucial role in catalysis, which also operates under technical pressures and temperatures that are beyond the normal laboratory conditions researched previously (Peden et al., 2002). This focus on actual operating conditions allows scientists to create catalysts that operate optimally under actual conditions, thereby enhancing their applicability in the industrial scenario. Second, the process of optimizing catalysts is more than just theoretical knowledge; it is a synergistic approach which merges experiment outcome with mechanistic insight. Most importantly, especially for electrochemical reactions like recycling carbon dioxide, creating catalysts with high selectivity and efficiency is critical. The progress in catalytic design needs to be oriented toward the reduction of energy barriers while achieving high current densities for target reactions. Current studies highlight the need for catalysts with the capacity to maintain performance under extended periods of use, and this involves not just enhancing stability but also that performance metrics are maintained under varying conditions (Luna et al., 2019). Finally, understanding the mechanistic processes of surface catalysis will enable the creation of renewable energy products that fall within the sustainability goals of the world. Overall, the future of surface catalysis is promising with challenge. New material science and technologies can drive enormous advances in catalytic performance and efficiency. While we pursue the elaborate interactions that exist at surfaces, it is important that interdisciplinary collaboration be fostered to address fundamental as well as applied problems. Progress in this field will not only result in the development of improved catalytic systems but will also play a vital role in addressing severe global issues, such as climate change and sustainable energy production. By focusing emphasis on empirical and theoretical research, the scientific community will be able to make sure that the conclusions arrived at contribute to the practical implementation of new catalytic solutions (Peden et al., 2002; Luna et al., 2019). Such co-operative effort will certainly pave the way for the next generation of catalysts that are vital to a greener world.

In surface catalysis study, it is necessary to summarize points of emphasis that highlight the complexity and the effect of surface interactions on the catalytic reaction. Catalysts' performance is typically determined by their surface properties and the excellence of their interaction with reactants. For instance, hydroxide catalysis has been a great method for enhancing bond stability utilized in gravitational wave detectors. Specifically, study of the shear strengths of hydroxide catalysis bonds formed on mono-crystalline sapphire indicates that they will maintain satisfactory strength at cryogenic temperatures with mean strengths above 23 MPa. This discovery not only emphasizes the importance of material selection in catalytic application but also indicates towards the ways in which surface direction may affect overall catalytic function and therefore become the foundation for future engineering applications of catalysis (Douglas et al., 2015). In addition, understanding catalytic mechanism at surfaces involves the study of the mechanisms by which molecular-level interaction controls catalytic performance. Monte Carlo simulation has been utilized in recent works to probe the dynamic interactions between structural motifs, such as the key-lock assembly mechanism. These studies have established that solvent influences can strongly affect the assembly and disassembly of catalytically active systems. Notably, the self-assembly of some structures can be triggered by means of entropic contributions, without any attendant direct attractive forces. This finding is not only important for enzyme catalysis but for use in general biochemical reactions where transport by proteins is crucial. This kind of research incites further investigation of the sensitive balance of molecules at surfaces, emphasizing their role in deciding catalytic pathways and efficiencies (Jiménez-Ángeles et al., 2010). Lastly, the delineation of key ideas in surface catalysis shows not just the significance of material properties but also the intricacy of molecular interactions involved. The outcomes in hydroxide catalysis in mechanical devices and the pivotal role of solvent dynamics in self-assembly procedures are reflective of the three-dimensional character of surface catalysis. All these elements elucidate the importance of maintaining a tailored approach toward material selection and interaction optimization in catalytic processes. As catalytic process discovery continues, it will be important to understand these fundamental concepts as a means of optimizing reactions, enhancing efficiency, and driving technological and industrial advancement in catalysis. Therefore, continued research in surface interactions and their effects is important to the development of future improvements in the field (Douglas et al., 2015; Jiménez-Ángeles et al., 2010).

## REFERENCES


1. Adam, M., Aggarwal, V., Amadori, C., Armstrong, J., Beckwith, R., Bellamy, L., et al. (2013). Heterogenization of ketone catalysts within mesoporous supports for asymmetric epoxidation.
2. Alammari, A., Allen, C., Cai, Z., Chen, Y., Conner, J., Farooque, M., Gong, X., et al. (2018). Graphene-based nanostructures in electrocatalytic oxygen reduction. http://arxiv.org/abs/1805.03149 (accessed on 3rd October 2025).
3. Andreussi, O., Giordano, L., Hwang, J., Katayama, Y., et al. (2019). An in-situ surface-enhanced infrared absorption spectroscopy study of electrochemical $CO_2$ reduction: Selectivity dependence on surface C-bound and O-bound reaction intermediates. http://arxiv.org/abs/1904.11299 (accessed on 3rd October 2025).
4. Aschauer, U., & Ouhbi, H. (2019). Nitrogen loss and oxygen evolution reaction activity of perovskite oxynitrides. http://arxiv.org/abs/1904.03077 (accessed on 3rd October 2025).
5. Bailleul, S., De Wispelaere, K., & Van Speybroeck, V. (2016). Towards molecular control of elementary reactions in zeolite catalysis by advanced molecular simulations mimicking operating conditions. https://core.ac.uk/download/55855326.pdf (accessed on 3rd October 2025).
6. Basséguy, R., Bergel, A., Etcheverry, L., Iken, H., et al. (2008). Local analysis of oxygen reduction catalysis by scanning vibrating electrode technique: A new approach to the study of biocorrosion. https://core.ac.uk/download/12039570.pdf (accessed on 3rd October 2025).
7. Bergel, A., Erable, B., & Féron, D. (2012). Microbial catalysis of the oxygen reduction reaction for microbial fuel cells: A review. https://core.ac.uk/download/12043853.pdf (accessed on 3rd October 2025).
8. Berger, D., Bhattacharya, S., Ghiringhelli, L. M., Levchenko, S., et al. (2017). Theoretical evidence for unexpected O-rich phases at corners of MgO surfaces. http://arxiv.org/abs/1706.05695 (accessed on 3rd October 2025).
9. Blake, G. A. (1988). Chemistry in dense molecular clouds: Theory and observational constraints. https://core.ac.uk/download/10211213.pdf (accessed on 3rd October 2025).
10. Bocquet, M. L., Boer, R., Stampfl, C., Christensen, M., Ertl, G., Greeley, J., et al. (2008). Ag-Cu alloy surfaces in an oxidizing environment: A first-principles study. http://arxiv.org/abs/0811.0864 (accessed on 3rd October 2025).

11. Bondzie, L. A., Kellogg, G. L., Hendriksen, B. L. M., Reuter, K., Taylor, K. C., Kaxiras, E., et al. (2003). Oxide formation at the surface of late 4d transition metals: Insights from first-principles atomistic thermodynamics. http://arxiv.org/abs/cond-mat/0304184 (accessed on 3rd October 2025).

12. Böttcher, A., Stampfl, C., Kreuzer, H. J., Skriver, H. L., et al. (1999). Towards a first-principles theory of surface thermodynamics and kinetics. https://core.ac.uk/download/154913245.pdf (accessed on 3rd October 2025).

13. Buckley, D. H. (2025). Application of modern surface analytical tools in the investigation of surface deterioration processes. https://core.ac.uk/download/pdf/42851354.pdf (accessed on 3rd October 2025).

14. Camellone, M. F., & Marx, D. (2013). On the impact of solvation on a Au/$TiO_2$ nano catalyst in contact with water. http://arxiv.org/abs/1303.5235 (accessed on 3rd October 2025).

15. Chen, M., Johnson, J. A., & Zhong, M. (2015). Light-controlled radical polymerization: Mechanisms, methods, and applications. https://core.ac.uk/download/85123358.pdf (accessed on 3rd October 2025).

16. Daturi, M., Frenkel, A. I., Gibson, E. K., Keturakis, C., et al. (2016). Dynamics of $CrO_3$–$Fe_2O_3$ catalysts during the high-temperature water-gas shift reaction: Molecular structures and reactivity. https://core.ac.uk/download/96882578.pdf (accessed on 3rd October 2025).

17. De Luna, P., Dinh, C. T., Kim, D., Li, Y., et al. (2019). Designing materials for electrochemical carbon dioxide recycling. https://core.ac.uk/download/287622524.pdf (accessed on 3rd October 2025).

18. Douglas, R., Haughian, K., Hough, J., et al. (2015). The effect of crystal orientation on the cryogenic strength of hydroxide catalysis bonded sapphire. https://core.ac.uk/download/42360364.pdf (accessed on 3rd October 2025).

19. Duda, R. D., & Renner, J. N. (2018). Catalysts for nitrogen reduction to ammonia. https://core.ac.uk/download/596190744.pdf (accessed on 3rd October 2025).

20. Faglioni, F., Gdownski, G. E., & Goddard, W. A. (2005). Energetics of hydrogen coverage on group VIII transition metal surfaces and a kinetic model for adsorption/desorption. https://core.ac.uk/download/4870705.pdf (accessed on 3rd October 2025).

21. Farra, R., Hevia, M., Knop-Gericke, A., et al. (2012). *In situ* surface coverage analysis of $RuO_2$-catalyzed HCl oxidation reveals the entropic origin of compensation in heterogeneous catalysis. https://core.ac.uk/download/210619064.pdf (accessed on 3rd October 2025).

22. Ford, Z. R. (2019). Effect of thiols for nitrogen reduction to ammonia. https://core.ac.uk/download/215464183.pdf (accessed on 3rd October 2025).
23. Ghiringhelli, L. M., Illas, F., Levchenko, S. V., et al. (2020). Ab initio data-analytics study of carbon-dioxide activation on semiconductor oxide surfaces. https://core.ac.uk/download/491525678.pdf (accessed on 3rd October 2025).
24. Grodzka, P. G., & Picklesimer, E. A. (2025). Improved catalysts by low-G processing. https://core.ac.uk/download/pdf/42877672.pdf (accessed on 3rd October 2025).
25. Güell, A. G., Zhang, G., Unwin, P. R., Patten, H. V., & Schmickler, W. (2016). Nanoscale electrochemistry of $sp^2$ carbon materials: From graphite and graphene to carbon nanotubes. https://core.ac.uk/download/293881802.pdf (accessed on 3rd October 2025).
26. Guo, L. (1999). The Synthesis and Characterization of New Higher Nuclearity Arene-Ruthenium-Sulfur Clusters: A Thesis Presented in Partial Fulfillment of the Requirements for the Degree of Master of Science in Chemistry at Massey University, New Zealand. https://core.ac.uk/download/162617233.pdf (accessed on 3rd October 2025).
27. Helmick, L. S., & Jones, W. R. (1992). Determination of the oxidative stability of perfluoropolyalkyl ethers and correlation with chemical structure. https://core.ac.uk/download/pdf/42805494.pdf (accessed on 3rd October 2025).
28. Heo, J., Jain, P. K., Kumari, G., et al. (2018). *In situ* formation of catalytically active graphene in ethylene photo-epoxidation. https://core.ac.uk/download/323308648.pdf (accessed on 3rd October 2025).
29. Jiménez-Ángeles, F., Odriozola, G., Lozada-Cassou, M., Michaelis, L., & Odriozola, G. (2010). Entropy driven key-lock assembly. http://arxiv.org/abs/1010.3000 (accessed on 3rd October 2025).
30. Jóźwik, P., Kobiela, T., Krupski, A., et al. (2015). Structure determination of Au on Pt (111) surface: LEED, STM and DFT study. https://core.ac.uk/download/29589400.pdf (accessed on 3rd October 2025).
31. Kim, K., Wagner, P., Wagner, K., & Mozer, A. J. (2022). Electrochemical $CO_2$ reduction catalyzed by copper molecular complexes: The influence of ligand structure. Energy & Fuels, 36, 4653–4676. https://doi.org/10.1021/acs.energyfuels.2c00400 (accessed on 3rd October 2025).
32. Kröger, V. (2007). Poisoning of Automotive Exhaust Gas Catalyst Components: The Role of Phosphorus in the Poisoning Phenomena. https://core.ac.uk/download/628859940.pdf (accessed on 3rd October 2025).

33. N/A. (2025). ECUT (Energy Conversion and Utilization Technologies Program). Biocatalysis Project. https://core.ac.uk/download/pdf/42839622.pdf (accessed on 3rd October 2025).
34. Peden, C.H.F., Weinert, C.M., Reuter, K., Scheffler, M., Cant, N.W., Blaha, P., et al. (2002). First-principles, atomistic thermodynamics for oxidation catalysis. http://arxiv.org/abs/cond-mat/0211602 (accessed on 3rd October 2025).
35. Reiher, M., Simm, G. N., & Vaucher, A. C. (2018). Exploration of reaction pathways and chemical transformation networks. http://arxiv.org/abs/1810.07490 (accessed on 3rd October 2025).
36. Reuter, K., & Wang, T. (2015). Structure sensitivity in oxide catalysis: First-principles kinetic Monte Carlo simulations for CO oxidation at $RuO_2$ (111). https://core.ac.uk/download/532709023.pdf (accessed on 3rd October 2025).
37. Salmeron, M. (2018). From surfaces to interfaces: Ambient pressure XPS and beyond. https://core.ac.uk/download/287621955.pdf (accessed on 3rd October 2025).
38. Steiner, M. G. (2017). Photocatalytic decomposition of phenol under visible and UV light utilizing titanium dioxide-based catalysts. https://core.ac.uk/download/84124263.pdf (accessed on 3rd October 2025).
39. Voecks, G. E. (2025). ECUT: Energy Conversion and Utilization Technologies Program. Heterogeneous Catalysis Modeling Program Concept. https://core.ac.uk/download/pdf/42849346.pdf (accessed on 3rd October 2025).
40. Walander, M. (2020). Towards Validated Catalytic Reactor Models. https://core.ac.uk/download/287066160.pdf (accessed on 3rd October 2025).
41. Wang, K., Zhang, J., Hsu, Y. C., Lin, H., Han, Z., Pang, J., Yang, Z., et al. (2023). Bioinspired framework catalysts: From enzyme immobilization to biomimetic catalysis. Chemical Reviews, 123, 5347–5420. https://doi.org/10.1021/acs.chemrev.2c00879 (accessed on 3rd October 2025).
42. Xiao, Y., Zhang, W., & Zhou, P. (2009). An investigation into catalysts to improve the low temperature performance of an SCR. https://core.ac.uk/download/9023356.pdf (accessed on 3rd October 2025).

CHAPTER 6

# COLLOIDS AND DISPERSED SYSTEMS

## Contents



## INTRODUCTION

The study of colloids and dispersed systems comprises a significant and complicated branch in the field of materials science and chemistry. Characterized as the dispersion of very small particles in a continuous medium, colloidal systems exhibit unique physical and chemical properties that are quite different from bulk particles. The sizes of colloidal particles are between 1 nanometer and 1 micrometer, which allows them to display unique properties that may be utilized across a broad spectrum of activities from medicine to food science. Research on such systems not only allows for the innovative design of new products but also enhances our ability to control existing goods, so it is important to research colloid interaction and stability. In colloid science, recent advances in synthesis and functionalization techniques have made it possible to achieve improved dispersion and stability in colloidal particles. For instance, recent research focuses on new methods such as grafting of functional groups on materials like graphitic carbon nitride (g-CN) for their better dispersibility in organic and aqueous media. Functionalization of such modified materials demonstrates the minimization of time required for particle redistribution using sonication, which maximizes dispersion stability. Such developments exemplify the need for modification of surface characteristics for achieving desirable functional features, which is most essential for forming high-performance colloidal systems, as demonstrated in studies on g-CN and its dispersible improved analogues (Antonietti et al., 2017). Furthermore, the research of exotic particle forms and their behavior in liquid crystals provides valuable understanding of the ordering of colloidal systems. Research on microbullet particles with distinct anisotropic forms provides insights into their ability to form the structure of colloidal assemblies in various conditions. When dispersed between nematic liquid crystals, the microbullets become able to engage in dynamic interactions driven by elastic forces and adopt ordered structures such as linear arrays or hexagonal arrays. By modulation of surface anchoring and external electric fields, researchers have been able to manipulate such organizations, which illustrates the intricate way in which particle shape and orientation control aggregate behavior. These findings emphasize the significance of carrying out thorough investigation of various colloidal structures and the ensuing properties towards advancing knowledge in dispersed systems (Beller et al., 2013).

## *Definition of Colloids*

The term colloid refers to a heterogeneous mixture made up of very fine particles dispersed in a continuous medium that does not settle out. Colloids come in numerous forms in the form of aerosols, emulsions, and foams, and each of them is characterized depending on the state of matter present. These very small particles, typically ranging from 1 nanometer to 1 micrometer in dimension, are colloidal and are invisible to the naked eye. In colloids, particle structure will have a profound effect on properties, including stability, viscosity, and potentially optical properties. This unique dispersion of fine particles allows colloids to exhibit properties that are different from true solutions and suspensions, making them essential in some industries such as materials science, food science, and pharmaceuticals. Understanding the dynamics and interactions in colloidal systems is important to being able to predict their behavior and to optimize their uses. Colloidal behavior is commonly researched through examining hydrodynamic medium and particle interactions, which is the most important for the maintenance of stability. An example is Smoothed Profile (SP), a numerical simulation approach in which scientists can simulate colloidal dispersion interaction in various fluid media. This method enables the study of complex fluid interactions by connecting the dynamics of rigid colloidal particles to continuum mechanics of the host fluid and providing insight into the behavior of multi-component colloidal systems (Chatterji et al., 2008). Sophisticated such simulations are very useful in the development of new colloidal formulations that could improve the effectiveness of products in industrial applications. Besides, colloids are influenced by other forces at the microscopic level, and they dictate their stability and interaction. The effective potential between particles of colloids can be attractive or repulsive depending on their surroundings, and it plays a significant role in their aggregation pattern and stability. For instance, by changing the interaction potential when colloids are suspended in a critical solvent, the behavior of these particles can be controlled in response to environmental conditions, which is very important in designing stable colloidal systems (Zaccarelli et al., 2012). Such dynamics highlights the key role of external conditions, such as temperature and concentration, in controlling colloidal formulation performance. Understanding these variables is key to the creation of the application of colloids in many areas of science and engineering (Figure 6.1).

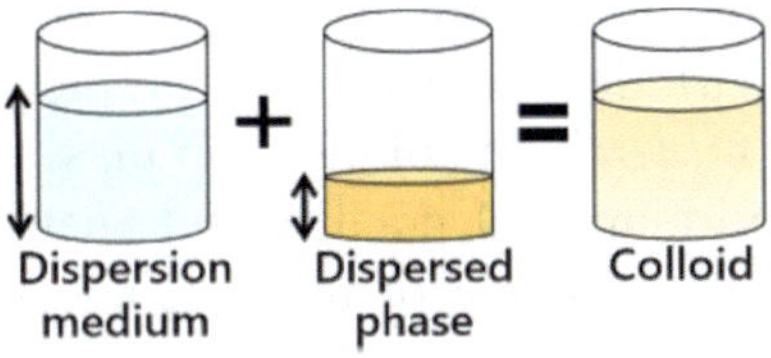


**Figure 6.1.** Dispersed medium vs. dispersed phase in colloids.

*Source:* https://chemistrytalk.org/what-are-colloids/.

## *Importance of Colloids in Various Fields*

The significance of colloids in various disciplines stems from their unique properties that bring forth new functions for uses in various fields. In materials science, colloids play a vital role in the production of nanomaterials and composites. They offer engineered surface functionalities that enhance mechanical, thermal, and optical properties, hence being crucial in the creation of advanced materials. For instance, the ability to manipulate particle size and distribution in a colloidal system allows researchers to achieve material properties demanded in high-performance products, such as aerogels and photonic crystals. Moreover, the versatility of colloidal systems means they will be incorporated into emerging technologies, thereby making them necessary building blocks for the development of future materials and applications. In biomedical science, colloids are important in drug delivery and imaging diagnostics. Colloidal vehicles such as liposomes and nanoparticles enhance therapeutic drug bioavailability by facilitating the transport and controlled release of drugs in the body. Targeted therapy lessens side effects and enhances therapeutic efficacy, addressing one of the largest challenges faced by pharmacotherapy. Apart from that, the unique optical characteristics of colloids enable them to be used for bio-sensing applications, which provide a means of early disease detection and monitoring. Research into the behavior of colloidal systems goes on, and they keep unearthing new means for problem-solving in bioengineering, accentuating the breakthrough contribution of colloids in increasing health benefits. Furthermore, in environmental science, colloids contribute significantly to learning and solving diverse problems, such as the control of pollution and soil remediation. Their role as contaminant carriers also confirms the necessity of colloidal transport and stability studies in natural systems. Engineered colloid applications to deliver and encapsulate contaminants, for example, can achieve maximum remediation efficiency. Research progress has also come up with new approaches to particle separation of microparticles with infinitesimal surface differences using liquid crystals and improvement of related separation processes critical for environmental purposes (Guo et al., 2018). Secondly, nematic droplets' response to electrical fields indicates how the research on colloids can make subtle remarks on their interactions in most environmental conditions (Auernhammer et al., 2009). Environmental technologies capable of minimizing pollution and increasing sustainability can be developed based on colloids' unique properties.

## *Overview of Dispersed Systems*

Dispersed systems are critically essential components of a wide range of scientific and industrial phenomena in which one phase occurs as a dispersion in another phase. Dispersed systems take numerous forms, including colloids, emulsions, and suspensions in which particles of one substance are finely dispersed in another. One of the very critical characteristics of dispersed systems is stability, which relies on the particle size, concentration, and on the nature of the dispersion medium. In colloidal systems, for instance, the particle-particle interactions are governed by steric and electrostatic forces stabilizing the dispersion and preventing aggregation. These interactions need to be understood when developing stable formulations ranging from pharmaceuticals to food technology. The understanding of dispersed systems allows scientists to take advantage of their intrinsic properties to drive progress in many disciplines. Physical and chemical characterization of dispersed systems is key to their manipulation and application. Various techniques have been applied to study the physical and chemical nature of dispersed systems, each contributing valuable information on particle shape, size, and stability. The techniques include microscopy, light scattering, and spectroscopy, which allow scientists to quantify parameters such as polydispersity and elemental composition. For instance, electron and optical microscopy techniques give details on the nanostructural characteristics of the nanoparticles, enhancing our understanding of their behavior when they are in colloidal form. A review puts it that different characterization techniques are grouped based on the properties they offer, demonstrating their role in sensing engineered nanoparticles (ENPs) and their suggested applications (Anderson et al., 2014) (Figure 6.2).

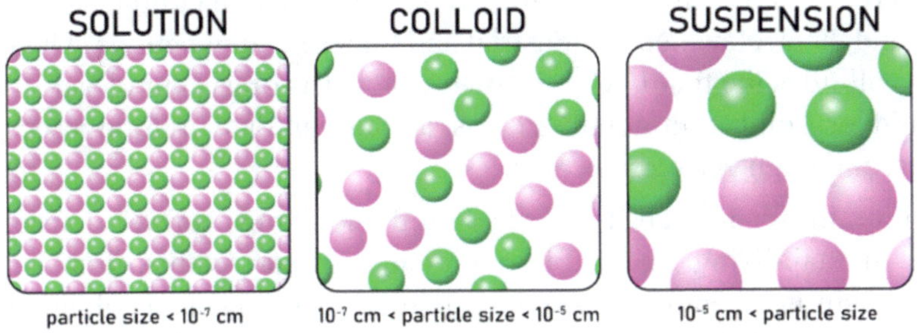


**Figure 6.2.** Dispersed phase.

*Source:* https://www.embibe.com/exams/dispersion-systems/.

This integrated approach facilitates dispersed systems to be highly suited for specific purposes and therefore improve their application. Colloids of transition metal sulfides are an indication of the diversity and variability that exist within dispersed systems because of their different properties compared to oxides. These materials exhibit a wide range of functional properties and potential uses, particularly in novel fields like bioimaging. Nevertheless, stability in metal sulfide colloidal suspension is a challenge of great magnitude, necessitating robust functionalization approaches. More comprehensively studied metal or metal oxide colloids have comparatively well-established surface chemistry, but that for metal sulfides is not well known. Bridging this gap in knowledge is critical toward further exploring their potential for use. Constant investigation as an ongoing endeavor on transition metal sulfides is important, as they may pave the way for new areas of high-end usage in various industries, towards the realization of the intricate interrelationship between particle chemistry and function (Dengo et al., 2017). This research can lead to significant breakthroughs in the dispersed systems field.

## 6.1. TYPES OF COLLOIDS

Colloids are classified into several types based on the nature of the dispersed phase as well as the dispersing medium. One of the principal classes is the sol, which is a liquid containing solid particles in dispersion. Sols are permanent and typically exhibit some unique optical properties, like light scattering, resulting in the Tyndall effect. Paint and ink are sols. In contrast, emulsions are colloidal systems where a liquid is distributed in another non-miscible liquid and are usually present in foods such as mayonnaise. Emulsion stabilization usually requires the application of surfactants to lower interfacial tension and enhance stability, again highlighting the complexity of colloidal systems. These categories reveal the multifunctionality of the utilization of colloids in everyday products and their relevance in industrial and scientific applications. The second principal colloid is the foam, whose structure consists of gas bubbles distributed in a liquid or solid continuum. Foams are ubiquitous in food processing, cosmetics, and engineering materials.

Stability in foams relies on the liquid medium's surface tension and presence of stabilizing agents, like proteins or surfactants. For instance, whipped cream is a common food foam that relies on mechanical agitation to incorporate air, while industrial foams may use specialized additives to enhance ruggedness. Furthermore, the texture and mouthfeel of foods like mousses and creams can strongly be attributed to the characteristics of foams. This mixing of gaseous phase and continuous phase emphasizes the sophisticated nature of colloid interaction and their significance in consumer products and complicated applications. Pickering emulsions represent a highly fascinating type of colloid wherein solid particles stabilize an emulsion with a different alternative to traditional surfactant-stabilized emulsions. Their success resides in the ability of solid particles to adsorb at the oil-water boundary and thus provide mechanical stability. The novel development has demonstrated the ability of intrinsically hydrophilic nanoparticles to stabilize oil-in-water emulsions, as presented in studies on dialkyl adipate oils.

The studies have established that the particles become hydrophobized in situ, forming stable emulsions without the need for conventional surfactants. The hydrogen bonding between the surfaces of particles and oil molecules are pivotal in this stability since they indicate how surface chemistry impacts colloidal behavior. Typically, Pickering emulsions are the innovative application of material properties towards bridging the stability and function of colloidal systems, creating new opportunities in food, pharmaceuticals, and cosmetics (Binks et al., 2016; Guo et al., 2018) (Figure 6.3).

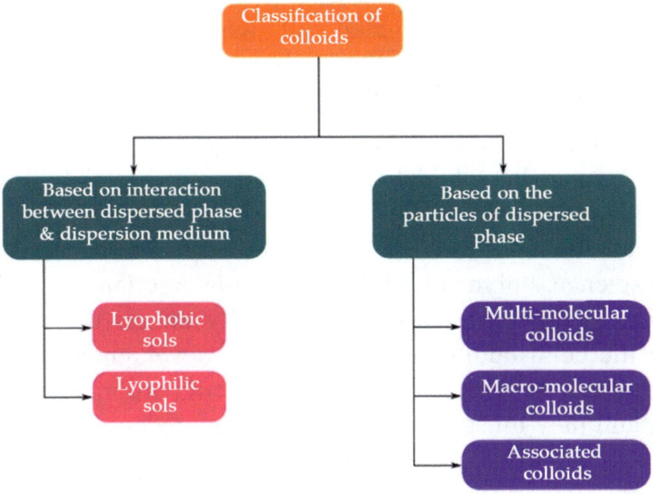


**Figure 6.3.** Classification of colloids.

*Source:* https://external-content.duckduckgo.com/iu/?u=https%3A%2F%2Fcache.careers360.mobi%2Fmedia%2Farticles%2Fuploads%2Ffroala_editor%2Fimages%2F2021%2F12%2F21%2F1640083962717.png&f=1&nofb=1&ipt=974be6f5d8a2ff01bb4001cf0e6f9ab26e29fefeee41679db7efbc2548f6371c.

## 6.1.1. Sols: Solid in Liquid

In colloids and dispersed systems, sols, i.e., solid particles dispersed in a liquid matrix, are of major relevance. Such colloidal systems possess unique features, primarily due to the extremely small size of the solid particles, typically between 1 nanometer and 1 micrometer. The movement of such particles allows for the sols to be evenly dispersed in the liquid and not precipitate by gravity, a process fueled by Brownian motion and fluid molecule interactions. This stable suspension not only enables the development of a pleasing appearance in items like paint and ink but also is suitable for enhancing functional properties of various applications in medicine, food technology, and materials science, hence imparting importance to understanding sol properties in colloid science. The functional properties of sols may greatly be influenced by the nature of solid particles and the liquid medium. For instance, size, surface chemistry, and solids concentration determine how these particles interact with the solvent molecules and thus affect viscosity, stability, and reactivity. Very small crystals such as those found in zeolites may be embedded within a liquid matrix to form composite materials with enhanced properties. The use of zeolite crystals, which are extremely porous, to a sol enables certain molecular sieving characteristics, in which big molecules are excluded and small molecules permeate easily, showing an example of practical application of sols in formulation of next-generation materials (Barrer et al., 1988).

This kind of interaction between solid and liquid phases shows the inherent complexity of balance in colloidal systems, with vast potential for development in various fields of science. Apart from that, research on sols also provides the solutions for the application of broader principles in material engineering and polymer science. Polymer-colloidal system interaction workshops provide the forum for scientists to explore how gravity affects these materials and thus provide the contributions towards product development and use (N/A). With advancements in science, applications of sols are more than traditional applications, and they must be reevaluated in new technologies, such as drug delivery systems or environmentally friendly coatings. The ability to size and surface control solids in liquids is a revolutionary advance, and new fields of applications might be leveraging these dynamics to create new solutions for diverse industries. This ongoing investigation of the nature and application of sols is crucial to the development of materials science.

## 6.1.2. Gels: Liquid in Solid

The concept of gels as liquid determines a solid matrix is the best description of the idea of the entangled interaction of gel and sol in colloidal dispersions. Gels are formed by trapping liquid phases in a solid-like matrix via cross-linking polymer chains forming a three-dimensional matrix. With this specific structure, gels retain water content along with exhibiting solid properties. Interestingly, one of the features of gels is their ability to tailor in various applications, ranging from biomedical fields, where they can be used as scaffolds in tissue engineering, to material sciences, where they are used in the development of hybrid structures of certain mechanical and optical properties.

The study of gels often demands sophisticated processing techniques, such as sonocatalytic methods, which have been shown to produce denser gels with homogenous porosity, significantly enhancing the performance of functionality and enabling possibilities in future materials (Fedriani et al., 2009). Understanding the behavior of gels involves being aware of their colloidal nature, which influences stability and mechanical properties (Figure 6.4).

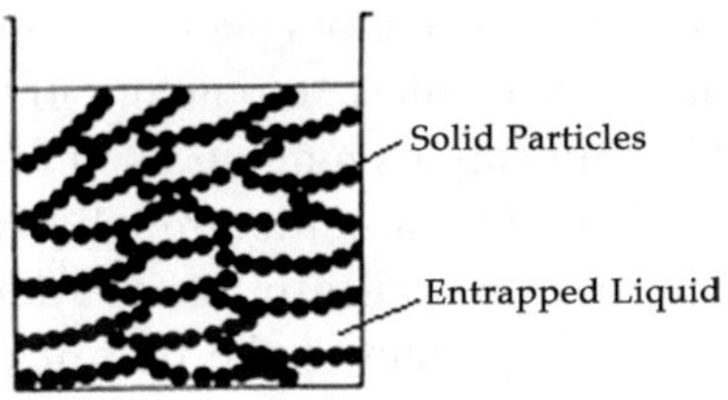


**Figure 6.4.** Gels: liquid in solid.

*Source:* https://thefactfactor.com/facts/pure_science/chemistry/physical-chemistry/gels/11922/.

Gels can be viewed as extremely complex agglomerates of dispersed phases that transform radically with time by their sustained interaction with the environment. Extremely charged, nanometer-sized particles within a gel have a new dynamic structure that is responsive to applied stresses and fluidity-enhancing upon aging, for instance. These particulate system relaxation times reveal the mechanism by which structural rearrangements contribute to the mechanical strength of the gels, demonstrating a highly fine-tuned behavior able to support stresses (Bandyopadhyay et al., 2004). This is the effect of aging significant to application in industries such as construction and foods, where the elasticity and stability of the gels are controlling factors. Besides that, gel research has long-term implications on the development of energy-efficient materials and sustainability of the environment. While scientists venture into synthesis and characterization of silica-based gels, advances in this area can lead to the production of sustainable materials with potential for sequestering harmful chemicals such as $CO_2$. Structural integrity of gels capable of dispersion of functional nanoparticles supports their versatility in multifunctional application. With ongoing advancements in fabrication technologies, including sol-gel techniques coupled with acoustic cavitation, future technological advancement in gel technology can potentially provide materials that not only perform optimally in their intended uses but also facilitate environmentally friendly approaches to material preparation (Fedriani et al., 2009).

### 6.1.3. Emulsions: Liquid in Liquid

Emulsions, by definition, are dispersions of one liquid within another and constitute an important component of colloidal systems. Emulsion formation is rooted in the intricate interaction between the dispersed phase and the continuous phase, typically requiring the addition of emulsifying agents to stabilize the interface between immiscible liquids. Emulsifying agents, such as surfactants or nanoparticles, reduce interfacial tension, enabling fine droplet formation.

Emulsion stability is a function of factors such as droplet size, concentration, and physicochemical character of the components involved. Such parameters must be well controlled in applications ranging from food science, where emulsions are utilized to provide texture and mouthfeel, to pharmaceuticals, where emulsions are used to improve drug absorption and delivery. Emulsion formation mechanism is therefore important in the optimization of their properties for industrial use. Recent works have recognized the use of nanoparticles to stabilize oil-in-water emulsions with minimal reliance on traditional surfactants. An emerging idea is founded on hydrophobization of naturally hydrophilic nanoparticles so that they are extremely effective at the oil-water interface. Work has demonstrated that solubilization of some oils in the aqueous phase contributes towards improved stability as the oils surrogate surfactants by participating in hydrogen bonding with the nanoparticles.

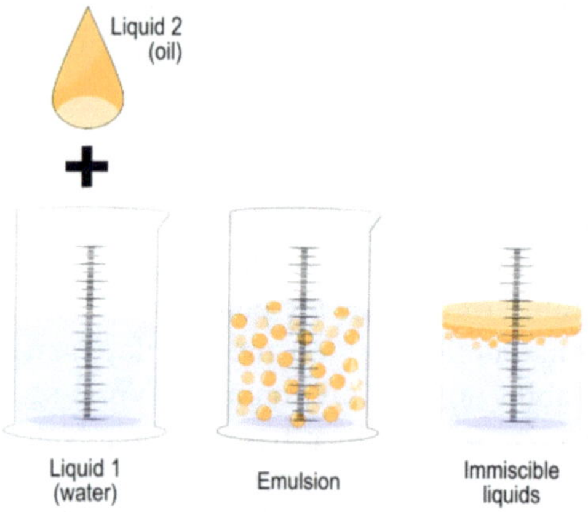


**Figure 6.5.** Emulsions.

*Source:* https://www.science-sparks.com/what-is-an-emulsion/.

For instance, emulsion stabilization using dialkyl adipate oils demonstrates this effect, where more stable emulsions are achieved with the use of shorter-chain oils due to increased interaction with the solid nanoparticle surfaces. These molecular interactions allow scientists to better improve the formulation and stability of emulsified systems for numerous applications, as evidenced by research currently being conducted (Binks et al., 2016). Moreover, rheological properties and drop size distribution of emulsions are greatly influenced by the selection of nanoparticles that is utilized for stabilization. Experimental works have established the effect of variations in wettability and the specific surface area of silica nanoparticles on the interfacial tension and resulting droplet formation in liquid-liquid systems. As opposed to classical processes with high-energy input for the formation of emulsion, low-energy stirring processes have been found to facilitate effective drop breakage and coalescence. This new knowledge aids in the building of comprehensive drop size models of nanoparticle-stabilized emulsions, thereby suggesting the relevance of addressing practical problems in manufacturing and product design. Therefore, the current research in this field has the promise of helping develop further understanding regarding emulsions, which leads to improved utilization in industrial as well as consumer products (Enders et al., 2019) (Figure 6.5).

## 6.2. PROPERTIES OF COLLOIDS

The properties of colloids are important to understanding their behavior in various applications, particularly in such areas as food science, drug production, and materials science. Among the unique features of colloids is that they can remain dispersed for extended times without settling, something that has been attributed to the colloidal particles' minute size, measuring from 1 nanometer to 1 micron. This stability is largely on the viscosity of the continuous phase and particle-particle interactions, such as Brownian motion, which suspends them. Knowing the properties becomes quite significant in developing new colloidal systems because they determine how the systems will respond to external forces and how effective stabilization mechanisms will be so that colloids will be able to work optimally in their applications. Besides, colloidal particle interaction with the medium is another dominant factor that determines the stability and functionality of colloidal systems. As an example, the wettability of particles will influence the performance of the particles to stabilize emulsions or foams. Recent research indicates that modifying nanoparticles' wettability and surface area can have a significant impact on the rheological properties of colloidal dispersions. Through the use of low-energy input pathways to create Pickering emulsions, researchers can control drop breakage and coalescence and further clarify drop size distributions in these systems (Enders et al., 2019). This would be more reliable in industrial application, where colloidal dispersion properties must be under controlled conditions in order to achieve desired effects. In addition, with the latest advancements, the versatility of colloidal systems has been enhanced, particularly in the formation of stable emulsions through new stabilization techniques. The use of inherently hydrophilic nanoparticles in oil-in-water emulsions is one of the prominent areas, where nanoparticles can be exposed to in situ hydrophobization in the presence of certain oils, thereby enhancing emulsion stability. It is a process that is dependent on hydrogen bonding between particle surface hydroxyl groups and ester oil, which illustrates the complexity of interactions in colloidal systems (Binks et al., 2016). Understanding such interaction not only deepens the fundamental science of colloids but also provides access to customized formulations that are capable of meeting the specific needs of numerous industries, suggesting the versatility and importance of colloidal science in contemporary science.

### 6.2.1. Tyndall Effect

The Tyndall effect is an important illustration of light scattering phenomena inherent in colloids and dispersed systems. It is the light scattering by particles in a medium and becomes operative when a beam of light is passing through colloidal particles in a suspension. Such scattering is particularly evident in systems whose particles are in the range of colloidal sizes (1 nm to 1 μm), as their size is comparable to the wavelength of visible light. Consequently, a beam of light traveling through a system of colloids can be seen, tracing out a bright path of light. This phenomenon has a practical application in solutions and colloidal suspension differentiation; clear solutions that lack the Tyndall effect are very different from the colloids, which cause light scattering and therefore reveal the presence of particulate matter. Beyond simple observation of the Tyndall effect are the complexities of the effect that dictate the understanding of phase behavior in colloidal systems. For example, colloid-polymer mixtures illustrate the degree to which the particle interactions reversibly alter light-scattering characteristics. These are particle size, concentration, and the nature of interaction that dictate not just the extent of the scattering but also phase separation and gelation potential in the medium. Scientists have used dynamical density functional theory to see the impact that these interactions have on the final configuration of a binary mixture and discovered that in certain set-ups, the interaction of the gravitational fields with the boundary conditions can cause inhomogeneous morphologies within the domains of phase separation (Malijevsk et al., 2013). The importance of the Tyndall effect as a diagnostic tool to discern and assess intricate colloidal behaviors is highlighted by these findings. Moreover, the Tyndall effect is crucial in describing phase transition mechanisms of colloid-polymer mixtures. Fluid-fluid phase separation or gelation is contingent on interactions between constituents—i.e., whether an attraction balance between polymers and colloids is accompanied by their respective magnitudes. During medium- to long-range attraction experiments, phase-separated state evolution indicates a complex landscape of dynamics in which the Tyndall effect is a pictorial map of the transitions (Bartlett et al., 2012). Scientists have observed typical dynamics in such processes, such as the dependence of particle mobility at various structural locations within the gel network, confirming the value of light scattering as a colloidal stability and structural integrity measure. The Tyndall effect, therefore, enriches pedagogical explanation of colloids both as a pedagogical tool and as an entry to more sophisticated investigation into dispersed systems (Figure 6.6).

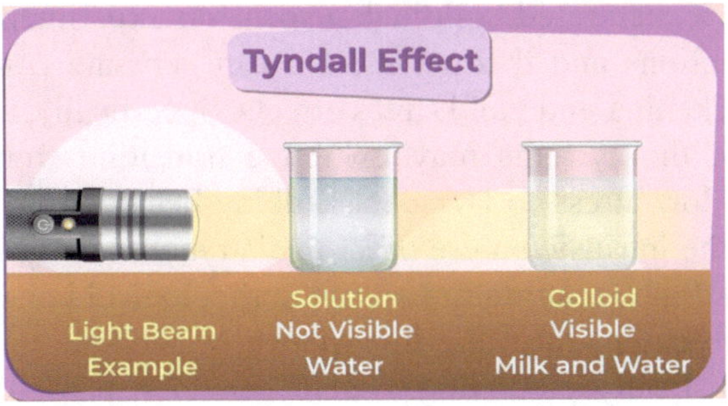


**Figure 6.6.** The Tyndall effect.

*Source:* https://media.geeksforgeeks.org/wp-content/uploads/20230327141025/Tyndall-Effect-1.jpg.

## 6.2.2. Brownian Motion

Brownian motion is an important phenomenon in colloidal and dispersed system research. It refers to the unsteady motion of suspended minute particles in a liquid resulting from frequent collisions with the molecules of the surrounding fluid, indicating the natural randomness of the system. Brownian motion is marked in colloidal dispersions at the microscopic level, where particles are faced with vast thermal agitation. This inherent dynamism has a bearing not just on the stability of colloidal systems but also on their macroscopic properties, such as viscosity and flow behavior. With an increase in particle density, the Brownian motion-mediated interactions can impart rich rheological characteristics, which are critical for predicting the manner in which these systems will respond under different conditions. Moreover, theoretical techniques of simulating Brownian motion also contribute significantly to our manipulation and control of colloidal dispersions. With the application of sound models of theory, researchers can predict the rheology and microstructure of colloidal systems subjected to external flow conditions. Predictability is priceless when it comes to the achievement of desired properties in real-world applications such as cosmetics, foodstuffs, and pharmaceuticals. It has been reported from the literature that flow behavior of colloidal dispersions can be regulated by a modulation of microscopic particle-particle interaction, particularly when interacting via Brownian motion. The understanding of statistical mechanics that controls such interactions leads to innovations that maximize the performance and stability of colloidal products, which have become a foundation in many industrial processes. Regarding high particle concentration, the effects of Brownian motion also complicate the dynamics of colloidal systems further. As demonstrated in concentrated colloidal suspension experiments, the interplay of particle interactions and thermal motion can generate complex behaviors such as shear-thickening and yield stress effects. Specifically, under fluctuating stress conditions, the systems may exhibit a transition from Bingham-like flow behavior at low stress to Herschel-Bulkley behavior at high stress. This dualism reflects the intrinsic nature of Brownian motion in governing both the microstructure and the resulting macroscopic rheological behavior of colloids. Current research demonstrates the importance of accurate modeling of these dynamics in realistically predicting system behavior, particularly in regimes that include strong particle interactions and fluid flow dynamics (Barrett et al., 2012; Batchelor et al., 2010) (Figure 6.7).

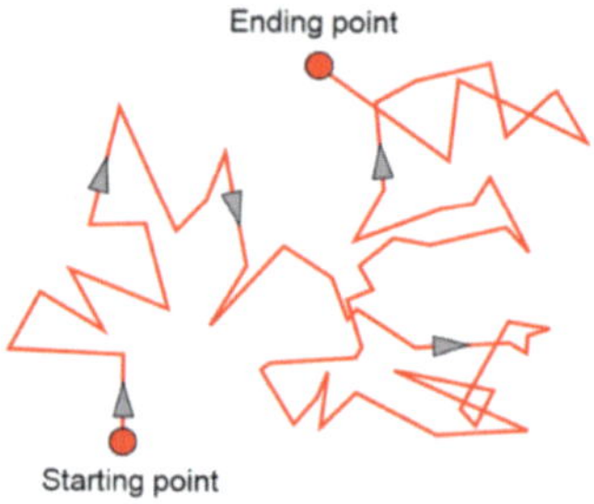


**Figure 6.7.** Brownian motion.

*Source:* https://chemistrytalk.org/brownian-motion/.

## 6.2.3. Stability and Flocculation

The flocculation and stability principles are quite significant in the behavior of disperse and colloidal systems. Stability in disperse colloidal systems usually refers to the ability of the particles to be dispersed uniformly without settling or agglomeration over a period of time. The situation is dictated by a number of factors, including particle size, surfactant content, and the ionic strength of the dispersing medium. A colloid system can be destabilized through interactions that are flocculating in nature, where particles come together to form large clusters. This kind of flocculation is induced by a change in environmental conditions, namely by changing the pH or the addition of an electrolyte, which will affect the electrostatic repulsion between particles. Understanding of these dynamics is essential to industries based on colloidal systems, as the physical properties of colloidal systems directly relate to their stability and their behavior under various applications. Flocculation mechanisms often rely on particle interactions in terms of specific potentials, e.g., the Yukawa potential. The Yukawa potential provides a mathematical framework by which to investigate the interaction between charged colloidal particles, particularly on collision at close proximity. Interestingly, it has been demonstrated in studies that a Yukawa potential can accurately describe the two-particle interactions down to distances on the order of the nematic coherence length in liquid crystals (Borštnik et al., 2003) (Figure 6.8).

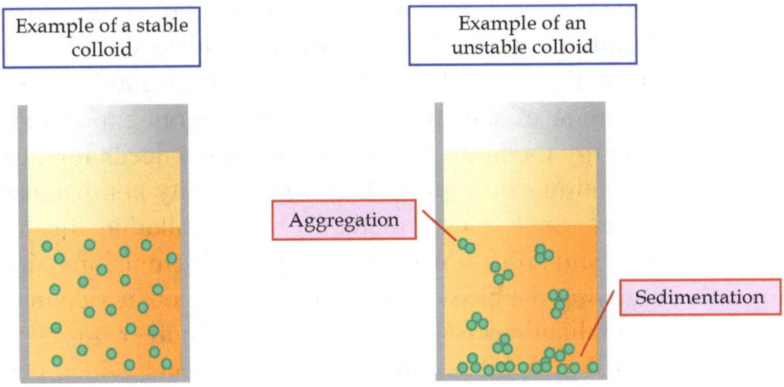


**Figure 6.8.** Examples of a stable and of an unstable colloidal dispersion.

*Source:* https://chem.libretexts.org/Bookshelves/Physical_and_Theoretical_Chemistry_Textbook_Maps/Supplemental_Modules_(Physical_and_Theoretical_Chemistry)/Physical_Properties_of_Matter/Solutions_and_Mixtures/Colloid.

By using such models, researchers are able to generate flocculation diagrams that depict the dependence of electrostatic parameters like surface charge density and Debye screening length. Such information derived from these diagrams allows for improved predictions and control of the flocculation phenomenon in most colloidal systems, particularly during phase transitions, which may contribute to the stability of the system dispersion. Aside from electrostatic concerns in classical fashion, the role of biopolymers, such as cellulose, in stabilizing emulsions has drawn considerable attention. Solubility of native celluloses in water is typically an issue for their application; however, chemical modification leads to derivatives that are soluble in water and exhibit amphiphilic behavior. Such molecular properties allow for the adsorption of modified celluloses at the oil-water interface, stabilizing the emulsions appropriately and preventing separation by flocculation (Costa et al., 2019). Research into the role of cellulose in emulsion systems emphasizes the need to examine molecular interactions in systems dispersed. By revisiting mechanisms of stability and flocculation, scientists can develop new formulations in food, pharmaceuticals, and other areas and manufacture products that leverage the desirable properties of cellulose and solve problems in the stability of colloidal systems.

## 6.3. METHODS OF PREPARATION

Preparation of colloids and disperse systems encompasses numerous various methods, each with certain advantages and suitability for practical use. One of the familiar techniques is using spontaneous emulsification, particularly convenient in food and pharmaceutical uses to encapsulate sensitive bioactive compounds. Based on the oil and water phases' self-assembly, the method facilitates the preservation of sensitive compounds without exposure to harsh conditions of high-energy methods. It also minimizes the needs for surfactants and co-surfactants, resulting in improved thermal stability compared to other low-energy approaches. Such a new method of encapsulation illustrates the ability of spontaneous emulsification in producing double emulsions, since a key feature towards improving the bioavailability of two or more active ingredients in a broad range of applications (Binks et al., 2016). Another important route of preparation is emulsion stabilization by Pickering agents, which makes use of solid particles to stabilize oil-in-water systems. Novel research illustrates how naturally hydrophilic nanoparticles can form stable Pickering emulsions on contact with oil, owing to in situ hydrophobization. This oil-particle surface interaction causes enhanced emulsion stability. By introducing a variation in chain length of the oil, researchers found that stability improved, with oil being thought of as a natural surfactant. The efficacy of this method has been confirmed by other confirmatory experiments, including contact angle determination of oil-water-solid, which illustrates the fundamental processes of colloidal system formation and stabilization (Binks et al., 2016). Aside from the aforementioned techniques, the most significant influence on colloidal and dispersed system preparation procedures comes from surfactant selection. The specific surfactant used can significantly alter both the particle diameter of the resulting droplets and ultimately the final stability of the resulting colloidal system. For instance, ionic surfactants such as Span 80 and polyglycerol polyricinoleate (PGPR) have similarly been reported to be useful in the formulation of micro-emulsions, with research confirming that a properly balanced surfactant-to-water ratio can produce optimal droplet sizes. As the proportion of surfactant is increased, a significant shift in shear viscosity is experienced, which directly influences the rheological properties of the resultant emulsions. This sensitive balance between surfactant level, particle size, and stability is critical when designing effective colloidal and dispersed systems for a variety of purposes (Jafari et al., 2016).

## 6.3.1. Dispersion Methods

The study of the dispersion methods in dispersed systems and colloids indicates numerous ways of achieving homogenously dispersed particles in a medium. Among these techniques, surface modification processes have been of special interest because of their capacity to stabilize dispersed systems. For instance, material functionalization as graphitic carbon nitride (g-CN) has demonstrated that amine groups, which allow pH-dependent water dispersibility, can be incorporated. This improvement is crucial as it enables low sonication time dispersion of g-CN, a drastic contrast with its unmodified counterparts. The ability to have high concentrations of modified g-CN in aqueous and organic dispersions is a breakthrough in the functionality of such materials, showing the direct impact of dispersion processes on the integrity and usability of colloidal systems (Antonietti et al., 2017) (Figure 6.9).

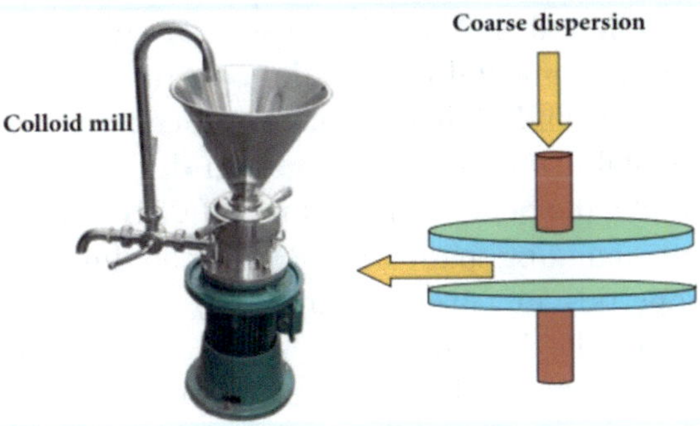


**Figure 6.9.** Colloid mill.

*Source:* https://www.brainkart.com/article/Preparation-of-Colloids_ 41346 /#googl e_vignette.

Concurrently, the application of different surface modification procedures to carbon nanotubes, namely single-walled carbon nanotubes (SWNTs), elucidates the range of responses to dispersion procedures. Experimental work has established that oxidization and surfactant or humic acid coating processes significantly enhance SWNT stability and water solubility in various aqueous media. Every procedure has specific characteristics regarding retention and mobility of modified SWNTs in porous media. Surprisingly, humic acid coating method has been noted to produce increased mobility than surfactant adhesion and oxidation, highlighting the central role of precise dispersion methods in enhancing the performance and behavior of colloidal particles in complex environments (Gao et al., 2012) (Figure 6.10).

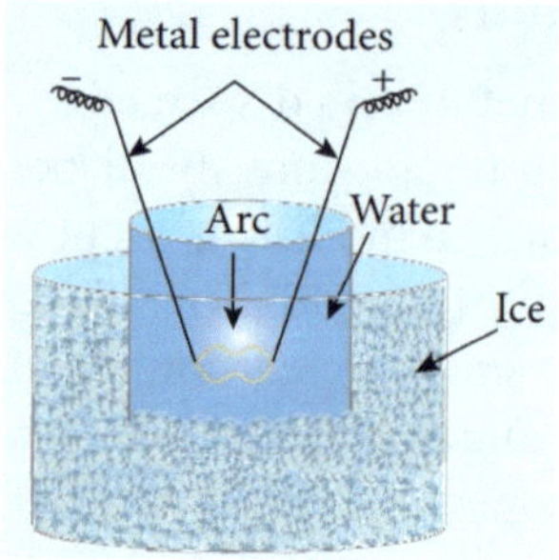


**Figure 6.10.** Electro dispersion.

*Source:* https://www.brainkart.com/article/Preparation-of-Colloids_41346/.

The effectiveness of such dispersion processes not only affects the short-term stability of colloidal systems but also their ultimate performance for purposes such as environmental remediation, pharmacy, and material science. An understanding of the underlying mechanisms of dispersion, e.g., electrostatic repulsion or steric stabilization, is important in designing improved colloidal systems. For example, the phenomenon of modified SWNTs at water interfaces accounts for how changing particle surfaces can influence their retention and transport properties in porous media and, in extension, their practical applications. Collectively, these processes of dispersion illustrate a fundamental concept of colloid science, which is the importance of controlled particle distribution in advancing both stability and function in numerous applications in colloids and dispersed systems (Figure 6.11).

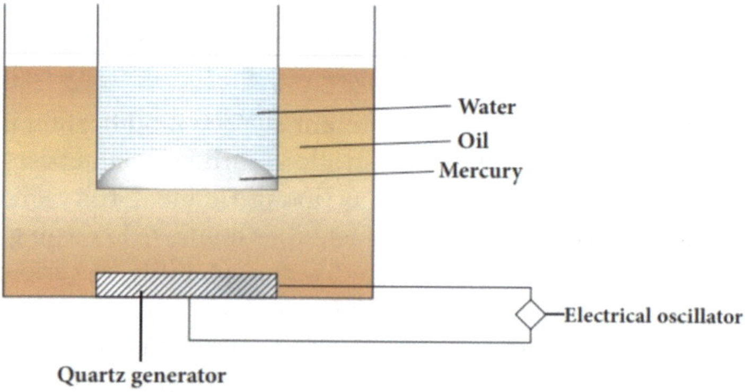


**Figure 6.11.** *Ultrasonic dispersion.*

*Source: https://www.brainkart.com/article/Preparation-of-Colloids_41346/.*

## 6.3.2. Condensation Methods

Condensation processes are of major importance in stabilizing and generating colloidal systems, invoking principles of molecular aggregation to promote the development of nanoscale and microscale materials. During condensation processes, phase change from gas to liquid or liquid-liquid phase leads to colloidal particles that undergo self-assembly. The process mechanism involves temperature and pressure fluctuations, which enable the crossing of energy barriers for clustering molecules. Through this method, not just particle size and distribution but also the presence of multiple components, i.e., surfactants or polymers, influencing the physicochemical properties of the resulting colloids can be obtained. Since the system is approaching equilibrium, these particles also exhibit unique behavior which has to be researched further, particularly on what happens to their behavior in different aqueous media, which can become integral to application in pharmaceuticals, materials science, and nanotechnology. When analyzing how efficient condensation processes are, it becomes apparent that they play a vital role in optimizing the performance of colloidal systems. For instance, it has been proven through research that changing the concentration of precursors during condensation can lead to extreme changes in the nature of the end product, such as shape, size, and stability.

Changing environmental conditions, such as humidity and temperature, also enables researchers to prepare particular colloidal dispersions. As in studies on geopolymers and composites, the material properties developed are largely determined by ratios of constituent elements in proportion to their condensation reactions. Addition of cellulose fibers into the systems has been observed with enhanced mechanical properties, attesting to the significance of investigation of the underlying condensation processes in the development of improved materials (Bernardo et al., 2018). Furthermore, the processes through which self-assembly occurs within condensation processes highlight both theoretical significance and practical relevance within physical chemistry. These effects are dynamical patterns that are common to many physical systems even with different microscopic details. Investigation of how condensation results in self-assembly, particularly under far-from-equilibrium conditions, illuminates the complexity of particle interactions. The ability to predict and control these pathways is critical to moving material science forward, and this is a consequence of the need for continued research into condensation methods. As our understanding grows, the potential applications in building new colloidal architectures—ranging from drug delivery systems right through to advanced coatings—can be maximized significantly, and this results in the creation of new technologies based on the energy of condensed colloidal systems (Jack et al., 2014).

### 6.3.3. Chemical Synthesis

Chemical synthesis is the central discipline in the formulation of colloidal systems because it provides the elementary processes necessary for creating nanoparticles and other colloidal substances. Chemical synthesis processes, particularly those that produce colloids by performing controlled reactions, are the finest illustrations of the intricate balance between kinetics and thermodynamics. As can be seen, the use of reducing agents in the synthesis of metal nanoparticles has gained traction due to the ability to tune the size and shape of the particles. Tight control over reaction conditions can lead to colloids with desirable characteristics, e.g., stability, and reactivity, which are important across applications that vary from catalysis to medicine. With ongoing research, the design of sophisticated synthesis methods will further expand our understanding regarding how to control colloidal properties to meet particular demands for function in disperse systems. Colloidal solution mechanical and chemical stability relies to a great extent on those processes involved in synthesizing them.

Through carefully controlled experimental procedures, particle size, shape, and distribution can be optimized to furnish optimal stability. One notable illustration is the chemical synthesis of nanoscale ferrihydrite particles, which highlighted their ability to facilitate microbial degradation of contaminants in aquifer remediation. These nanoparticles possess characteristic transport properties dependent on ionic strength and velocities of flow, and they reveal how chemical synthesis can have direct effects on the relevance of colloids to environmental remediation. Understanding these parameters is essential to improve methods used in the synthesis of colloids that perform optimally in dispersion systems and maximize their use in environmental applications (Bosch et al., 2012). Additionally, recent research in colloidal synthesis typically explores varying parameters used in the determination of physical properties of final particles.

With mechanisms comparable to the colloidal gold synthesis, researchers have mimicked the preparation of monodisperse silver colloids with iso-ascorbic acid and silver-amine complexes. The focus in this two-step formation is on controlling reaction kinetics for desired particle stability and size. As described, manipulation of parameters such as temperature and solvent viscosity is critical in determining the synthesis pathways and hence the properties of the resulting colloids. Such outcomes are exemplary of the growing sophistication of chemical synthesis methodologies in being capable of delivering well-defined colloidal systems with the potential to accomplish disparate functions in disparate fields (Goia et al., 2008).

## 6.4. APPLICATIONS OF COLLOIDS

The applications of colloids extend to numerous fields, which says much about their utility and significance in industrial and scientific processes. Functionalized colloids such as grafted graphitic carbon nitride (g-CN) are now key components in materials science due to their enhanced dispersibility in media.

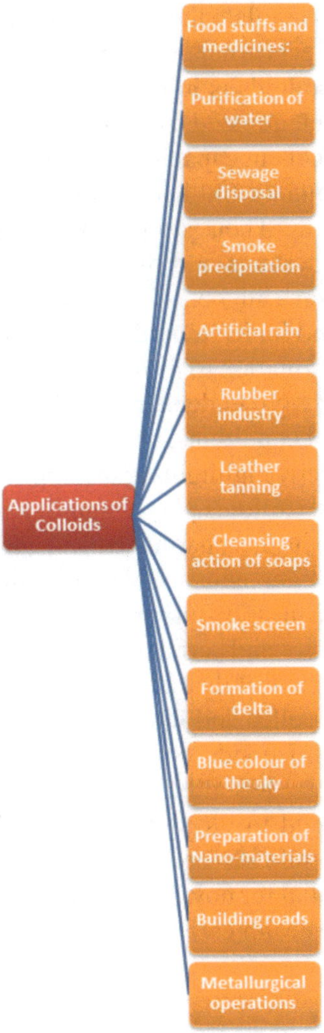


**Figure 6.12.** Applications of colloids.

*Source:* https://www.askiitians.com/iit-jee-chemistry/physical-chemistry/applications-of-colloids.aspx.

The new process in using functionalized olefinic molecules in grafting allows one to retain g-Cs inherent features while maximizing its potential applications. This improved dispersibility results in higher stability and facilitates higher concentration achievement in aqueous and organic dispersions, thus making g-CN a promising material for catalysts and sensors in chemical processes (Antonietti et al., 2017). Optimization of the dispersion properties provides scientists with the ability to regulate the physical properties of materials, which paves the way for advancements in nanotechnology and green chemistry. Another significant application of colloids is in the area of environmental science, and that is in the cleaning of contaminated environments. Ferrihydrite nanoparticles at the nanoscale display excellent capacity to catalyze microbial degradation of certain environmental pollutants and hence are used to carry out in situ clean up of contaminated aquifers. Their unique size and properties render them injectable into subsurface systems and dispersible in water, forming reactive zones where biodegradation can be effectively stimulated. Studies indicate that subsurface flow rate and ionic strength have a strong influence on particle mobility, imparting key information to formulate optimal remediation strategies (Bosch et al., 2012). Through understanding how these colloidal systems behave under environmental conditions, practitioners can design more effective remediation processes, thereby highlighting the applied significance of colloids to address ecological issues. Colloids play a significant role in drug formulation and drug delivery systems in the pharmaceutical industry. Colloidal systems like emulsions and suspensions enhance the solubility and bioavailability of most pharmaceutical molecules, ensuring an effective therapeutic response. Through adjustment of the size and surface characteristics of colloidal particles, the formulators are able to control the rates of drug release, which translates to better patient outcomes. Micelles and liposomes, for instance, are site-specific colloidal carriers that minimize side effects and maximize therapeutic effect. Moreover, colloidal chemistry has allowed for the development of smart drug delivery systems that respond to physiological conditions, making treatment protocols more accurate. While researchers continue to create novel colloidal drug preparations, the potential for novel drugs is vast, affirming the significance of colloids in medicine and health care (Figure 6.12).

## 6.4.1. Food Industry

The food industry has become increasingly aware of the significant role of colloids and dispersed systems to food quality and stability. The systems, such as emulsions, suspensions, and foams, play a fundamental role in food formulation since they affect the texture, appearance, and acceptability by consumers. The application of particles of calcium carbonate ($CaCO_3$) in stabilizing foam bubbles is one instance of applied colloid science. As discussed in a study paper, once these particles are hydrophobized with food-grade emulsifiers, they act well as foam stabilizers in food applications under normal pH conditions. This is not only useful to contribute a better texture but also assists with the growing trend towards clean label ingredients because consumers more and more demand products consisting of recognizable and safe materials. Thus, understanding interactions within these spread systems is paramount to food businesses which desire to innovate and optimize their products in a competitive business landscape. Resuming the investigation of emulsions, one can observe that such colloidal systems are priceless with regard to preserving as well as delivering flavors in foods. Flavor materials, which influence the consumer acceptance to a significant extent, are volatile in nature and can further deteriorate during processing and storage. The food industry can prevent flavor loss by encapsulating such flavor compounds in emulsions. Emulsions, however, also suffer from physical stability problems like an increase in particle size that renders them less effective with time (Figure 6.13).

**Figure 6.13.** Colloids in food.

*Source:* https://library.fiveable.me/_next/image?url=https%3A%2F%2Fstorage.googleapis.com%2Fstatic.prod.fiveable.me%2Fsearch-images%252F%2522Foams_in_food%253A_whipped_cream_meringue_bread_dough_stability_colloidal_dispersions_texture_nutrition%2522-mini_raspeberry_pavlovas.jpg&w=828&q=75.

Nanoemulsions' emergence overcame these challenges in the aspects of higher stability and smaller droplet size, solidifying the promise of enduring flavor enhancement. High-pressure homogenization techniques further solidify the establishment of stable emulsions, underpinning the pivotal role of colloidal science in producing high-quality food products based on consumers' demands (et al., 2016). Lastly, the intersection of consumer health and colloid science is also a significant issue for the food industry. The use of effective food-grade emulsifiers, such as sodium dodecanoate (SD) and sodium stearoyl-2-lactylate (SSL), is one great illustration of how effective ingredient choice affects not only stability but also health profile. With an increased emphasis on nutrient delivery and reduced unhealthy fats in products, the food industry is posed with achieving balance between consumer demands and technological challenges offered by colloidal systems. Improved surfactant technology enables manufacturers to produce healthy products without sacrificing sensory characteristics. Therefore, food technology, health, and consumer protection are a critical intersection of progress and colloidal knowledge in food items, and the need for ongoing research and development here is underscored.

## 6.4.2. Pharmaceuticals

The colloid-dispersed system interaction is of special significance in pharmaceuticals, particularly in the formulation of drug delivery systems. Among the most notable trends here is the application of Pickering emulsions, which are stabilized by solid particles instead of traditional surfactants. In the case of oil-in-water Pickering emulsions, hydrophilic character of the nanoparticles may be tailored by in situ hydrophobization, it is suggested. This approach stabilizes the emulsions by establishing hydrogen bonds between the particles and the oil molecules, leading to increased stability and effectiveness in drug formulations. These developments not only enhance the functionality of emulsions but also reinforce the safety and efficacy of the overall drug preparation, illustrating the way colloidal science can drive the development of sophisticated drug delivery systems (Binks et al., 2016). In addition to Pickering emulsions, solid lipid nanoparticles (SLNs) have proved to be a worthy innovation in the pharma industry, particularly for controlled drug release. SLNs are unique due to their size-dependent properties, with full control over drug release rates and targeted delivery mechanisms. Using drug encapsulation in lipid matrices, SLNs offer sustained release profiles and enhanced bioavailability while reducing drug side effects. This aspect proves to be especially valuable in therapeutic applications with localized action of drugs, such as the treatment of skin diseases. As the pharmaceutical world continues to evolve, SLNs constitute a promising field of research in treating complex diseases, thus redefining the traditional therapeutic modalities and offering better outcomes for the patients (Conway et al., 2014) (Figure 6.14).

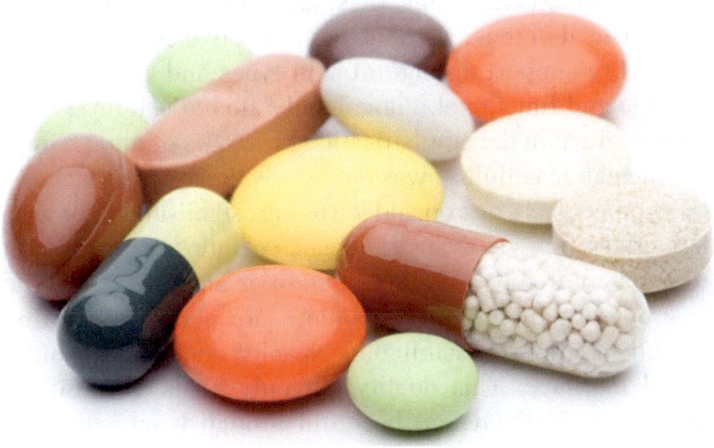

**Figure 6.14.** Colloids plays a crucial role in the pharmaceutical industry.

*Source:* https://files.askiitians.com/cdn1/images/20141020-114051500-1683-capture.png.

The impact of colloidal and dispersion systems on drugs extends beyond mere formulation milestones; it also extends to the general therapeutic environment. Through the union of sophisticated approaches such as nanoparticle stabilization with hydrophilic components, researchers can overcome the solubility and distribution problem usually plaguing drug delivery systems. The potential of these technologies to streamline drug formulation implies that drugs will remain effective even as they meet stringent standards of safety. Furthermore, with ongoing research into the colloidal synergies and drug delivery systems, new treatments for previously challenging diseases are also on the agenda. Thus, by acquiring knowledge of the fundamental principles of colloids and dispersed systems, not just does it establish a foundation for pharmaceutical development, but also paves the way for future healthcare development (Binks et al., 2016; Conway et al., 2014).

### 6.4.3. Cosmetics

The intersection of cosmetics and colloid and dispersed system science demands critical examination of the ingredients upon which these products are based. The majority of cosmetic products are highly reliant on lipid ingredients, which may be either solubilized or insolubilized. Such amphiphilic lipid molecules spontaneously self-assemble into complex nanostructures when they are dispersed in water. This self-assembly behavior not only enhances the stability of the products but also allows their properties to be modified to accommodate various physical forms, i.e., creams, gels, and pastes. Control over these lipid systems is very critical in the development of effective cosmeceutical products, which often need to support a range of skin types and consumer preferences and remain stable and functional. Functionally altered cellulose has now been an important ingredient in the emulsification processes of utmost importance to cosmetics. While native cellulose was previously annoying in its non-solubility in water, development has gone on and led to the establishment of water-soluble cellulose derivatives. These advances enhance cellulose's performance so that it can function as an effective rheology modifier and emulsifying agent. The amphiphilic properties of modified cellulose are particularly valuable; they possess an excellent emulsion-stabilizing capacity since they are able to adsorb at the water-oil interface. This duality not only helps to stabilize cosmetic products but also to make them more beautiful through their texture and sensorial qualities. Cellulose application in cosmetics is thus a classic illustration of the intricate relationship between material science and cosmetic formulation. The application of nanostructured lipid systems also shows the complexities of colloids and dispersed systems in cosmetics. The lipid nanostructures can be utilized to encapsulate active ingredients that are hydrophilic, hydrophobic, or amphiphilic in nature and enhance their delivery and performance in topical formulations. Preparing nano-hybrid materials that combine lipid systems with other nanomaterials such as silica nanoparticles and hydrogels, formulators are able to prepare novel cosmetics with novel properties and functionalities. The range of such materials allows one to create products that not only ensure visual attractiveness but also offer enhanced performance benefits, such as increased skin penetration and longer release of active substances. That is why incorporation of innovative nanotechnology in the cosmetic industry is opening up new opportunities for novel approaches that re-create consumer perception and channel efforts toward efficacy and safety.

## 6.5. COLLOIDAL STABILITY

The concept of colloidal stability is pivotal to the characterization of the disposition of disperse systems, particularly in environmental and industrial processes. Stability refers to the ability of colloidal particles to stay dispersed, against such processes as sedimentation, aggregation, or coagulation. Stability is influenced by several conditions, such as particle size, surface charge, and ionic strength of the medium. For instance, smaller particles tend to be more stable since their large Brownian motion opposes gravitational forces that attract them towards each other. Electrostatic repulsion is also significant; where particles possess equal surface charges, they repel one another and thus maintain stable dispersion. In metastable or intermediate stability systems, slow aggregation kinetics may be present, indicating that such factors as membrane filtration design must be optimized to achieve both critical flux and the dynamics of colloidal interactions (Aimar et al., 2010). One must hence learn about these principles in order to optimize processes across industries from pharmaceuticals to environmental remediation. Colloidal stability is especially valuable when used for membrane filtration operations. In this process, stable colloids are responsible for maintaining optimal flux rates and preventing fouling, thereby enhancing filtration system efficiency. Fouling involves particles adhering to the filter membranes, leading to a gradual decrease in flow rates and filtration efficiency.

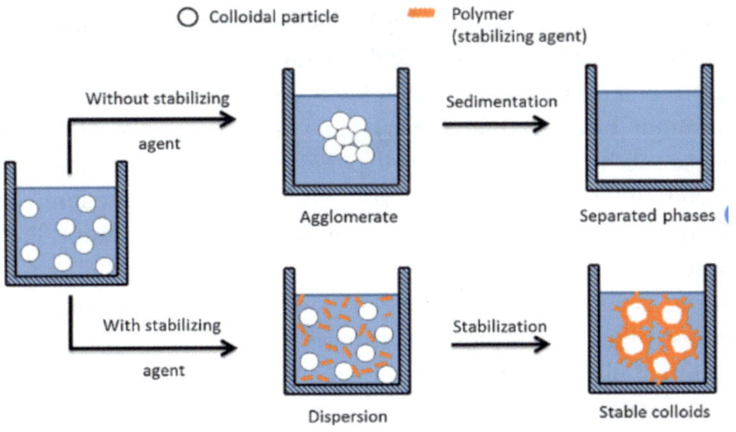


**Figure 6.15.** Schematic presentation of colloidal stability, with and without stabilizing agent.

*Source:* https://www.researchgate.net/figure/Schematic-presentation-of-colloidal-stability-with-and-without-stabilizing-agent_fig1_297591140.

From studies, it is clear that the interaction between factors affecting colloidal stability can reveal mechanisms for ultrafiltration or microfiltration flux decline. As an example, it should be noted that even if a critical flux defines the limiting flow for fixed colloids, operational conditions that promote slow particle coagulation acquire greater significance (Aimar et al., 2010). From these data, one can form design and operation conditions for filtration processes to facilitate better fouling control and overall system life and efficiency. In environmental science, colloidal stability is critical to the transport and remediation of pollutants. An example of application of this use is the use of nanoparticles, such as ferrihydrite, in groundwater cleanup technologies.

Ferrihydrite nanoparticles enhance microbial degradation processes by developing reactive zones in contaminated aquifers. However, the distribution of these nanoparticles in subsurface systems largely relies on factors like flow rate and ionic strength, which have direct influence on their stability and retention in groundwater systems. Experimental studies indicate that at typical ionic strengths, these nanoparticles can move a few meters in aquifers, whereas enhanced ionic strengths can decrease their mobility by a significant amount. By focusing our knowledge on these colloidal dynamics, we can better optimize groundwater cleanup measures, using mobility as a tunable parameter to enhance the effectiveness of remediation actions (Figure 6.15).

## 6.5.1. Factors Affecting Stability

Stability of the dispersed system and colloids is based on different factors that are dynamic in nature and have to do with the nature of the components involved, as well as their interaction at the molecular level. Among the significant constituents is the surfactant employed in the system since it governs the equilibrium between forces at the interface of the dispersed particles. For instance, a suitable surfactant can enhance colloidal stability significantly by reducing interfacial tension, thereby inhibiting particle deposition and sedimentation. Particle size and distribution may be drastically affected by the alteration of surfactant concentration, as seen in surfactant-to-water ratio experiments, where increased concentrations of surfactant initially reduced droplet size but ultimately led to destabilization at elevated concentrations. Such interactions emphasize the key role of surfactant properties and content in formulating and stabilizing stable colloid systems (Jafari et al., 2016). Another significant parameter influencing colloid stability is associated with the ionic strength of the surrounding medium. The presence of electrolytes can increase or decrease stability, primarily by screening electrostatic repulsion between particles with the same charge. Low ionic strength conditions see strong double layers that envelop dispersed particles, thus maintaining stability based on repulsive forces. With increased ionic strength, however, particles can be seen to suffer less repulsion, which helps in aggregation and sedimentation. This effect is expressed in research focusing on nanoscale ferrihydrite particles, whereby high salt concentrations exhibited a linear relationship with particle mobility and retention within aquifer systems, with the effects thus influencing remediation efforts. Thus, being aware of ionic strength is essential in order to forecast and manage the stability of colloidal dispersions in various processes from environmental remediation to food technology (Bosch et al., 2012). Temperature should not be disregarded either, as thermal conditions will alter the kinetic energy of particles and thereby their interaction dynamics. High temperatures tend to produce enhanced molecular motion, tending to increase particle dispersion and decrease viscosity in some colloidal systems. Yet high temperatures might also encourage destabilization by introducing increased Brownian motion, possibly resulting in aggregation and phase separation. Cooling, on the other hand, may stabilize some systems by decreasing kinetic energy but could also cause crystallization or gelation in others. This balance of the favorable and unfavorable temperature influences on colloid stability must be regulated in a sensitive way in real systems. Basically, temperature profoundly affects the stability of colloids and dispersed systems, and necessitates an tailored solution depending on the precise conditions and system needs.

### 6.5.2. Role of Surfactants

Surfactants, or surface-active agents, play a vital role in stabilizing colloids and structuring dispersed systems. They function by reducing surface tension at the interface of two immiscible liquids, e.g., water and oil. Such a reduction in surface tension promotes dispersion of the particles and thus more uniform mixtures. When added to a colloidal system, surfactants position themselves at the interface and form a monolayer preventing droplet coalescence or particle agglomeration. Surfactants are also essential in uses like emulsions of foods and drugs, where a stable distribution of the ingredients is required to meet efficacy and shelf life. Surfactants can also charge colloidal particles and influence their stability in a dispersion. For instance, studies have proved that the charge of sterically-stabilized colloids is a function of the surfactant applied and various surfactants result in either a negative or positive charge on particulate matter. All classes of surfactants interact with particles in colloidal systems differently, affecting their characteristics in the colloidal system. As depicted in the literature, reverse micelles may charge colloids upon introduction into non-polar solvent systems, thereby stabilizing them and making them transposable under various environments. The relationship between surfactant properties and charge of the colloid demonstrates a complex interaction that can be used in industrial operations (Bartlett et al., 2007). The applications of surfactant use extend beyond stabilization; they also affect the mobility of colloids in porous media and other locations. As an example, research has indicated that surface alteration by surfactant coating significantly enhances the mobility of carbon nanotubes in granular porous media. Various surfactant modification methods have different impacts on the retention and transport of particles, highlighting surfactants central role in facilitating the transport of colloidal systems. These findings point to the critical selection of surfactants with respect to desired effects in actual applications, particularly in material science and environmental engineering. Ultimately, surfactants are critical regulators of dispersed system behavior, proving their critical influence on physical and chemical properties (Gao et al., 2012).

### 6.5.3. Techniques for Stabilization

Colloidal system stabilization is a critical parameter that ensures that particles are evenly dispersed in a medium and prevent particle agglomeration or settling. One of the most cited techniques involves the use of surfactants, which lower the tension at water and oil interfaces, ultimately decreasing the particle coalescence tendency. Surfactants are further sub-divided into anionic, cationic, nonionic, and amphoteric surfactants, each of which possesses unique properties that can be specifically tailored to find use in food, pharmaceuticals, and cosmetics. The choice of a surfactant is crucial; it must be not only effective at stabilizing the system but also have to be compatible with other ingredients. The intricate interplay of particle size and surfactant concentration also affects stability, and hence optimization of such factors is of paramount importance in achieving desired performance characteristics. The use of biopolymers, such as cellulose derivatives, is a prevalent technique of stabilization. Biopolymers are also instrumental in maintaining greater stability of the emulsion. Native cellulose, due to its amphiphilic character, is capable of adsorbing efficiently on the oil-water interface and inhibiting emulsion droplet coalescence. The drawback with native cellulose is its insolubility in water, which generally makes it necessary to chemically modify it to render it water-soluble and applicable in colloidal systems. Modified cellulose is a rheology modifier, which has the capability to alter the dispersion viscosity and functionally stabilize foams and emulsions. Recent studies have rather indicated towards cellulose structural anisotropy, emphasizing its potential as a new stabilizing agent that not only facilitates emulsion formation but also enhances performance through mechanisms akin to the Pickering effect, significantly broadening its range of application across industries (Costa et al., 2019). Nevertheless, additional stabilization advances have also emerged through the utilization of functionalized materials, such as grafted graphitic carbon nitride (g-CN). This new strategy involves surface functionalization of the g-CN to improve its water and organic solvent dispersibility. By adding functional groups within the framework of the g-CN, researchers have observed remarkable stability and dispersion improvements, which are critical towards optimizing the efficiency of deployed colloidal systems. The ability to disperse higher concentrations of the functionalized g-CN compared to its unfunctionalized precursor enhances the role of surface chemistry in stabilization processes. This method not only facilitates the incorporation of g-CN into various formulations but also ensures efficient dispersion in various media, thus emphasizing the importance of material modification towards the attainment of stable colloidal formulations (Antonietti et al., 2017).

# SUMMARY

In summarizing the colloids and dispersed systems discussion, it is apparent that the multifaceted behavior and properties of these systems are the mainstay to a myriad of applications in scientific as well as industrial areas. The dynamic interaction between the particles and the medium permits various phenomena such as optical features, stability, and mass transport. Better understanding of these interactions not only reveals the fundamental principles that control colloidal stability and behavior but also gives more rigor to methodology used in real-world scenarios, such as drug delivery systems, food industries, and cosmetics. Thus, the realization of colloids impact on diverse sectors emphasizes the demand for continuous research, hence the innovations and technology advancements in material sciences. Moreover, recent advances in imaging instrumentation and analysis, such as neutron grating interferometry, provided researchers with novel means to investigate microstructures of colloidal systems more effectively. With the possibility of recording simultaneously transmission, phase differentiation, and dark-field imaging, scientists are able to obtain essential information on the local structure and dynamics of dispersed systems. This type of high-resolution imaging capability picks up detail like anisotropy and microstructural heterogeneity that influences profoundly the nature and the utilization of colloidal systems. Such findings have practical implications beyond theoretical interest and create opportunities for designing and optimizing colloidal materials for particular purposes, justifying the relevance of advanced analytical techniques in this field (Betz et al., 2016). In electrochemical applications, direct electron transfer between immobilized enzymes and other nanostructured materials is another major characteristic of colloids and dispersed systems. Investigations of such types of interactions, such as the electrochemical response of hemoglobin with modified electrodes, emphasize the potential for the development of highly sensitive biosensors and energy conversion devices. The effective demonstration of appreciable electrochemical activity and stability of such systems based on nanomaterials not only demonstrates their applicability in analytical chemistry and biochemistry but also highlights the broader applicability of colloids toward enabling environmentally friendly technologies. Collectively, these findings highlight the multidisciplinary importance of colloids and dispersed systems towards substantiating their role in advancing current science and technology (Arabnezhad-Khanooki et al., 2012).

The field of colloids and dispersed systems is one of multifaceted interaction of components with a significant influence on their behavior and uses. Central to the systems is the interfacial property concept, with the latter playing a central role in emulsion stabilization. Of interest, recent studies show that particle-stabilized emulsions can gain specific characteristics by the process of nucleation and subsequent coarsening of minority phase domains when cooled. While the interfaces are covered by particles, interfacial energy is reduced and, consequently, stabilizes the emulsion with efficient barriers against coalescence. The significance of this study underscores the significance of interfacial tension dynamics as the system goes through temperature changes, highlighting the intricate relationship between temperature conditions and emulsion stability (Schofield et al., 2005). It is the foundation upon which more efficient colloidal systems are developed in various industrial processes. In investigating the dynamics of colloidal dispersion, the effects of gravitational fields are more prominent. The binary colloidal system offers complex behaviors toward sedimentation, which generates unique morphological patterns within the system. With the processes of the colloidal mixture transitioning through fluid state to fluid state, liquid-liquid and gas-liquid phase separations become evident, with multiple equilibria occurring under different conditions. This phase separation interaction has profound impacts on the dynamic density distribution, dictating how the system evolves over time. The findings indicate that in certain scenarios, the symmetry of the final density profiles cannot match with external potentials to conventional expectation from statistical mechanics (Malijevsk et al., 2013). This makes understanding dispersed systems more complex and proves the extensive impact external conditions exert on colloidal activity. Finally, the consolidation of these results explains the multi-dimensionality of dispersed systems and colloids. By understanding the forces of particle stabilization, the alteration of interfacial tension, and gravitational forces' impacts, scientists can better manipulate and tailor these systems for operational uses, such as food and medicine. Exploiting the understanding of phase behavior holds promise for product innovation, realizing peak performance through better stability and function. Moreover, transitioning from the lab bench to the outside world highlights the need for interdisciplinarity in colloid science. By continued exploration of such intricate systems, the scientific community can exploit hitherto untapped potential that influences both scientific insight and industrial processes in colloid science.

# REFERENCES


1. Aimar, P., & Bacchin, P. (2010). Slow colloidal aggregation and membrane fouling. https://core.ac.uk/download/12042069.pdf (accessed on 3rd October 2025).
2. Alexandridis, P., Alexandridis, P., Balazs, A. C., Bartlett, P., Bockstaller, M. R., Boucenna, I., Cipelletti, L., et al. (2012). Structure of nanoparticles embedded in micellar polycrystals. http://arxiv.org/abs/1205.4746 (accessed on 3rd October 2025).
3. Antonietti, M., Kumru, B., & Schmidt, B. (2017). Enhanced dispersibility of graphitic carbon nitride particles in aqueous and organic media via a one-pot grafting approach. https://core.ac.uk/download/212793821.pdf (accessed on 3rd October 2025).
4. Arabnezhad-Khanooki, T., Farasat, A., Mazdapour, M., et al. (2012). A novel hydrogen peroxide biosensor based on modified electrode with hemoglobin and zinc oxide nanoparticles. https://core.ac.uk/download/52204783.pdf (accessed on 3rd October 2025).
5. Auernhammer, G. K., Ullrich, B., Vollmer, D., et al. (2009). Frequency dependent deformation of liquid crystal droplets in an external electric field. http://arxiv.org/abs/0812.0745 (accessed on 3rd October 2025).
6. Bandyopadhyay, R., Borthwick, M. A., Harden, J. L., et al. (2004). Evolution of particle-scale dynamics in an aging clay suspension. http://arxiv.org/abs/cond-mat/0410376 (accessed on 3rd October 2025).
7. Barrer, I., Breck, D., Brinker, C., Jacobs, J., Janata, J., Seiyama, T., & Ward, M. (1988). Formation and characterization of inorganic membranes from zeolite-silica micro composites. https://core.ac.uk/download/12165051.pdf (accessed on 3rd October 2025).
8. Barrett, K. E. J., Petekidis, G., Hunter, R. J., Isa, L., Larson, R. G., Lee, B., Ballesta, P., et al. (2012). Wall slip and flow of concentrated hard-sphere colloidal suspensions. http://arxiv.org/abs/1201.4305 (accessed on 3rd October 2025).
9. Bartlett, P., Faers, M. A., Royall, C. P., et al. (2012). Phase separation dynamics in colloid-polymer mixtures: The effect of interaction range. http://arxiv.org/abs/1209.3973 (accessed on 3rd October 2025).
10. Bartlett, P., Kemp, R., Roberts, G. S., et al. (2007). Electrostatic charging of non-polar colloids by reverse micelles. http://arxiv.org/abs/0712.2768 (accessed on 3rd October 2025).
11. Batchelor, G. K., Berthier, L., Bird, R. B., Born, M., Bouchaud, J. P., Cates, M. E., Cates, M. E., et al. (2010). Nonlinear rheology of colloidal

dispersions. https://core.ac.uk/download/20651718.pdf (accessed on 3rd October 2025).

12. Beller, D. A., Cavallaro, M., Gharbi, M. A., et al. (2013). Micro-bullet assembly: Interactions of oriented dipoles in confined nematic liquid crystal. http://arxiv.org/abs/1210.1881 (accessed on 3rd October 2025).
13. Bernardo, E., Dlouhy, I., & Taveri, G. (2018). Mechanical performance of glass-based geopolymer matrix composites reinforced with cellulose fibers. https://core.ac.uk/download/168406423.pdf (accessed on 3rd October 2025).
14. Betz, B., Horisberger, M., Schulz, M., Böni, P., Mühlbauer, S., & Reimann, T. (2016). The new neutron grating interferometer at the ANTARES beamline–Design, principle, and applications. http://arxiv.org/abs/1602.08846 (accessed on 3rd October 2025).
15. Binks, B. P., & Yin, D. (2016). Pickering emulsions stabilized by hydrophilic nanoparticles: *In situ* surface modification by oil. https://core.ac.uk/download/151161333.pdf (accessed on 3rd October 2025).
16. Binks, B. P., Koman, H., Muijlwijk, K., et al. (2016). Food-grade Pickering stabilization of foams by *in situ* hydrophilization of calcium carbonate particles. https://core.ac.uk/download/151161467.pdf (accessed on 3rd October 2025).
17. Borštnik, A., Borštnik, A., Borštnik, A., Derjaguin, B. V., Bauer, C., Prieve, D. C., Terentjev, E. M., et al. (2003). Interaction and flocculation of spherical colloids wetted by a surface-induced corona of paranematic order. http://arxiv.org/abs/cond-mat/0301256 (accessed on 3rd October 2025).
18. Bosch, J., Meckenstock, R., Sethi, R., et al. (2012). Transport of ferrihydrite nanoparticles in saturated porous media: Role of ionic strength and flow rate. https://core.ac.uk/download/pdf/11424714.pdf (accessed on 3rd October 2025).
19. Ceolín, M., Correa, N. M., Falcone, R. D., et al. (2018). Structural characterization of biocompatible reverse micelles using small-angle X-ray scattering, $^{31}P$ nuclear magnetic resonance, and fluorescence spectroscopy. https://core.ac.uk/download/288802561.pdf (accessed on 3rd October 2025).
20. Chatterji, A., Malevanets, A., Zick, A. A., Ladd, A. J. C., Likos, C. N., Peskin, C. S., Beenakker, C. W., et al. (2008). Simulating (electro) hydrodynamic effects in colloidal dispersions: Smoothed profile method. http://arxiv.org/abs/cond-mat/0601322 (accessed on 3rd October 2025).

21. Conway, B. R., & Kakadia, P. G. (2014). Solid lipid nanoparticles: A potential approach for dermal drug delivery. https://core.ac.uk/download/30730310.pdf (accessed on 3rd October 2025).
22. Costa, C., Edlund, H., Filipe, A., et al. (2019). Emulsion formation and stabilization by biomolecules: The leading role of cellulose. https://core.ac.uk/download/288958416.pdf (accessed on 3rd October 2025).
23. de la LLana, A., Anderson, I. S., Capitán Vallvey, L. F., et al. (2014). A general perspective of the characterization and quantification of nanoparticles: Imaging, spectroscopic, and separation techniques. https://core.ac.uk/download/186617942.pdf (accessed on 3rd October 2025).
24. Dengo, N., Gross, S., & Vittadini, A. (2017). Functionalization of colloidal transition metal sulphides nanocrystals: A fascinating and challenging playground for the chemist. https://core.ac.uk/download/141699569.pdf (accessed on 3rd October 2025).
25. Esquivias Fedriani, L. M., Morales Flórez, V., Piñero, M., et al. (2009). Nanostructured sonogels. https://core.ac.uk/download/322844972.pdf (accessed on 3rd October 2025).
26. Gao, B., Morales, V. L., Tian, Y., et al. (2012). Effect of surface modification on single-walled carbon nanotube retention and transport in saturated and unsaturated porous media. https://core.ac.uk/download/228177112.pdf (accessed on 3rd October 2025).
27. Goia, D. V., Halaciuga, I., Privman, V., et al. (2008). Synthesis of silver colloids: Experiment and computational model. http://arxiv.org/abs/0810.0004 (accessed on 3rd October 2025).
28. Guo, Y., Lavrentovich, O. D., Peng, C., et al. (2018). Sorting and separation of microparticles by surface properties using liquid crystal-enabled electro-osmosis. http://arxiv.org/abs/1808.10559 (accessed on 3rd October 2025).
29. Jack, R. L., & Whitelam, S. (2014). The statistical mechanics of dynamic pathways to self-assembly. http://arxiv.org/abs/1407.2505 (accessed on 3rd October 2025).
30. Jafari, S. M., Maghsoudlou, Y., Makhmal-Zadeh, B. S., et al. (2016). Crocin loaded nano-emulsions: Factors affecting emulsion properties in spontaneous emulsification. https://core.ac.uk/download/52205581.pdf (accessed on 3rd October 2025).
31. Jimenez-Martinez, C. J., Davila-Ortiz, G. D., Tellez-Medina, I. T. D., O., et al. (2016). Application of high-pressure homogenization to improve stability and decrease droplet size in emulsion-flavor systems. https://

media.neliti.com/media/publications/238581-application-of-high-pressure-homogenizat-78afca62.pdf (accessed on 3rd October 2025).

32. Kaz, D. M., Manoharan, V. N., McGorty, R., et al. (2016). Contact-line pinning controls how quickly colloidal particles equilibrate with liquid interfaces. http://arxiv.org/abs/1607.08154 (accessed on 3rd October 2025).
33. Kulkarni, C. (2016). Lipid self-assemblies and nanostructured emulsions for cosmetic formulations. https://core.ac.uk/download/73961481.pdf (accessed on 3rd October 2025).
34. Malijevský, A., Archer, A. J., Evans, R., Henderson, J. R., & Rowlinson, J. S. (2013). Sedimentation of a two-dimensional colloidal mixture exhibiting liquid-liquid and gas-liquid phase separation: A dynamical density functional theory study. https://core.ac.uk/download/288377920.pdf (accessed on 3rd October 2025).
35. Schofield, A. B., Binks, B. P., Herzig, E. M., Evans, D. F., Cates, M. E., Clegg, P. S., Pamplona, D. C., et al. (2005). Colloid-stabilized emulsions: Behaviour as the interfacial tension is reduced. http://arxiv.org/abs/cond-mat/0508088 (accessed on 3rd October 2025).
36. Zaccarelli, E., Fisher, M. E., Sciortino, F., Hansen, J. P., & Gnan, N. (2012). Tuning effective interactions close to the critical point in colloidal suspensions. http://arxiv.org/abs/1209.2254 (accessed on 3rd October 2025).

CHAPTER 7

# INTERFACES IN LIQUIDS AND SOLIDS

## Contents



## INTRODUCTION

In materials engineering and science, interface analysis in liquids and solids is vital to the understanding of an immense variety of phenomena that span from phase transitions to the behavior of composite materials. Interfaces dictate the physical and chemical properties of the substance, from molecular interactions to macroscopic properties such as strength and stability. As advancements are made in computational methods and experimental procedures, the internal mechanisms of these interfaces may be explored to further levels. In particular, the problem of rigid solids interacting with fluids has been one of concern because it involves complex interactions that determine material properties at microscopic levels. This increasing focus on interfaces demands a wide foundation for the understanding of the associated theories, methodologies, and applications at play related to the topic of interfaces in liquids and solids. To illustrate the applicability of this field of research, various models and approaches, such as phase field theory (PFT) and vortex particle methods, can be applied to simulate interfacial dynamics. For example, the PFT has been effectively utilized to determine interfacial properties and nucleation barriers in binary eutectic systems, proving that it can handle complex interfacial properties. Meanwhile, advanced computational methods like those discussed in Allard et al. (2006) provide realistic bi-phase flow simulations, stressing the importance of correct modeling in the visualization of interactions between solids and fluids. By a subtle interplay of simplicity and accuracy, these techniques provide entry to deeper insight into material behaviors that are of vital importance in a broad array of applications ranging from manufacturing processes to biological functions, further emphasizing the significance of interface research. In addition to that, knowledge of the interface phenomenon has functional significance in a broad spectrum of disciplines, from nanotechnology to materials science, and even environmental studies. Examination of how the interface responds to changing conditions—such as varying temperature or composition—can yield significant information about central processes such as nucleation and stability. For instance, the work reported in Gyula et al. (2007) illustrates how temperature and composition influence equilibrium interfacial properties, making predictions that are consistent with experimental measurements of nucleation and undercooling. The complete comprehension of these interactions not only improves theoretical knowledge but also enhances the optimization and design of materials for specific applications. Hence, the study of interfaces between liquids and solids emerges as a necessary drive in developing scientific understanding and practical advances in technology and material science.

## *Definition of Interfaces in Materials*

The concept of interfaces in materials is central to the comprehension of intricate interactions of different phases, particularly in liquids and solids. An interface is the boundary that separates different phases, e.g., liquid and solid, through which different physical and chemical phenomena take place. Here, at the boundary, properties are very different from bulk phase properties, leading to unusual behaviors that can be very different from the two phases. An example is interfacial tension between a solid and a liquid, which may influence processes of nucleation and how solid phases precipitate out from liquid solutions. An understanding of these interfaces allows scientists and engineers to manipulate material properties to apply to areas such as materials science, nanotechnology, and chemical engineering, where controlled interfaces are employed to enhance performance and functionality. Interfacial properties also become relevant to thermodynamic properties like free energy and phase stability. The temperature- and composition-dependent interfacial equilibrium properties are the outcome of temperature- and composition-dependencies, which can be characterized using various theoretical models, such as classical nucleation theory (CNT) and phase field theory (PFT). For example, PFT accounts for interfacial free energies from empirical evidence obtained from molecular dynamics simulations and experimental evidence. It outlines how variations at the interface can lead to varying nucleation processes under a set of conditions, which reveals the complexity of phase transitions and stability on the nanolevel. This is extremely crucial when developing material with an optimum interface to create improved mechanical, thermal, and electrical properties in practical applications, as in the binary eutectic system Ag-Cu (Gyula et al., 2007). Moreover, interfaces are central to probing material behavior under new conditions, e.g., microgravity. Under no gravity, it is possible to realize modified interface interactions so that material processing can be studied irrespective of conventional earth-based limitations. NASAs foray into microgravity materials science demonstrates this possibility, as it calls for investigations of interfacial dynamics in various classes of materials, including metals, polymers, and ceramics (Holland et al., 2025). These studies not only push forward the theoretical understanding of interfaces but also propel the development of new materials that find applications in numerous industries. Studying the definition and implications of interfaces in this regard puts researchers in the best position to benefit from controlled interfacial interactions to develop material science breakthroughs.

## *Importance of Studying Interfaces*

Studies of interfaces between liquids and solids form a core part of understanding overall appreciation for the properties and behaviors of material in chemical, physical, and materials sciences. Interfaces have a tendency to demarcate drastic changes in physical and chemical properties with effects such as adhesion, corrosion, and catalysis. Through the study of such interfacial properties, researchers can develop new materials with improved performance for electronic to biomedical device applications. For instance, the phase field theory (PFT) allows one to calculate nucleation barriers and equilibrium interfacial characteristics in complex systems, such as the binary eutectic system Ag-Cu, which provide information about the thermodynamic stability of the interfaces (Gyula et al., 2007). Understanding such behaviors also facilitates the control and optimization of material interfaces for specific purposes, a classic example of which is the support for heterogeneous catalysts that demonstrate the paradigmatic nature of studying such areas. In addition, interface behavior is very important within the framework of quantum fluids, where the properties are such that they deviate from what is observed in classical liquids. One prominent example includes the measurement of neutron reflection from a helium surface, where one finds that the superfluid has a smoother interface than the normal liquid. This effect suggests the presence of a surface layer with definite properties, potentially related to a greater Bose-Einstein condensate fraction near the helium surface (Ganshin et al., 2008). Such a result refers to the complexity of interfacial interactions in quantum fluids and their capability to revise the existing understanding of fluid mechanics and phase transitions. The study of these interfaces not only generates new basic scientific knowledge but also generates new prospects for technological invention through the control of quantum properties in engineered systems. Interface generalizations extend far beyond science; they are critical to the solution of practical problems. In fields such as environmental science and engineering, understanding about liquid-solid interfaces helps enhance control of pollution, water treatment, and material recycling. Increased knowledge of interfacial tension and interactions helps design better adsorbents and catalysts, significantly optimizing energy efficiency and resource utilization. Furthermore, in nanotechnology, control at the nanolevel over interfaces is crucial for the development of sophisticated materials with desired functionalities. Thus, the need to comprehend interfaces is not only evident in enhancing scientific principles but also in enhancing practical uses of solutions to important global challenges, hence rendering interfaces a prominent area of research in the interaction of liquids and solids.

## 7.1. THE NATURE OF INTERFACES

The nature of interfaces is crucial in determining the behavior of materials, particularly in the areas of liquids and solids. Such interfaces, described as boundary areas between two or more phases or materials, dictate not just the mechanical properties of the materials but also their functional capability in various applications. For example, in the case of lubricants in contact with solid surfaces, the interfacial physical properties can strongly affect friction and wear behavior. It is through the understanding of these interfacial phenomena that researchers can develop means of optimizing material performance in applications like adhesives and coatings. By the analysis of hydrodynamic interactions among soft, deformable interfaces, elastic deformation due to viscous stress is revealed to be a key factor in determining material response within overall contact, showing a convenient correlation between material design and interface property (Dhong et al., 2017). Furthermore, molecular arrangement of the material is also a key factor in determining the nature of interfaces. Water is a good example, which shows distinct properties emerging at the liquid-solid interface. The adhesive and cohesive properties of waters facilitate the formation of complex interactions with other solids, which further modulate its function in chemical as well as physical processes. Thus, water structure based on its unique hydrogen bonding and tetrahedral geometry directly affects the stability and properties of interfaces formed through interaction with other materials. An understanding of why these properties induce waters anomalous behavior, e.g., ice being less dense than liquid water, is central to the comprehension of larger geochemical and biological systems. Such an understanding emphasizes the importance of molecular orientation and structuring in setting the nature of interfaces, influencing not only nature but also industrial operations (Brini et al., 2017). Besides, interface research also entails investigating the challenge posed by complex interfacial phenomena in actual applications. Under dynamic conditions to which materials are subjected, parameters such as substrate layering, viscoelasticity, and heterogeneity at surfaces become crucial in predicting performance at the interface. Such complicating factors have the potential to introduce unforeseen changes in material properties, which affect performance in actual systems. For example, the ion-water interface has fascinating entropic behavior and heat capacities, which are significant for uses ranging from biochemical processes to the conceptualization of next-generation materials. Thus, overcoming such challenges through enhanced theoretical modeling and experimental validation is still relevant to conceptualize advanced materials that exploit some interfacial features for new applications (Dhong et al., 2017; Brini et al., 2017).

### 7.1.1. Characteristics of Liquid Interfaces

The behavior of liquid interfaces is very important in determining how fluids will behave in various applications, ranging from industrial applications to biological processes. Liquid interfaces are primarily described by their surface tension, which is a result of the cohesive forces within the liquid molecule. This tension creates an energy barrier at the interface and defines the wetting properties of liquids on solid surfaces. Such properties as temperature, impurities, and chemical composition of the fluid are capable of altering these properties significantly. For instance, addition of surfactants may reduce surface tension for better wetting and spreadability of the fluid, which is particularly significant in uses such as coatings and medications. It is necessary to learn these properties in order to be able to maximize processes that utilize the manipulation of liquid interfaces, which are indicative of their overall significance in theoretical and practical terms. Furthermore, the interaction of liquid interfaces with solid surfaces is complex and consists of numerous aspects. When they come in contact with a solid, adsorption and chemical bonds are established, and wettability of the surface is altered. The degree to which a liquid spread on a solid is typically quantified by the contact angle, and this will determine whether a surface is hydrophobic or hydrophilic. The characteristic is not merely founded upon the property of the liquid; it will be determined by the roughness and chemical composition of the surface as well (Figure 7.1).

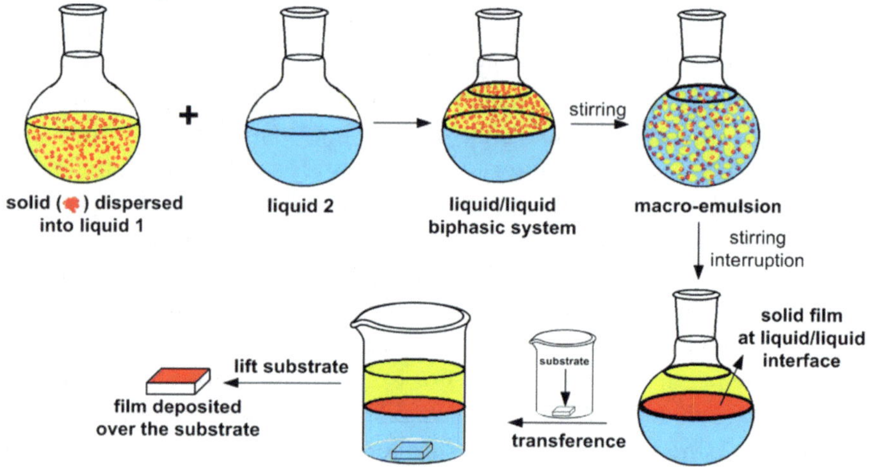


**Figure 7.1.** Liquid-liquid interfaces.

*Source:* https://external-content.duckduckgo.com/iu/?u=https%3A%2F%2Fpubs.rsc.org%2Fimage%2Farticle%2F2021%2FMH%2Fd0mh01676d%2Fd0mh01676d-f1_hi-res.gif&f=1&nofb=1&ipt=c8a54bb789ae0de3b8b664f327030c329f412a5fc7aadfd2cbd4810365410b31.

For example, $CO_2$ and Nd:YAG laser processing has the capability to dramatically modify the wettability of materials like $Al_2O_3/SiO_2$ ceramics through modifying their surface morphology and characteristics. This flexibility is a reflection of how important it is to control solid-liquid interaction for the achievement of desired functionality in material used in different applications (Lawrence et al., 1998). Moreover, a study of wettability modification through laser treatments in metals such as carbon steel offers significant facets of the interaction of liquid interfaces and solid materials. The use of high-power diode lasers (HPDL) has shown potential for creating extreme changes in surface characteristics, such as roughness and surface energy, which consequently enhances the wetting characteristic of treated metals.

These changes enable improved adhesion and coating processes, important in engineering and production. Research involving techniques such as scanning electron microscopy and X-ray photoemission spectroscopy also emphasized the connection of surface treatment and wettability, underlining how intentional changes may ensure maximum solid-liquid interaction (Agarwala et al., 2000). Thus, comprehensive understanding of liquid interface properties is most vital in the formulation of technologies in materials science and engineering.

### 7.1.2. Characteristics of Solid Interfaces

Solid interfaces have characteristics that are of essential significance in a variety of applications and control phenomena such as adhesion, wettability, and surface energy. The characteristics of solid interfaces are defined by their well-defined surface morphologies, chemical nature, and the atomic structure at the surface where solid phases coexist. Physical and chemical interface properties can significantly affect the interactions between solids and liquids, particularly in processes like coating, sintering, and corrosion. For instance, differences in surface roughness will alter the wettability of a solid, influencing the spreading or beading of liquids on its surface. Experiments have shown that differences in the microstructure of solids can lead to extremely large changes in its surface energy, which in turn dictates the observed fluid dynamics at such interfaces, emphasizing the significance of understanding and controlling these properties in practical applications. Further studies on solid interface properties indicate how treatments can enhance their properties in a very important way. HPDL treatments, for example, have been found to change the wettability of carbon steel through inducing changes in surface roughness and chemistry. The process entails surface melting followed by modification of polar components of surface energy, as evidenced in experiments on carbon steel exposed to HPDL radiation. This treatment effectively increases improved adhesion for subsequent coatings such as enameling, proving surface re-engineering optimizes material performance in technical use. Improved understanding of these processes facilitates innovative approaches in the tailoring of surface properties to create needed functional outcomes. The interaction between solid interfaces and external treatments not only controls their wettabilities but also suggests potential application across a range of materials, including ceramics. Laser processing, such as excimer and Nd:YAG laser processing, has been shown to have a dramatic effect on the surface properties of $Al_2O_3$/$SiO_2$ ceramic-based materials, significantly reducing contact angles to enhance wettability. Partial vitrification and increased relative surface oxygen content results from these treatments, thereby facilitating coating adhesion and material life. The variations of contact angles following laser treatment highlight the importance of understanding the interaction between material properties and laser wavelength. Thus, this comprehensive understanding of solid interfaces facilitates advances in material science with improved performance in engineering applications with multiple improvements while emphasizing the demands for surface modification with proficiency on specialization based on particular working requirements (Figure 7.2).

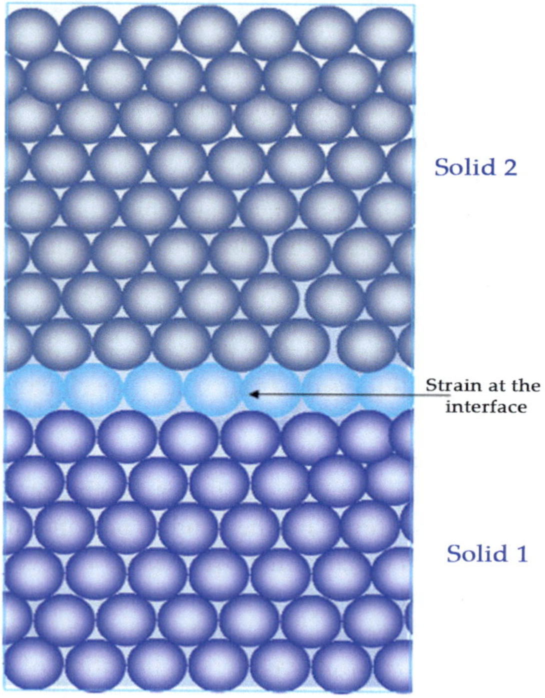


**Figure 7.2.** An illustration of solid–solid surface and interface scenario.

*Source:* https://www.sciencedirect.com/topics/chemistry/solid-interface.

### 7.1.3. Comparison Between Liquid and Solid Interfaces

The difference between solid and liquid interfaces is of prime significance to material behaviors in many applications. At solid interfaces, atoms are put into a highly ordered form with well-defined structural features that determine their physical properties. Solid interfaces, for instance, have abrupt boundaries, where there is a sudden change in structure from one crystalline phase to another. In contrast, liquid interfaces are also generally disordered, featuring a fluidic and dynamic molecular arrangement. This distinction endows solid interfaces with the capacity to provide stable supports for reaction and material processing, while liquid interfaces promote mass transport and intensify mixing processes. Interfacial behavior observed in these two phases is inseparably interconnected with their respective molecular dynamics and thermodynamic properties, with far-reaching consequences in materials science and engineering. In addition, mechanisms of nucleation at solid and liquid interfaces are quite distinct. Phase field theory (PFT) has given a method to compute equilibrium interfacial properties and nucleation energy in solid-liquid systems, such as the binary eutectic system Ag-Cu. The temperature and composition-dependent free energies, computed with CALPHAD, are characterized by a complex interplay between phases. Worth mentioning is that the PFT reveals that the critical fluctuations of concentrations of the Cu-rich and Ag-rich phases compete near the eutectic composition, which governs the interface stability. This forms the foundation under which heterogeneous nucleation ought to occur upon undercooling, providing a thermodynamic foundation upon which predictive modeling of solidification as well as synthesis of materials can be conducted, as attested by experimental results (Gyula et al., 2007). Conversely, liquid-solid transition processes employing density functional theory (DFT) reveal deep insights into the manner in which quantum mechanical phenomena dictate interfacial behaviors at a more microscopic level. Employing a weighted-density approximation that has been adjusted, researchers can investigate the interaction between solid phases, i.e., Bose and Fermi solids, and the corresponding liquid phase. The phase mapping relates static structure factor with input information from the liquid state to find transition dynamics. Notably, this approach allows for the broad comprehension of the quantum processes at interfaces between solids and liquids and illustrates how changes in energy states can influence the stability and configurational behavior at phase transitions. Despite quantitative differences with simulations, such advances are significant markers of the ongoing improvement in the account of interfaces in different states of matter and hence of the fundamental knowledge that will have to be in place to enable future innovation (Denton et al., 1997).

## 7.2. INTERFACIAL TENSION

Interfacial tension is a fundamental property that characterizes interaction between two distinct phases, particularly at the liquid-solid interfaces. Interfacial tension arises from the imbalance of intermolecular forces at the interface, where molecules in different phases are under varying degrees of attraction. Interfacial tension is a fundamental concept used in a wide range of scientific fields, ranging from materials science to biology, since it influences phenomena such as adhesion, wetting, and nucleation. For instance, in metallurgy, control of interfacial tension has a bearing on the solidification of alloys and subsequently on their mechanical properties. Further, phase field theory (PFT) provides predictive insight into interfacial tension evolution, detailing how temperature and composition are crucial in determining equilibrium interfacial free energies. This highlights the need for advanced models that are able to accurately depict the behavior of interfacial tension in complex systems, and the significance it plays both theoretically and in terms of application (Gyula et al., 2007). Another characteristic of interfacial tension is its central role in spreading and wetting phenomenon, most notably in ferrofluids, or fluids that undergo radically different behaviors to externally applied magnetic fields. The unusual properties of ferrofluids are due to their behavior at the interface, which affects the interaction of such fluids with solid surfaces at the nanoscale. Ferrofluids contain magnetic nanoparticles that can change the contact angle and hence the way these liquids spread on surfaces, making the classical models of wetting difficult.

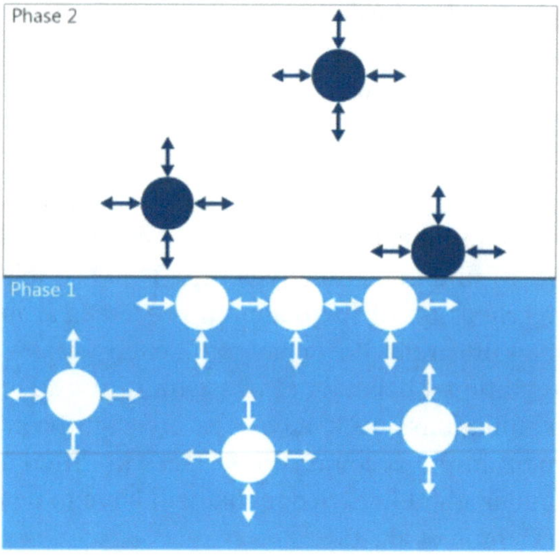


**Figure 7.3.** *Molecular environment in the volume phase and at the interface.*

*Source: https://www.kruss-scientific.com/en/know-how/glossary/interfacial-tension.*

The magnetic particle behavior when it interacts with solid surfaces can be counterintuitive, thus studies in ferrofluids as an essential field of research for applications in magnetic actuation and microfluidics. Exploring these interfacial properties enlightens one on practical applications and enables the development of better technologies (Backholm et al., 2018). In addition to its implication in specific applications, interfacial tension is also a policy for the thermodynamic stability of composite materials. Equilibrium of interfacial energies at phase transitions is important in determining the stability and morphology of phases that form. For example, the concept of the Tolman length, as a descriptor of deviation of the curvature effect on interfacial tension, has a direct correlation with the nucleation phenomena observed in various material systems. As more sophisticated studies are being carried out towards the intricate interplay between interfacial tension and material response, it is clear that proper understanding of these concepts is crucial to advancements in material design and synthesis. Therefore, future work on interfacial phenomena will not only expand our knowledge of existing materials, but also result in new applications across a variety of industries, justifying the importance of interfacial tension within the disciplines of material science and engineering (Figure 7.3).

### 7.2.1. Definition and Significance of Interfacial Tension

Interfacial tension is a fundamental physical property which arises at the interface between two immiscible phases, say liquids and gases or solid-liquid interfaces. It is the force per unit length along the interface and arises due to cohesive forces between molecules in each phase and the adhesive forces at the interface. This effect is a key factor in determining numerous macroscopic behaviors, such as emulsion stability, droplet formation, and capillary flow. In interfacial phenomena, changes in temperature, concentration, or surfactant addition can significantly change interfacial tension, thereby determining mass transfer rates and system behavior.

Therefore, interfacial tension knowledge is essential in the optimization of processes in chemical engineering, materials science, and environmental science, showing its significance in theoretical and real-life cases. The importance of interfacial tension is not limited to the mere definition because it has a fundamental role in numerous industrial operations. For example, in formulations of products such as emulsions, foams, and paints, the control of interfacial tension is essential to achieve certain stability and performance characteristics. Low interfacial tension may help in mixing immiscible liquids, leading to more stable emulsions with a longer storage life. Additionally, in slurry reactors, interfacial tension influences mass transfer characteristics, including volumetric gas-liquid mass transfer coefficients required for reaction efficiency. The inclusion of solid particles in a slurry has been proposed in recent studies to enhance mass transfer operations, which implies interfacial tension as the key parameter to be used in reactor efficiency optimization and the development of new solutions for chemical synthesis and energy production (Beenackers et al., 1993) (Figure 7.4).

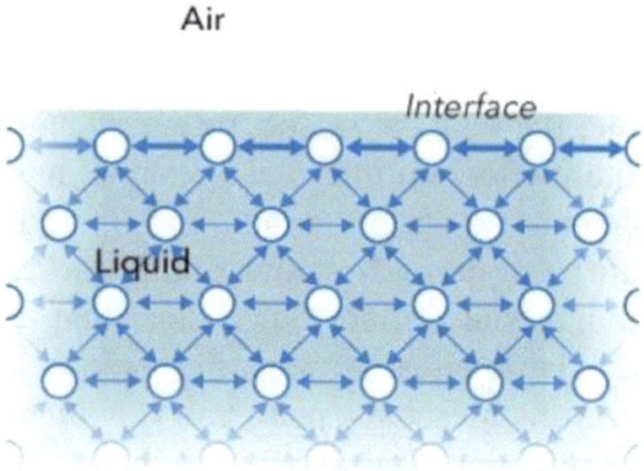


**Figure 7.4.** Intermolecular forces between molecules in the bulk and the interface.

*Source:* https://www.nanoscience.com/wp-content/uploads/2023/10/Forces-on-liquid-molecules.jpg.

Besides, study of interfacial tension is significant in the pursuit of expanding our understanding of fluid dynamics and thermophysical phenomena in other environments, including microgravity. Fluid behavior, including interfacial dynamics, in low-gravity environments varies significantly from that occurring on Earth and demands first-hand experimentation to unscramble the complications. The investigation of interfacial tension at these unusual conditions will assist us to understand capillarity, convection, and multiphase heat transfer, the most important issues in designing space-effective systems. Therefore, interfacial tension is a base property that not only impacts our understanding of simple fluid behavior but also underlies the development of beneficial technologies, and it is thus a crucial area of study in contemporary science and engineering (Abramson et al., 2025).

## 7.2.2. Factors Affecting Interfacial Tension in Liquids

The interfacial tension process in liquids is highly influenced by a variety of factors, the most noticeable being temperature, concentration, and surfactant levels. When temperature increases, molecules possess higher kinetic energy, causing interfacial tension to drop. Such a shift can contribute a lot to how materials behave at their interfaces, e.g., the type of liquids spreading on surfaces or if they can mix or not. In addition, changes in solute concentration can modify the interplay of cohesive and adhesive forces and hence impact interfacial properties. Surfactants have a special utility in reducing interfacial tension by disrupting interfacial molecular cohesion and thereby making stabilization mechanisms available for various applications from emulsification to detergency. Regulation of such variables is crucial to make processes optimal in areas of application including material science, where tight regulation of interfacial properties is critical for product performance. Another variable that affects interfacial tension is the molecular composition of the liquids present. The hydrophobic or hydrophilic nature of the molecules within the liquid controls the strength of intermolecular interactions at the interface. Polar solvents, for instance, experience greater interfacial tension when mixed with nonpolar solvents since there are no optimum interactions, resulting in total phase separation. Microgravity conditions explored as documented in the United States Microgravity Laboratory programs illustrate how reduced gravity forces allow for easier examination of fluid behavior at interfaces that would otherwise be disrupted by convection-driven turbulence. In such studies, specially designed experimental conditions allow for an insight into molecular dynamics, thus offering detailed insights into the role of different molecular characteristics on interfacial tension and accompanying phenomena, including stability and reactivity of mixtures in various applications (Accardi et al., 2025). A consideration of the effect of gravitational forces on fluid interfaces is also important, particularly in cases where standard laboratory environments prove inadequate to ensure a comprehensive understanding of complex phenomena. In microgravity experiments, absence of gravity-induced convection allows for a cleaner study of interfacial behavior where surface roughness, molecular orientation, and temperature gradients can be systematically varied to study their effects on interfacial tension. NASAs activities, with the aim of having a permanently manned space station, put a high value on such experiments in advancing materials science and fluid dynamics. Through the deconstruction of microgravity research functional needs, scientists are able to more effectively plan their experiments to research interfacial tensions in-depth. Not only does this research enrich academic study, but it also drives innovation within commercial uses, with direct repercussions for industries based on the principles of fluid dynamics and material science (Holland et al., 2025).

### 7.2.3. Role of Interfacial Tension in Solid Interfaces

The interplay between interfacial tension at solid interfaces is to blame for a key role in a variety of physical processes, governing stability and interfacial interactions at the microscopic level. Interfacial tension, which characterizes the energy required to enhance the surface area of a phase, governs much of wetting, adhesion, and nucleation behavior in solid materials. For instance, in alloy systems, interfacial tension in the solid-liquid state can control the shape and material properties as temperature and composition changes modify equilibrium conditions. These phenomena are explained by theoretical models such as phase field theory, which allows us to comprehend these dynamics by means of describing how interfacial behavior responds under various circumstances, contributing to the understanding of phase transformations in solid materials. Such elementary insight is essential to process optimization in materials science and engineering, particularly in tight control applications of interfacial phenomena (Gyula et al., 2007). Interfacial tension also dominates the behavior of ferrofluids, a type of colloidal suspension of magnetic nanoparticles. During their interaction with solid surfaces, the interfacial tension altered by these materials dictates their spreading characteristics and wetting properties, both of which are crucial in magnetic actuation technologies and microfluidic devices.

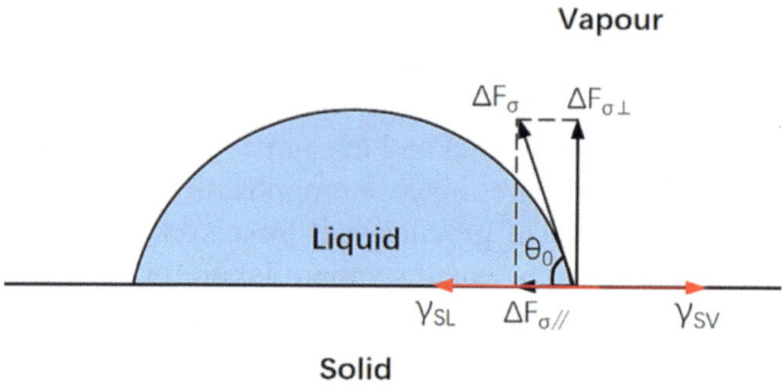


**Figure 7.5.** Surface tension and interface tension between solid and liquid.

*Source:* https://www.researchgate.net/figure/Surface-tension-and-interface-tension-between-solid-and-liquid_fig4_361243928.

The relationship between magnetic forces and interfacial dynamics can presage new applications, and that the manipulation of ferrofluid behavior through the tuning of interfacial tension can be the forerunner to new ways of technological innovation. In addition, magnetic nanoparticle behavior at solid surfaces has unique characteristics in the nanoscale that reveal the complexities of contact angle variations in defiance of traditional wetting theory. This necessitates strong experimental research in determining the mechanisms of ferrofluid spreading and wetting in solid materials (Backholm et al., 2018). Additionally, interfacial tension at solid interfaces is not only confined to different areas such as biotechnology and nanotechnology. For instance, control of interfacial properties can enhance biomaterials' compatibility and integration with living tissues or allow the stability of drug delivery systems. Manipulation of interfacial tension enhances adhesion, stability, and separation phenomena, which are significant for applications such as tissue engineering and regenerative medicine. Furthermore, with materials evolving at the nanoscale dimension, interfacial tension becomes increasingly crucial in accomplishing the properties of nanocomposites and coatings. Solving for interfaces is not only vital to the development of basic science but also to innovation in a wide variety of applications where material property can be specifically controlled through altering interfaces. Learning and applying these interactions will continue to be a fundamental priority in material engineering development (Gyula et al., 2007) (Figure 7.5).

## 7.3. SURFACE ENERGY

Surface energy is a phenomenon behind the physics of liquid and solid interfaces. It can be defined as residual energy at the surface of a material in comparison to its bulk by virtue of intermolecular forces. These forces are responsible for a surface desiring minimum area, which explains phenomena such as wetting and the tendency of droplets. The high surface energy in solids implies that the surface atoms possess characteristics different from the bulk atoms. Surface energy plays a crucial role in the prediction of a variety of phenomena including adhesion, stability of films, and mechanical properties of materials under external stresses. This fundamental knowledge allows researchers to tailor material properties to enhance performance in countless applications, ranging from coatings to nanotechnology and biomaterials. Surface energy is a major influencer of the dynamics of flow and interface behavior in liquids. For instance, in liquids where a liquid is flowing in nanometric channels, solid-liquid interface interactions dominate the controls on flow rates and permeability. These studies have clarified the hydraulic flow of hydrocarbon and water liquids and hydrocarbons through nanoporous materials such as Vycor glass monoliths in a wide range of conditions. For hydrophilic surfaces, the observed hydraulic permeabilities compare favorably with those predicted on the basis of a sticking boundary layer that complicates the velocity profile at the interface. On the other hand, in observing hydrophobic surfaces, the result indicates that the lack of such a boundary layer has a significant impact on the permeability of the liquid. These findings highlight the importance of surface energy in governing the behavior of fluids at solid surfaces and impacting engineering applications and material design drastically (Busch et al., 2015). Furthermore, nanometer- and micrometer-scale surface energy has opened up new possibilities for interrogating amorphous materials, with glasses being no exception. Recent work emphasizes the importance of the deformation potential energy landscape as compared to surface energy, and how changes of temperature and stress state can induce very significant differences in material behavior at interfaces. Especially for amorphous solids, phenomena like shear banding are intimately connected with the disorder in materials, and this influences their surface characteristics as well. By integrating multiscale modeling approaches, researchers are able to obtain critical information regarding the complex relationship between mechanical properties and surface energy and ultimately pave the way towards innovation in material interface control in a wide range of applications, from coatings and composite materials (Rodney et al., 2011) (Figure 7.6).

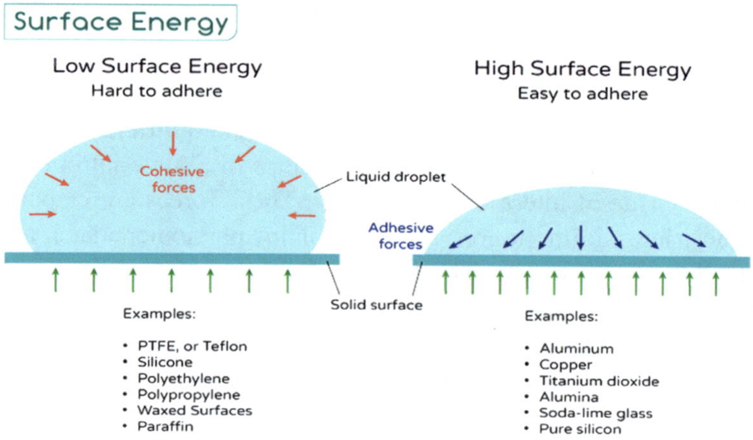


**Figure 7.6.** Surface energy.

*Source:* https://www.sciencefacts.net/surface-energy.html.

## 7.3.1. Concept of Surface Energy in Solids

The concept of surface energy in solids is at the core of understanding various physical processes and phenomena at the liquid-solid interfaces. Surface energy refers to the quantity of excess energy at the solid surface relative to its bulk, arising due to unsatisfied bonds at the interface. Surface energy is behind determining the stability, wettability, and adhesion properties of materials. For instance, enhanced surface energies generally lead to enhanced adhesion properties, which are significant in applications ranging from coatings to composite materials. Surface-energy changes may influence interactions between solids and liquids, and hence wetting and spreading, which are significant in materials science, engineering, and nanotechnology. The study of surface energy is thus essential for material performance optimization in several applications. Its significance is also enhanced by the character of nucleation effects, particularly in those systems with metastable phases. Nucleation is the formation of new phases within the parent phase, and the properties of surface energy contribute much to it. When a metastable phase—such as a liquid—encounters a stable solid, whether or not the liquid can wet the surface of the solid will have a precipitous effect on the nucleation barrier (Figure 7.7).

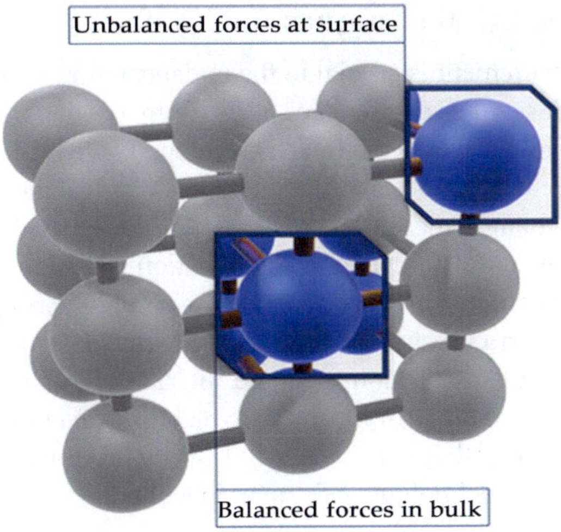


**Figure 7.7.** A basic depiction of bonding in a solid, showing balanced interactions in the bulk and unbalanced interactions at the surface.

*Source:* https://www.ossila.com/pages/a-guide-to-surface-energy.

As has been posited in recent theoretical frameworks, the overestimation of the nucleation barrier by classical nucleation theory can be attributed to the exclusion of the metastable phase contributions. These aspects make the relationship between surface energy and nucleation cumbersome, leading to complex phase transformation kinetics that require powerful theoretical tools such as density functional theory (DFT) for accurate modeling and prediction. Additionally, advances in experimental methods have begun to unravel the intricate relationship between surface energy and multiple material properties under varied conditions, for example, those of microgravity. In microgravity, forces due to gravity do not play a role, and hence one can track phase transitions and surface interactions with less external contamination.

For instance, NASAs work in developing materials science experiments in such conditions underscores the importance of understanding surface energy in the development of new materials with enhanced properties. By enabling improved studies of phase behavior and surface interactions in a controlled environment, scientists can utilize the outcomes to improve manufacturing processes and materials that are better able to withstand extreme conditions, thereby furthering the applications of solid interfaces in liquids and solids in industries across the board, including aerospace and electronics (Holland et al., 2025; Bagchi et al., 2013).

### 7.3.2. Measurement Techniques for Surface Energy

Surface energy measurement is crucial to the understanding of interfacial behavior in solids and liquids, where it dictates uses from coatings to the stability of nanocomposites. Various ways have been developed to calculate surface energy, including the Wilhelmy plate technique, sessile drop technique, and advancing and receding contact angle measurements. Each of these has different advantages depending on the material and distinct application situation. For instance, the sessile drop method can be employed to study the wettability of a surface from the curvature of a droplet on a substrate, while the Wilhelmy plate method is more suited for the measurement of the tension of liquid interfaces. Such variety in methods gives researchers a broad array of means by which to extract nuanced information about the surface energy that can be utilized in applications as wide-ranging as material science to nanotechnology. Advances in technology in the measurement of surface energy are also applicable in emulsion stabilizing studies, particularly with regard to the role of particulate adsorbents in interfaces. Recent developments have demonstrated significant advances in knowledge on the stabilization of oil-water emulsions by particles, particularly with the use of new contact angle measurements that quantify the interaction of a single particle at interfaces. This level of analysis adds strength to the foundation of theoretical models and increases predictive capability in terms of emulsion and foam stability. Furthermore, the observation of particle behavior at non-aqueous interfaces, as highlighted by recent work, points toward the necessity of adapting existing measurement techniques to appropriate alternative systems, thereby expanding the overall understanding of multiphase interactions (Binks et al., 2017). The uses of surface energy measurement techniques even extend to antiwetting surfaces, and in this instance, the ones being applied on membrane distillation processes. As membrane technologies become increasingly important in desalination and wastewater treatment, having the surface energy characteristics of the membranes can assist in solving the issue of fouling and wetting. Anti-fouling character created by engineered antiwetting interfaces can be highly impacted by their surface energy profiles. Through the application of hierarchical architectures to develop long-lasting antiwetting membranes, scientists are not only increasing efficiency but also extending the operational lifespan of such technology. The interaction between surface energy and membrane performance reveals an extremely crucial research niche that calls for worldwide measurement approaches to follow developments in fluid separation technologies (Kwon et al., 2019).

### 7.3.3. Implications of Surface Energy in Material Properties

The importance of surface energy in material properties is huge, particularly concerning understanding how surface properties regulate interface interactions between solids and liquids. Surface energy is defined as the excess energy of a material on its surface compared to its bulk, and it plays a crucial role in regulating many traits such as wettability, adhesion, and friction. For example, substances with low surface energy are also hydrophobic, and that influences their application in coatings and films. It is desirable to know such interactions when designing materials for application in applications such as adhesives and microsensors, where overall performance is dictated by the surface interactions. The performance of such interfaces is also made more complex by ambient conditions and temperature, which suggests that a nuanced consideration of surface energy is crucial in further advancing material science for beneficial use. The interdependency between elastic deformation and surface energy introduces additional complexity into our understanding of the properties of materials in elastohydrodynamics. When two surfaces approach each other in a viscous liquid, elastic deformation by hydrostatic pressure becomes significant. This type of effect has bearing on several force measurement techniques, including the surface forces apparatus (SFA) and atomic force microscopy (AFM) (Dhong et al., 2017). Surface energy not only influences how materials adhere to one another but also how they respond to dynamic forces under these conditions. The regulation of surface interactions in soft, deformable materials highlights the importance of surface energy in accounting for elastic and viscous contributions to material response and therefore influencing biomaterial and advanced coating design strategies. The latest results indicate the importance of surface energy contrasts among interfaces, particularly when considering particle-laden fluids. Experiments with scaled-particle theory have also demonstrated that thermal motion of adsorbed particles can be able to affect the saddle-splay modulus of a fluid interface to a large extent, illustrating that surface energy is able to affect material stability and behavior in unexpected ways (Bertrand et al., 2008). It has very significant implications in the design of colloidal systems and emulsions, where surface properties determine interactions and stability. By integrating the effects of surface energy into material design paradigms, researchers can better control the properties of interfaces, which leads to new applications for materials in energy storage, drug delivery, and environmental processes. Such knowledge not only enhances theoretical understanding but also gets transformed into practical solutions in many disciplines.

## 7.4. INTERFACIAL PHENOMENA

Knowledge of interfacial phenomena is extremely critical in a host of science and engineering applications since solid-liquid interfaces significantly influence material properties and behaviors. The standard fluid dynamic model has traditionally adhered to the no-slip boundary condition, which assumes that fluid velocity at a solid-liquid interface equals solid velocity. Despite this, emerging research has indicated that such a model does not generally account for the additional complexity afforded through surface wettability and roughness properties. For instance, how the existence of hydrophobic surfaces makes possible hydrodynamic slip, thereby changing flow properties in unexpected ways. Recent advances in interface dynamics show how variations in surface properties can result in novel hydrodynamic effects, as demonstrated in milestone works in the past decade, in favor of shifting toward more accurate models for fluid interaction at interfaces (Belyaev et al., 2011) (Figure 7.8).

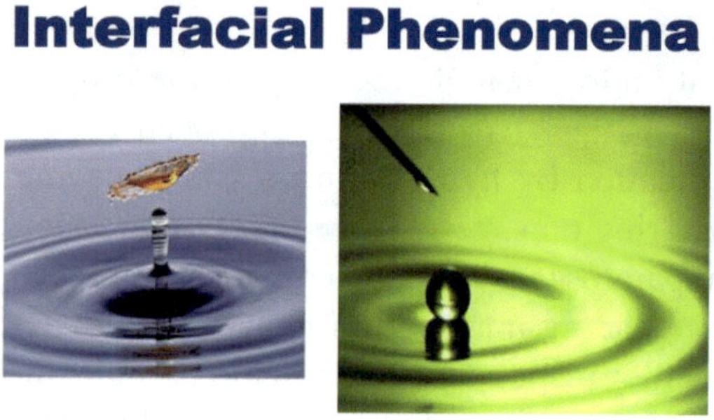


**Figure 7.8.** Interfacial phenomena.

*Source:* https://image5.slideserve.com/9660955/interfacial-phenomena-l.jpg.

In addition, interfacial effects are of theoretical interest, but they have practical implications for industrial processes. Surface treatments, for example, can be applied in an effort to change the wetting characteristics of materials, and illustrate technological use in coatings and adhesives. A good example is HPDL surface treatment of carbon steel, which has been shown to significantly alter wettability by varying surface roughness and energy characteristics. Such treatments can be used to enhance material-to-material adhesion, as indicated by enhanced enameling on treated carbon steel surfaces. Severe tests, including methods like scanning electron microscopy and energy-dispersive X-ray analysis, authenticate such alterations and their effect on actual applications, underlining the requirement of understanding interfacial phenomena in engineering applications (Agarwala et al., 2000). Lastly, the connection between material properties and interfacial phenomena surpasses conventional applications, influencing emerging technologies like nanotechnology and microsystems. At nanoscale systems, the importance of interfacial interactions is further increased due to increased surface area-to-volume ratios. These lead to emergent behaviors, like superfluidity and decreased hydrodynamic drag, which are absent in bulk materials. These multifaceted interactions at these interfaces usually comprise a mix of molecular forces, such as van der Waals forces and hydrogen bonding, which govern the physical properties of nanoparticles or thin films. With continuous research illuminating these interactions, it opens up possibilities for novel and innovative applications, from advanced drug delivery vehicles to new composite materials, reinforcing the supremacy of interfacial phenomena in the study of liquids as well as solids.

### 7.4.1. Wetting and Its Effects on Interfaces

Wetting is among the most dominant phenomena that determine the character of interface interactions, which contributes significantly towards the behavior of solids and liquids. Wetting is the manner in which the liquid wets the surface of a solid, which is basically determined by the balance between adhesive and cohesive forces. For a well-wetted surface, the contact angle must be low, indicating a strong adhesive interaction between the liquid and the solid. Conversely, wetting-resistant surfaces have high contact angles, which result in a beading effect on drops. Such phenomena point to some of the numerous applications from coating technologies to hydrophobic material engineering. Thus, knowledge of the nuanced dynamics of wetting is critical in the quest for innovation in a broad area of fields like nanotechnology and material science. The influence of surface roughness on wetting cannot be overstressed since it can radically alter interfacial interactions. Hypersurfaces are capable of enhancing hydrophobic or hydrophilic characteristics depending on their roughness and the material used. For instance, the lotus effect describes how micro- and nanostructured surfaces can be used to improve water repellency and hence reduce drag when fluid flows. This can be extremely significant in applications such as self-cleaning surfaces and efficient fluid transport systems. Additionally, surface roughness may introduce complication into flow dynamics, countering the traditional perception of hydrodynamic slippage in solid-liquid interfaces. The wettability-roughness relationship is thus a fertile field of research today and calls for innovative investigations in designing interface behavior (Belyaev et al., 2011). Capillary condensation also illustrates the intricate connection between wetting and interfacial effects, particularly in confined geometries. In nanoporous systems, the capillary forces become relevant, determining the adsorption properties and changing the effective properties of the confined fluid. The capillary condensation physics not only dictates the properties of liquids in small quantities but also governs such effects as static and dynamic friction at solid surfaces. Techniques such as atomic force microscopy (AFM) have considerably advanced our understanding of the role that surface roughness plays in these interactions, culminating in the formulation of material degradation mechanisms such as stress-corrosion cracking (Charlaix et al.,

2009). Understanding these capillary effects is important for the optimization of nanotechnology applications, reinforcing the necessity for stringent exploration of wetting behavior at the nanoscale.

### 7.4.2. Adsorption Processes at Interfaces

Adsorption at interfaces is the key to understanding liquid-solid interactions and their influence on a variety of scientific and industrial processes. Adsorption refers to the adhesion of molecules in the fluid phase to a solid surface that gives rise to a concentration gradient decreasing with distance over the interface. Adsorption efficiency can be regulated by temperature, pressure, and adsorbate and adsorbent properties. For instance, the topological and chemical surface properties of the solid surface, such as its roughness, control how molecules behave at the surface. The advancements in experimental approaches, as emphasized in most papers, allow for deeper investigation of adsorption mechanisms and eventually catalysis optimization in catalysis, sensing, and material fabrication. Grasping the thermodynamics of adsorption is crucial to understanding its intricacies at interfaces. Research shows that the Gibbs adsorption isotherm offers the basic framework for comprehending how molecules of a liquid fill a surface and the dynamics of energy involved in the interaction. The mathematical formulation describes the connection between the concentration of adsorbates in the bulk and their coverage on the interface, providing insight into the spontaneous character of adsorption. Moreover, processes like capillary condensation in geometrical confinement are crucial to the adsorption phenomenon (Figure 7.9).

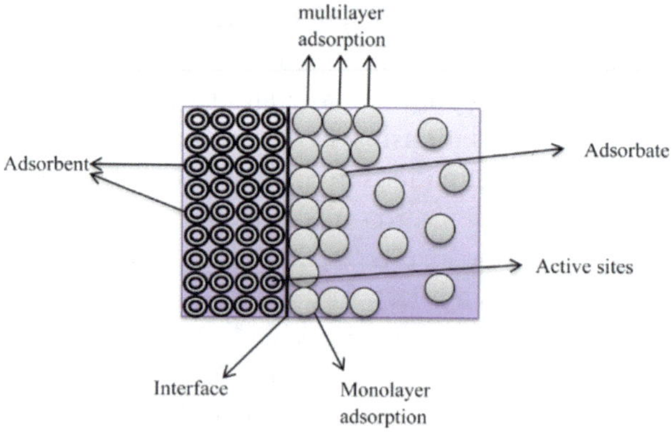


**Figure 7.9.** Schematic diagram of the adsorption process.

*Source:* https://www.researchgate.net/figure/Schematic-diagram-of-the-adsorp-tion-process_fig3_343224678.

Research brings to light the interplay between these thermodynamic principles and applications at the level of real situations, particularly in nanotechnologies, where control of surface interactions can lead to revolutionary breakthroughs in materials science and other fields. Moreover, the dynamic nature of adsorbed layers adds further complications in interface adsorption. The layers may not be static; they can be rearranged as a reaction to external stimuli such as variations in environmental conditions or mechanical stresses. As has been observed, interfaces in nanoscale systems are particularly vulnerable to such alterations, triggering multiple chemical and corrosive processes that change surface properties, including stress-corrosion crack growth (Holland et al., 2025). Such evolving processes highlight the importance of real-time monitoring techniques, which are progressively more advanced through instruments such as atomic force microscopy (AFM). With these approaches, researchers can discover profound insights into interface dynamics, leading to increased performance and lifespan of materials that rely on adsorption phenomena (Holland et al., 2025).

### 7.4.3. Capillarity and Its Relevance in Liquids

Capillarity, a fundamental phenomenon of liquid behavior at interfaces, is highly significant to describe the nature of liquid behavior in restricted environments. Capillarity results from the interaction between the cohesive forces within the liquid and the adhesive forces between the liquid and the surrounding surfaces. In actual applications, capillary action is responsible for numerous processes like water movement in porous material and nutrient transportation within living things. Besides, capillarity plays a very important role in nanofluidic technologies and materials design as demonstrated by pioneering nanotechnologies. The differences in capillary forces need to be grasped to unlock their possibilities across various fields of science, placing it in context for the study of interfaces where the liquid and solid phases meet face-to-face. Indicators illustrate that capillary actions become dominant in narrow confines, leading to phenomena such as capillary condensation of liquids in small regions. This is of particular concern to nanostructured material applications where pore size may be able to dramatically alter liquid behavior. For instance, the thermodynamics of thin film adsorption and capillary condensation is essential for producing materials utilized in nanotechnology, as described in recent literature. The measurement techniques of capillary forces such as surface force apparatus (SFA) and atomic force microscopy (AFM) demonstrate how roughness of the surface and nanoscale alterations are capable of influencing forces measured and associated physical properties. Capillarity effects influence their mechanical and chemical stability in nanomaterials and come to their surface, which makes it crucial to understand the effects at particular interfaces (Charlaix et al., 2009). Another area where capillarity plays an important role in liquid behavior is in the transport of liquids through channels of nanometer scale dimensions. From experimental studies, it is evident that water and hydrocarbons in nanoporous silica exhibit miscellaneous flow behavior with respect to surface material properties. More precisely, hydrophilic rather than hydrophobic surface conditions control permeability and flow rates, with hydrophilic surfaces enhancing water transport as a function of a boundary layer increasing flow efficiency. Conversely, hydrophobic surfaces have blocked flow until enormous pressure is introduced, and they have the reverse effect of surface tension and capillary forces in hydrophilic and hydrophobic regimes. These insights inform the design of flow at liquid-solid interfaces and provide insight into how capillarity controls fluid transport in confined systems, relevant to applications ranging from filtration to energy storage (Busch et al., 2015).

## 7.5. CHALLENGES IN INTERFACE RESEARCH

Perhaps one of the most important problems in interface research is an understanding of the complex interactions at the interface between liquids and solids. These interactions are crucial to establishing material physical and chemical properties in applications such as coatings and adhesives. Elastic deformation of soft solids can affect importantly these interactions, especially when subjected to viscous forces, like in dynamic force measurements with techniques like the surface forces apparatus (SFA) or atomic force microscope (AFM) (Dhong et al., 2017). The hydrodynamic interactions are sophisticated in nature and demand an overall theoretical model that includes characteristics like viscoelasticity and surface heterogeneity. Additionally, correctly modeling how these properties respond on surface forces is central to forming our understanding of soft hydrodynamic interactions and hence is a serious challenge for researchers seeking to optimize material performance at the microscopic scale. A second challenge is presented by understanding nanofluids, which have attracted attention as they exhibit higher thermal conductivity using low concentrations of nanoparticles. Despite the encouraging possibilities of nanofluids for heat transfer, researchers have yet to explain the mechanisms behind their effectiveness. One perspective claims that structuring of the nanoparticle in the liquid improves thermal properties but, in turn, raises viscosity, thereby making them challenging to implement (3rd Micro and Conference NF (MNF2011) et al., 2011). The compromise between optimal thermal conductivity and lowest viscosity is of importance for potential practical application across a wide range of engineering disciplines. To address these issues in a meaningful way, though, is crucial for the development of new-generation nanofluid technologies that can leverage their unique characteristics without compromise to efficiency. Lastly, the interfacial physics that take place during contact of soft matter present significant research challenges. Understanding how soft, deformable interfaces respond when loaded with varying loading and environmental conditions is at the center of several technological applications, including biomaterials and microsensors. Challenges such as substrate layering and poroelasticity complicate the challenges by complicating predictive modeling of the behavior of interfaces. Researchers must deal with how such phenomena influence material adhesive and mechanical properties, including how they respond to variable loads (Dhong et al., 2017). Thus, the establishment of our understanding of these soft hydrodynamic interactions is an endeavor requiring a multidisciplinary collaboration involving theoretical, experimental, and computational approaches. A unified perspective is the door to unlocking the promise of soft materials in future innovations, and it drives the need for sustained research into the challenges of interface study (Figure 7.10).

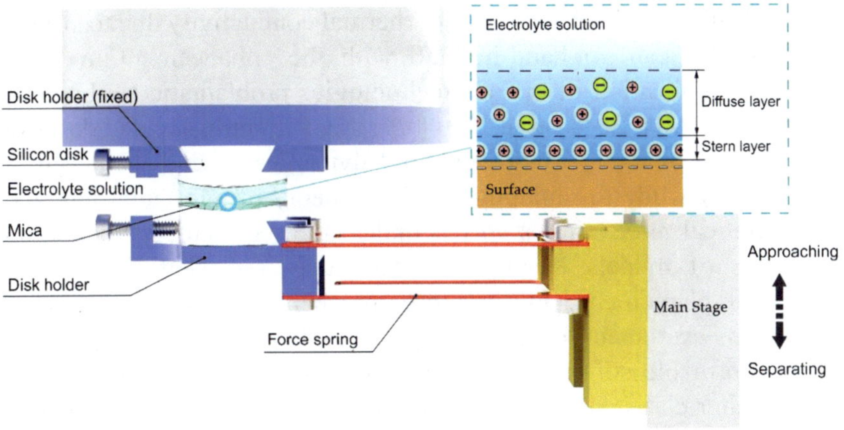


**Figure 7.10.** Schematic diagram of the surface forces apparatus (SFA) system. The electric double layer (EDL) structure formed on a mica surface (both the stern layer and the diffuse layer) is illustrated in the upper right corner.

*Source:* https://www.researchgate.net/figure/Schematic-diagram-of-the-surface-forces-apparatus-SFA-system-The-electric-double-layer_fig1_347574655.

## 7.5.1. Difficulties in Measuring Interfacial Properties

Measurement of interfacial properties comprises a difficult array of issues that impact our understanding of interfaces in liquids and solids directly. The initial principal challenge arises from the inherent fluctuations of the interface itself, which may change with external factors like temperature, pressure, and the addition of impurities. Fluctuations complicate experimental setup for measuring properties like interfacial tension and adhesion. Besides, the nanoscale nature of most interfaces often necessitates high-level measurement techniques, such as atomic force microscopy (AFM) or surface forces apparatus (SFA). These techniques, while useful, are not free from caveat; they can produce artifacts if interfacial conditions are not well controlled, thus restricting reproducibility of data. Consequently, the nuanced and dynamic nature of interfaces necessitates a multilevel measurement strategy that balances carefully accuracy against utility. In addition to the physical intricacy of measurement, the theoretical frameworks of interfacial properties are lacking or inadequate. The majority of existing models are having a hard time explaining the nanoscale complexities, and as such, scientists are having a difficult time correlating experimental outcomes to significant interpretations.

For example, when dealing with nanofluids, suspensions of nanoparticles in a fluid in the form of colloids, the structure-property relationship is complex. The theoretically noted enhancement of thermal conductivity due to structuring by nanoparticles can go hand-in-hand with the enhancement in viscosity, making their application in various technologies problematic (3rd Micro and Conference NF (MNF2011) et al., 2011). Moreover, multi-faceted descriptions of the influence of nanoparticles on fluid dynamics around them can create more ambiguity while formulating a unified theory of interfacial interactions. These theoretical uncertainties correspond to the requirement for continuous improvement of models in order to accurately represent the complexity of interfacial phenomena. Furthermore, mechanical response at interfaces complicates measurement from the perspective of materials design. Interactions between deformable soft materials and fluids introduce elastohydrodynamic effects that alter classical measurements of interfacial properties. When either of the two interacting surfaces at an interface is elastic, deformation under viscous stress must be considered, and this adds layers of complexity to data analysis (Dhong et al., 2017).

Investigation of the impact of this elastohydrodynamic coupling is called for in applications in adhesives, coatings, and biomaterials, where accurate measurement of interface dynamics is essential to material performance. Some of the most pressing issues of modern relevance include describing how viscoelasticity, heterogeneity at the surface, and layering in the substrate affect measured interactions, and so it is essential that scientists develop new methods that describe these sophisticated behaviors with accuracy. It is this integration of mechanical and fluid attributes that makes a multi-disciplinary understanding of interfacial properties necessary and ultimately furthering our understanding of material interfaces in theory and practice.

## 7.5.2. The Complexity of Multi-Phase Systems

Multi-phase systems are characterized by the richness of their dynamics and the challenge to model the behavior of the various phases that interact at their boundaries. Such systems, ubiquitous as they are in industrial operations and natural phenomena, consist of the coexistence of solids, liquids, and gases with varying characteristics each of which contributes to the system as a whole. For instance, in the field of pharmaceuticals, the behavior and structure of amorphous solids hold immense promise to influence the function of drug delivery systems. A clear and thorough understanding does not yet exist, and as such, these materials' potential uses are limited. Such intricacies in such systems call for one to explore advanced approaches to examining interfacial behaviors, such as terahertz time-domain spectroscopy, which can explain molecular behavior in non-destructive manners and, in so doing, help to bridge the knowledge gap regarding amorphous solids (N/A, 2016). Adding complexity to the analysis of multi-phase systems, particularly in geochemical systems, are diverse rheological conditions emanating from the variation in phase composition. Igneous processes, for example, are dominated by a multifaceted interaction among the solid, liquid, and gas phases, each of which affects flow dynamics in magma bodies. The earlier models used to concentrate mainly on two-phase interactions, such as porous transport of melt or crystal settling, which fail to cover the entire spectrum of behaviors seen in complex systems. Advances in model development have created frameworks that account for multi-phase interactions on an all-encompassing level, borrowing ideas from Mixture Theory and Non-equilibrium Thermodynamics. Such approaches are essential in unraveling the different flow modes and reactions dominating during material transport in igneous systems, accessing information that was otherwise lost with oversimplification of models (Keller et al., 2019). Indeed, the challenges offered by multi-phase systems are not merely conceptual in nature; they demand rigorous experimental approaches for validating the theories in question. As the material properties such as viscosity and density may be quite different based on the phase compositions, understanding of the behavior at the interface becomes critical to making an estimate of the systems overall performance. For instance, in drug delivery fields, control over the molecular mobility of amorphous solids can inherently enhance the bioavailability of drugs that are poorly soluble. The addition of advanced experimental tools that yield precise information about the dynamics involved thus remains crucial. The study of such multi-phase systems not only advances science but also has major ramifications for industrial operations, environmental systems, and technological innovation (N/A, 2016; Keller et al., 2019).

## 7.5.3. Future Directions in Interface Research

The field of solid and liquid interfaces is one of those new fields with a huge potential as a research field in the future. As the hydrodynamic slippage physics at solid-liquid interfaces continues to develop, researchers are beginning to realize that the traditional no-slip boundary condition so deeply rooted in fluid mechanics is not perfect. This new understanding reveals how alterations in surface properties, and also wettability and roughness, can lead to nonconventional flow behaviors such as hydrodynamic slip. Investigating these effects may unlock the intricate relationships that govern flow behavior, which will revolutionize the material and system design for a variety of applications from coatings and filtration to microfluidics. The effect of evolving surface properties on hydrodynamic slip is used to emphasize an even more nuanced need for boundary conditions, and heralds that interface science is on the verge of a revolutionary era that can reshape decades of traditional theory (Belyaev et al., 2011). Furthermore, opportunities for interface research under microgravity conditions are another new frontier in interface science. The unique microgravity conditions not only provide a unique environment to observe fluid dynamics but also provide a deeper understanding of the fundamental physics of phase interaction. Microgravity provides the freedom to carry out complex research in fluid dynamics, phase segregation, and materials science, which becomes simply impossible under Earth's gravitational forces. With these conditions, scientists can observe phenomena such as capillary action and fluid interfaces with unprecedented precision. This is an exciting opportunity to develop new processes and materials that exploit microgravity science. Secondly, microgravity research on the International Space Station provides a continuous stream of data that can inform terrestrial applications, mirroring the critical nexus between space exploration and applied science (Rogers et al., 1997). Finally, an important area of research in the interface domain is the convergence of advanced computational simulation and modeling techniques. With the advances in high-performance computing, researchers are now able to create sophisticated virtual models of interfaces with various physical and chemical interactions at the molecular level. The models can help predict performance under different conditions, thereby guiding experimental activity and enabling the design of new materials with specific properties. With a blend of experimental outcomes and sophisticated simulations, scientists are more able to comprehend complex interfacial processes such as surfactant behavior, adsorption dynamics, and interfacial reactions. This melding of computation and experiment not only allows for research streamlining but also optimizes the potential for innovation in technologies such as drug delivery devices, nanofabrication, and energy storage devices. Thus, the interface studies of the future will be a fascinating integration of computational, theoretical, and experimental techniques (Belyaev et al., 2011).

## SUMMARY

While exploring the complex interactions at interfaces between liquids and solids, one realizes that interactions at such interfaces are crucial in numerous scientific as well as engineering applications. The knowledge that can be learned by observing interfaces between liquids and solids is that how furthering the science of computational fluid dynamics, especially with techniques like vortex particle simulation, can go a long way in understanding the dynamics of multiphase flows better. The new approaches described, including the application of vortex methods, demonstrate increased efficiency within complexity and turbulence present within these interfaces. By the validation of these approaches through empirical observation, scientists are able to further define how such interfaces will behave under diverse conditions, which creates an understanding that forms the foundation both for theoretical research and practical applications within material science and fluid mechanics (Allard et al., 2006).

In addition, the reconsideration of classical boundary conditions, the no-slip condition in this context, emphasizes the evolving comprehension of fluid behavior along solid-liquid interfaces. Recent studies debunk traditional assumptions by drawing attention to the influence of hydrodynamic slip under specific conditions such as changing surface roughness and wettability. This is no mere curiosity but has far-reaching relevance in fields from microfluidics to lubrication technology. Recognition that hydrodynamic slippage has the ability to create emergent behavior as well—such as superfluidity and drag reduction—opens up new avenues of study and uses. Indeed, ongoing research into these phenomena is certain to yield significant contributions, further integrating fluid dynamics' theoretical underpinnings with its applied uses (Belyaev et al., 2011). In conclusion, solid and liquid interface research is a dynamic and rapidly developing field that combines theoretical, computational, and experimental approaches. As researchers refine their methods and expand their knowledge of interfacial behavior, the potential for innovative applications in nanotechnology, materials science, and fluid mechanics becomes increasingly tangible. The interdependence of elementary theories and contemporary advances in technology is a driving force for progress. Finally, the insights derived in this area may not just advance our scientific knowledge but also translate into groundbreaking impact on numerous industries, which will substantiate the worth of continued research into the intricate aspect of interfaces (Allard et al., 2006; Belyaev et al., 2011).

The research of interfaces in liquid and solid phases is a critical area of research within materials science that offers insight into physical molecular interactions. One significant area of such research comprises liquid-solid phase transitions. Density-functional theory has been used by researchers to compute ground-state energies of quantum hard-sphere solids, showing the way theoretical models flow into practical applications. The weighted-density approximation, appropriately adapted, offers a means of comparing quantum solids and their respective liquid phases, particularly emphasizing the unimportant role that exchange interactions play in Fermi solids. These theoretical advances emphasize the continued need for improvement of liquid-state input data, particularly by superior alternatives to standard approximations, thus pointing towards the complexities surrounding successful modeling of such transitions (Denton et al., 1997). Moreover, the microgravity environment offers a unique chance to push the frontiers of material interface research. NASAs vision towards achieving a permanently manned space station is geared towards pushing the boundaries of materials processing in this unique environment. Microgravity enables experiments unencumbered by the conventional gravitational forces that have made liquid-solid interactions problematic. The focus on areas like combustion science, electronic materials, and fluids is an indication of the stations multidisciplinary research strategy. With 11 classes of research named as user requirements, this project is focused on equipping the space station with hardware that enables extensive study of these interfaces, possibly revealing groundbreaking findings in material properties and behavior (Holland et al., 2025). The cross-feeding between experimental procedure and theoretical models is a sign of the dynamic nature of liquid-solid interface studies. As scientists continue to advance computational techniques such as the Wu-Feenberg cluster expansion, which addresses antisymmetry effects in Fermi liquids, differences between theory and experiment could be closed. Methodological improvements are all that are required for agreement in detail between simulation and observation in nature. This enhanced understanding will most likely lead to further innovations in new materials with applications in many fields of science and technology. The success of an extremely efficient pressurized laboratory module aboard the space station will be key to making these experiments feasible, thereby reconciling theoretical know-how and practical innovation in liquid and solid interface studies (Denton et al., 1997; Holland et al., 2025).

## REFERENCES

1. Abramson, H. N., Angrist, S. W., Catton, I., et al. (2025). Fluid Physics, Thermodynamics, and Heat Transfer Experiments in Space. https://core.ac.uk/download/pdf/42890762.pdf (accessed on 3rd October 2025).
2. Accardi, D., McMahan, T., Mikatarian, J., et al. (2025). The First United States Microgravity Laboratory. https://core.ac.uk/download/pdf/42810667.pdf (accessed on 3rd October 2025).
3. Affatigato, M., Backhaus, M., Bordia, R. K., et al. (2012). Challenges in ceramic science: A report from the workshop on emerging research areas in ceramic science. https://core.ac.uk/download/33113281.pdf (accessed on 3rd October 2025).
4. Agarwala, A., Agathopoulos, S., Bourell, D., Cassie, F., Chidambaram, D., Fowkes, F., Girifalco, L., et al. (2000). Carbon steel wettability characteristics enhancement for improved enameling using a 1.2 kW high power diode laser. https://core.ac.uk/download/53228.pdf (accessed on 3rd October 2025).
5. Allard, J., Cani, M. P., Coquerelle, M., et al. (2006). A vortex method for bi-phasic fluids interacting with rigid bodies. https://core.ac.uk/download/50439047.pdf (accessed on 3rd October 2025).
6. Backholm, M., Latikka, M., Ras, R. H. A., et al. (2018). Wetting of ferrofluids: Phenomena and control. http://arxiv.org/abs/1810.05581 (accessed on 3rd October 2025).
7. Bagchi, B., Santra, M., & Singh, R. S. (2013). Nucleation of a stable solid from melt in the presence of multiple metastable intermediate phases: Wetting, Ostwald step rule and vanishing polymorphs. http://arxiv.org/abs/1304.1663 (accessed on 3rd October 2025).
8. Beenackers, A. A. C. M., & van Swaaij, W. P. M. (1993). Mass transfer in gas-liquid slurry reactors. https://core.ac.uk/download/pdf/11461337.pdf (accessed on 3rd October 2025).
9. Belyaev, A. V., & Vinogradova, O. I. (2011). Wetting, roughness and hydrodynamic slip. http://arxiv.org/abs/1012.5642 (accessed on 3rd October 2025).
10. Bertrand, M. J., Binks, B. P., Hansen, J. P., Helfrich, W., Nelson, D., Lishchuk, S. V., & Safran, S. A. (2008). Saddle-splay modulus of a particle-laden fluid interface. http://arxiv.org/abs/0812.1133 (accessed on 3rd October 2025).

11. Binks, B. P., Cabrerizo-Vilchez, M. A., Fernandez-Rodriguez, M. A., et al. (2017). Particles adsorbed at various non-aqueous liquid-liquid interfaces. https://core.ac.uk/download/151161743.pdf (accessed on 3rd October 2025).

12. Brini, E., Dill, K. A., Fennell, C. J., et al. (2017). How water's properties are encoded in its molecular structure and energies. https://core.ac.uk/download/323073023.pdf (accessed on 3rd October 2025).

13. Busch, M., Greulich, S., Gruener, S., et al. (2015). Hydraulic transport across hydrophilic and hydrophobic nanopores: Flow experiments with water and n-hexane. http://arxiv.org/abs/1512.03908 (accessed on 3rd October 2025).

14. Cazabat, A. M., Dietrich, S., Oshanin, G., et al. (2012). Precursor films in wetting phenomena. http://arxiv.org/abs/1205.1541 (accessed on 3rd October 2025).

15. Charlaix, E., & Ciccotti, M. (2009). Capillary condensation in confined media. http://arxiv.org/abs/0910.4626 (accessed on 3rd October 2025).

16. Denton, A. R., Brandow, B. H., Campbell, C. E., Choudhury, N., Dalfovo, F., Dalfovo, F., Denton, A. R., et al. (1997). Density-functional theory of quantum freezing: Sensitivity to liquid-state structure and statistics. http://arxiv.org/abs/cond-mat/9703165 (accessed on 3rd October 2025).

17. Dhong, C., Frechette, J., Pilkington, G., et al. (2017). Elastic deformation during dynamic force measurements in viscous fluids. http://arxiv.org/abs/1809.10850 (accessed on 3rd October 2025).

18. Ganshin, A., Fermi, E., Fomin, I. A., Ivanov, Y. B., Kirichek, O., McClintock, P. V. E., Dalgliesh, R. M., et al. (2008). Neutron reflection from the liquid helium surface. https://core.ac.uk/download/69756.pdf (accessed on 3rd October 2025).

19. Holland, L. R., Uhran, M. L., & Wear, W. O. (2025). Accommodation Requirements for Microgravity Science and Applications Research on Space Station. https://core.ac.uk/download/pdf/42842190.pdf (accessed on 3rd October 2025).

20. Keller, T., & Suckale, J. (2019). A continuum model of multi-phase reactive transport in igneous systems. https://core.ac.uk/download/228164365.pdf (accessed on 3rd October 2025).

21. Kwon, Y. N., Lee, H. K., & Ray, S. S. (2019). Review on blueprint of designing anti-wetting polymeric membrane surfaces for enhanced membrane distillation performance. https://core.ac.uk/download/286607990.pdf (accessed on 3rd October 2025).

22. Lawrence, J., & Li, L. (2001). Wettability characteristics of polyethylene (PE) modified with $CO_2$, Nd:YAG, excimer and high-power diode lasers. https://core.ac.uk/download/53078.pdf (accessed on 3rd October 2025).
23. Lawrence, J., Li, L., & Spencer, J. T. (1998). Surface modification of an $Al_2O_3/SiO_2$-based ceramic treated with $CO_2$, Nd:YAG, excimer and high-power diode lasers for altered wettability characteristics. https://core.ac.uk/download/52944.pdf (accessed on 3rd October 2025).
24. Meyyappan, M., Neyts, E. C., & Ostrikov, K. (2013). Plasma nanoscience: From nano-solids in plasmas to nano-plasmas in solids. http://arxiv.org/abs/1306.6711 (accessed on 3rd October 2025).
25. Mitchell, J., Strange, J. H., & Webber, J. B. W. (2008). Nuclear magnetic resonance cytophotometry. https://core.ac.uk/download/91939.pdf (accessed on 3rd October 2025).
26. N/A. (2016). Direct Measurement of Molecular Mobility and Crystallization of Amorphous Pharmaceuticals Using Terahertz Spectroscopy. https://core.ac.uk/download/77411088.pdf (accessed on 3rd October 2025).
27. Pohl, H. A., Zhu, H., Rowlinson, J. S., Liu, J. M., & Dong, S. (2007). Di electrophoresis model for the colossal electro resistance of phase-separated manganite's. http://arxiv.org/abs/0705.3296 (accessed on 3rd October 2025).
28. Rodney, D., Tanguy, A., & Vandembroucq, D. (2011). Modeling the mechanics of amorphous solids at different length and time scales. http://arxiv.org/abs/1107.2022 (accessed on 3rd October 2025).
29. Rogers, M. J. B., Vogt, G. L., & Wargo, M. J. (1997). Microgravity: A Teacher's Guide with Activities in Science, Mathematics, and Technology. https://core.ac.uk/download/pdf/10472848.pdf (accessed on 3rd October 2025).
30. Tóth, G. I., Kelton, K. F., Korn, G. A., Gránásy, L., Powell, G. F. L., Powell, G. F. L., & Rowlinson, J. S. (2007). Phase field theory of interfaces and crystal nucleation in a eutectic system of fcc structure: I. Transitions in the one-phase liquid region. https://core.ac.uk/download/338512.pdf (accessed on 3rd October 2025).

CHAPTER 8

# ELECTROCHEMICAL INTERFACES

## Contents



## INTRODUCTION

Electrochemical interface research constitutes an integral nexus across several fields, including materials science, chemistry, and engineering. The interfaces determine the performance and efficiency of electrochemical systems, which are used in devices from batteries to fuel cells and sensors. The physical and observed phenomena taking place at these interfaces can directly influence charge transfer kinetics, electrolyte interactions, and device stability. An in-depth understanding of the electrochemical activity at these interfaces is necessary in order to design more efficient technologies, particularly as there is an increasing demand for sustainable energy solutions. Research in this field typically engages both theoretical modeling and experimental approaches in exploring the complex interactions that occur at the nano- and micro-scale, thus revealing information that drives innovation in energy conversion and storage systems. New methods now make it possible for researchers to modulate electrochemical interfaces to some degree of precision previously thought impossible. For instance, the use of two-dimensional (2D) materials in semiconductor interfaces is a promising way of enhancing electrochemical performance due to their unusual electronic properties. Research on inserting dipoles into semiconductor surfaces presents a means of improved control over double-layer structures at electrochemical interfaces. Experiments taking into account the open-circuit voltage of various material pairs, such as p-Si/Gr (graphene) and p-Si/h-BN (hexagonal boron nitride), show sensational changes in voltage response corresponding to adjustments in electrochemical potential. These findings show how atomic-scale modifications can yield unprecedented performance gains in terms of metrics, which makes well-selected choice of materials essential in electrochemical applications (Lewis et al., 2020). Further demonstrating the intricacy embedded in electrochemical systems, studies of lithium-ion electrodes have shown how microstructural design influences lithium plating dynamics. Utilizing a stochastic model for creating three-dimensional (3D) microstructures overcomes the shortcomings that exist in conventional imaging techniques and contributes enormously to our knowledge of porous electrodes. By coupling a microscopic, 3D spatially-resolved model of electrochemical response, scientists can study lithium plating and stripping in great depth. The complexity posed by such models, coupled with the computational cost of modeling realistic microstructures, calls for novel solutions like model order reduction schemes. These innovations provide one workflow that streamlines engineering tasks, eventually resulting in the mass production of more efficient lithium-ion cells (Feinauer et al., 2018). The continuing research into these interfaces underscores the pivotal role they will play in providing next-generation technological breakthroughs.

## *Definition of Electrochemical Interfaces*

Electrochemical interfaces are the prime areas where chemical reactions occur under the influence of an electric field so that energy is converted from a chemical to an electrical form. They tend to be located at the boundary between an electrode and an electrolyte and are a fundamental part of various electrochemical devices such as batteries, fuel cells, and sensors. One of the characteristic aspects of electrochemical interfaces is their complexity, which may be due to parameters such as the composition of materials employed, surface topography, and operational conditions. The interface properties regulate electron transfer and ionic transport rate in reactions, and therefore it is imperative that they are known for the sake of enhancing performance parameters in electrochemical devices. The electrochemical interface performance can be explained with the help of advanced techniques such as conducting atomic force microscopy and AC impedance spectroscopy. The techniques enable quantitative investigation of nanoscale reactions, and researchers can quantitatively study reactions at the nanoscale in order to disentangle the kinetics of processes such as oxygen reduction in interfaces, for example, Pt|CsHSO_4. It is discovered that these interfaces exhibit two predominant kinetic processes; one overpotential-dependent and the other being somewhat invariant. Both the processes exhibit near-ideal capacitive behavior, exhibiting the specific characteristics of nanoscale interfaces in which standard averaging methods may fail. The data obtained from these techniques can become a significant factor in the understanding of charge-transfer processes, ultimately directing the design of more efficient electrochemical systems (Haile et al., 2010) (Figure 8.1).

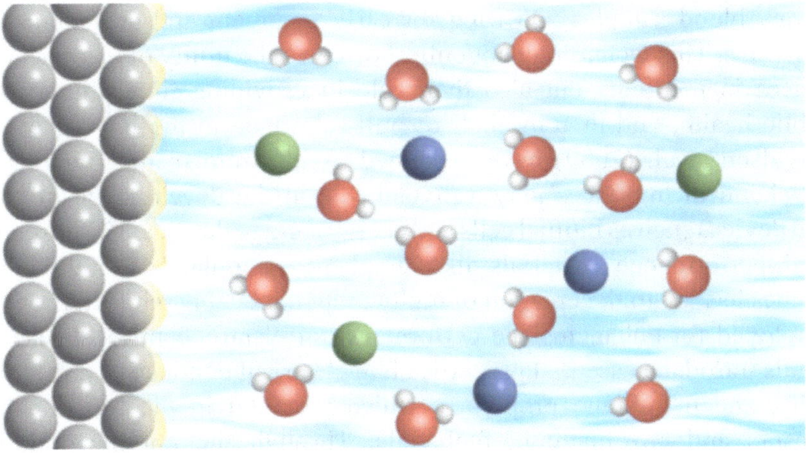

**Figure 8.1.** Electrochemical interfaces.

Source: https://www.nature.com/collections/aecijaiigd.

Furthermore, new simulation workflows have been developed to explore the intricate interaction between three-dimensional electrode microstructures and their electrochemical activity. This follows the difficulties in characterization of porous electrodes and their performance under a variety of operating conditions, such as in lithium-ion cells. By synthesizing 3D microstructures via parametric stochastic modeling, researchers can overcome the tomographic imaging restrictions and enable in-depth analysis of lithium plating and stripping processes. Integration of spatially-resolved physical models in the simulations sheds light on the transient nature of electrochemical systems, but the models' complexity demands high computational resources. But computational advances pave the way toward greater understanding of electrochemical interfaces and how they perform in next-generation energy storage devices (Feinauer et al., 2018).

## *Importance in Various Fields*

The significance of electrochemical interfaces is widespread, primarily in the domain of sensor science and medical diagnostics. The major advancement here is the inclusion of carbon nanomaterials such as carbon nanotubes (CNTs), which have been found to possess great promise for enhancing electrochemical reactivity. By virtue of their capability to facilitate electron transfer reactions, these materials have been key to the development of biosensors capable of detecting numerous biomolecules with high sensitivity. As an example, a CNT-based sensor has been used for quick diagnosis of hepatitis B virus (HBV) from raw blood samples, allowing for early treatment and detection of fatal liver disease. This technology not only showcases the utility of electrochemical interfaces in medicine but also their role in providing timely interventions in public health, and thereby their overall relevance in the biomedical field, as Davidson et al. (2009) have explained. Aside from medical applications, electrochemical interfaces play a pivotal role in the development of energy-related devices such as biofuel cells and energy storage devices. The introduction of high-performance materials like graphene nanoplatelets and halloysite nanotubes has led to the creation of multicomponent nanoarchitectures that can be employed for enhancing the performance of electrochemical devices. The materials introduce essential properties like electrical conductivity, mechanical stability, and membrane properties, which are vital in the functionality of biosensors and enzymatic biofuel cells. Through the optimization of the electrochemical interfaces using innovative hybrid systems, scientists are in a position to deliver enhanced energy efficiency and stability, paving the way for sustainable energy conversion and storage. The adaptability of the materials tests their ability to handle energy challenges, as documented in Aranda et al. (2019). Electrochemical interfaces also have impacts on environmental monitoring and the detection of pollutants. The development of electrochemical sensors allows in situ monitoring of environmental contaminants, thus enabling the sustaining of ecological safety and well-being. For instance, electrochemical biosensors may be designed to identify heavy metal or biochemical markers in water networks, representing an important tool for environmental policymakers and researchers. The sensitivity and precision of these sensors enable effective tracking of compliance with environmental regulation and the detection of toxic substances in ecosystems. Hence, the applicability of electrochemical interfaces extends from scientific inquiry into the development of social benefits through improved practice in protecting the environment. Such a potential highlights the need for more study and development of electrochemical technology in safeguarding our natural resources and a sustainable planet.

## 8.1. FUNDAMENTAL CONCEPTS OF ELECTROCHEMISTRY

The field of electrochemistry is in effect concerned with the interaction at the interface between an electrode and an electrolyte where significant redox reactions occur. This interaction is crucial to knowing and applying electrochemical systems, such as batteries and fuel cells. These responses are governed by kinetics and thermodynamics, and the Nernst equation plays an important role in determining the electrode's equilibrium potential under varying conditions. This link between chemical potential and ionic concentration demonstrates the manner in which variations in ion activity directly influence electrochemical performance. Further, the design and architecture of electrochemical cells are permeated with these fundamental principles, dictating the material engineering for efficiency in conductivity and reaction rates. Therefore, the knowledge of these underlying mechanisms is paramount to advancing technology for energy storage and conversion applications. Apart from thermodynamics, the kinetics of electrochemical reactions plays an important role in dictating the efficiency of processes at the electrochemical interface. The rate of electron transfer and the kinetics of surface reaction control the overall rate of electrochemical reactions. The electrode material, surface area, and roughness determine important characteristics in allowing higher rates of reactions. The employment of advanced materials, such as nanostructured electrodes, has expanded very broad avenues for enhancing the kinetics of reactions. Furthermore, the knowledge derived from investigation of reaction mechanisms at the atomic level allows us to engineer better electrocatalysts, which are pivotal in enabling efficient complex redox reactions. For this reason, the combination of kinetics and material choice improves energy efficiency as well as promotes more sustainable electrochemical technologies. Interface control in electrochemical applications includes not just electron transfer but also strategic design of materials to enhance their performance in fields such as energy storage. For instance, the focus in current studies has been on aqueous battery zinc anodes, addressing concerns of dendrite growth and corrosion that undermine their long-term performance and efficiency (Shin et al., 2020). Understanding these problems lays a framework for the development of improved anode materials by engineering, with implications towards improved cycle stability and performance. Furthermore, in the realm of carbon dioxide reduction, efficient electrocatalysts have a key role in favoring target reaction pathways towards desirable products, especially higher-energy density compounds (Zheng et al., 2019). This underscores the necessity of integrating fundamental electrochemical principles with advanced material science to innovate technology that can meet the world's energy needs sustainably. The synergy of these disciplines will remain integral to future advancements in electrochemical interfaces (Figure 8.2).

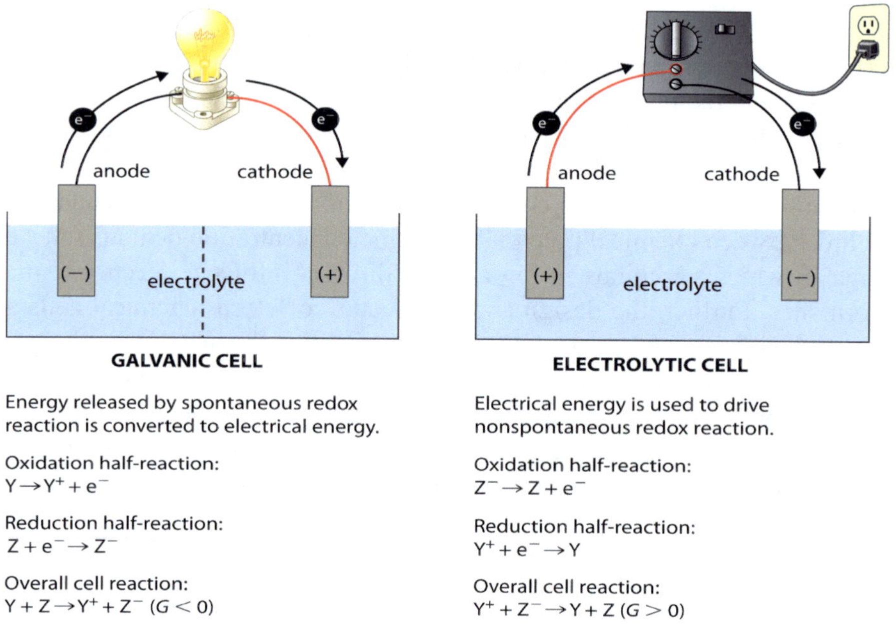


**Figure 8.2.** Electrochemical cells.

Source: https://saylordotorg.github.io/text_general-chemistry-principles-patterns-and-applications-v1.0/s23-electrochemistry.html.

## 8.1.1. Basic Principles of Electrochemical Reactions

Electrochemical reactions are fundamentally characterized by electron transfer between reactants, typically occurring at the interface between an electrolyte and an electrode. Oxidation and reduction reactions are the characteristic parameters of the system at such electrochemical interfaces, in which one species is losing electrons (oxidation) and another gaining electrons (reduction). The kinetics governing these reactions depends on a variety of variables, including temperature, concentration, and nature of the electrode material. Moreover, the electric double layer (EDL) is an important consideration, with charged species that locate at the interface determining ionic as well as electronic conduction. Not only does such a layer act as a boundary to charge separation but also undergoes alteration by applied voltage, which can lead to extreme changes in electrochemical operation and efficiency, particularly in more recent battery systems. A more profound understanding of electrochemical reactions necessitates research into the likely barriers that influence reaction kinetics in these interfaces.

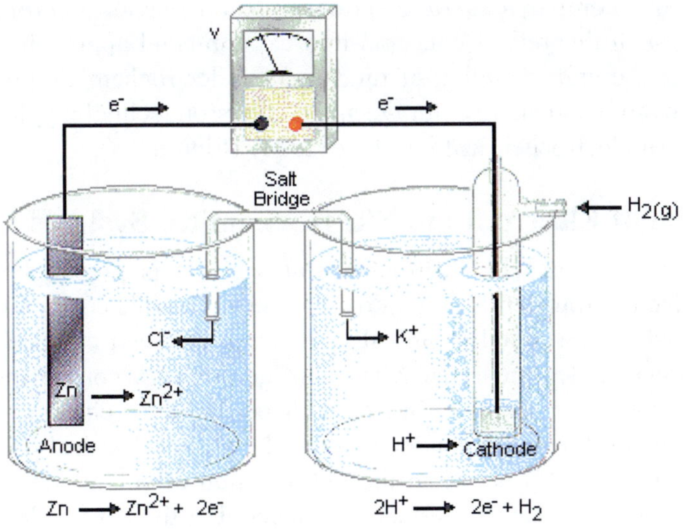


**Figure 8.3.** Electrochemical reactions.

*Source:* https://chemed.chem.purdue.edu/genchem/topicreview/bp/ch20/electro.php.

Overpotentials are caused by differences between the theoretical and actual voltage required to catalyze an electrochemical reaction. This is especially crucial in the optimizing of efficiencies in processes such as fuel cells. Through the application of advanced techniques such as electrochemical impedance spectroscopy, scientists are able to quantify these overpotentials, providing information about the electro-catalytic efficiencies of various materials. For instance, in a study where nickel and platinum electrodes are compared, the dynamics of reaction were found to radically differ depending on the properties of the material of the electrodes. Thus, an understanding of how interaction occurs between these potentials can lead to the creation of electrochemical systems with better functions, reinforcing the intrinsic nature of electrochemical interfaces in this research area. Modern research relies on theoretical models like Density Functional Theory (DFT) to describe atomic-level interactions defining electrochemical interfaces. Using DFT, researchers can predict the impact of voltages applied on charge distributions and dipole moments at the interface, which informs electrode material design to improve their efficiency. For example, in gold substrate thin films research on investigation, manipulation of surface dipole has remained a shining light in the advancement of electrode passivation, which ultimately reflects in enhanced performance. Real electrode potential difference and self-consistency needs on lithium content further determine the complexities of electrochemical behavior interpretation. By integrating such theoretical concepts into experimental approaches, scientists can advance the understanding of fundamental electrochemical processes and inspire innovations in energy storage and conversion technology, justifying the importance of electrochemical interfaces today (Figure 8.3).

## 8.1.2. Role of Electrodes in Electrochemical Systems

The role of electrodes in electrochemical systems is pivotal because they facilitate electron transfer during redox reactions. The role is pivotal for battery operation, where sustained chemical conversion between cathode and anode relies on effective electron transfer. The configuration and composition material of electrodes can decisively influence their performance characteristics, such as efficiency and longevity. Understanding the electrochemical processes that occur within electrodes is key in the optimization of energy storage devices. Existing research indicates interfaces engineered at the nanoscale possess the potential to enhance lithium-ion battery performance to a significant level. For instance, regulation of lithium-ion transport at the microstructure level can steer the direction of ion migration, thus resulting in efficient and long-lasting batteries (Dayeh et al., 2015). Apart from energy storage devices, electrodes have a significant function in electrochemical sensors, where the surface properties of the electrodes determine sensitivity and selectivity of analyte detection. Carbon-based electrode materials, including carbon nanotubes and nanocrystalline diamond, have proven promising because of their unique electrochemical characteristics. They possess high conductivity and high surface area, which favors improved reaction kinetics and high rates of electron transfer, making them very suitable candidates for sensing applications. Their inclusion in biosensors has proven to be beneficial, enabling biomolecules to be detected with sensitivity at low concentrations. The application of carbon nanotube-modified electrodes, for example, has contributed to enhanced real-time medical diagnosis, with the capability of detecting diseases early, like hepatitis B (Davidson et al., 2009). Moreover, expanding the knowledge base regarding electrode material characteristics and their interaction with electrolytes plays an important role in flexible electrochemical system design. Enhanced nanofabrication techniques enable the realization of advanced layered materials that enhance electrochemical interfaces, thus enhancing the overall sensor and battery performance. With sustainability being a priority, researchers are looking for environment-friendly and low-cost materials for the fabrication of electrodes. As the demand for energy-effective technology rises, optimization of electrode design will be the key to the next generation of electrochemical devices. The current research into novel materials and their fundamental properties will continue to generate innovation in electrochemical interfaces as well as expand their application in various fields (Figure 8.4).

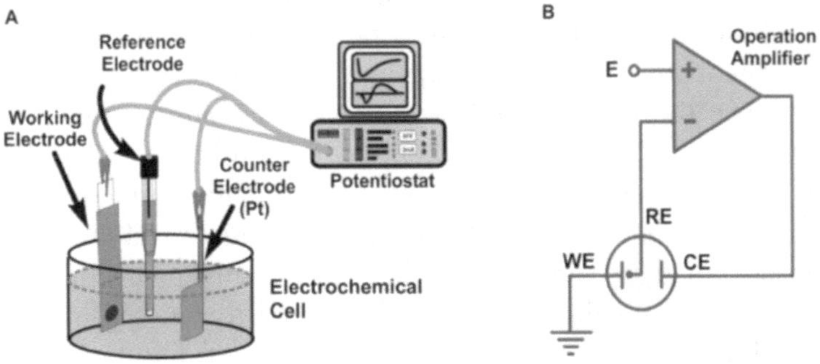


**Figure 8.4.** (A) Electrochemical cell with three electrodes connected to a potentiostat; and (B) Electronic sketch illustrating the mode of operation of a typical potentiostat.

*Source:* https://www.researchgate.net/figure/A-Electrochemical-cell-with-three-electrodes-connected-to-a-potentiostat-B-Electronic_fig1_353321941.

## 8.1.3. Importance of Electrolyte Solutions

Electrolyte solutions are key to the function of electrochemical interfaces, enabling ion transport that controls charge transfer processes. In batteries, for instance, the choice of the electrolyte directly influences energy efficiency and system performance of the electrochemical system. Electrolyte solutions not only facilitate ion transport between electrode material but also stabilize interfaces needed for battery function. As defined by the Joint Center for Energy Storage Research, low-cost, high-performance electrolyte solutions can significantly revolutionize next-generation batteries to achieve breakthroughs in energy storage technology (Crabtree et al., 2015). These breakthroughs can facilitate sustainable energy systems with increased efficiency and lifespan in batteries, while emphasizing the pivotal role of electrolytes in contemporary energy solutions. In addition to their function as transport media, electrolyte solutions affect thermodynamic properties of electrochemical systems. Reaction rates, thus capacity and cycle life of batteries, are affected by the electrolyte ion-elektrode surface interactions. The porous electrode theory shows how composition and concentration of the electrolyte lead to intercalation processes in battery material. Classical concentrated solution theory offers a tool for describing the behavior of concentrated electrolytes under varying electrochemical conditions, that is, in the context of developing the emergent properties critical to efficient charge transfer (Bazant et al., 2017). With insights into these complex relationships, researchers can optimize electrolyte compositions to make stable and efficient systems available for advancement in energy storage technology. The role of electrolyte solutions is not restricted to traditional lithium-ion technologies; they play a critical role in the evolution of new battery chemistries. With development shifting towards beyond-lithium-ion technologies, the function of the electrolyte is more critical. Electrolyte solutions must be engineered to accommodate different materials, such as those with phase separation characteristics. Innovation in novel electrolyte chemistries is key to making novel battery architectures viable and improving their electrochemical performance. Scientists are advancing the frontiers of electrolyte science with the help of sophisticated theoretical techniques and computational models to enhance the efficiency of future energy storage devices (Crabtree et al., 2015; Bazant et al., 2017). These multidisciplinary approaches mirror the dynamic nature of electrolyte solutions, with a focus on their multifunctional functions in the construction of electrochemical interfaces and energy technologies.

## 8.2. TYPES OF ELECTROCHEMICAL INTERFACES

Electrochemical interfaces are very significant in many applications, particularly in the development of sensors and energy storage devices. Among all kinds of interfaces, carbonaceous materials have attracted attention due to their unique properties. Carbon nanotubes (CNTs), for instance, are highly conductive and have high surface area, which enhances their electrochemical reactivity. Their ability to perform electron transfer reactions renders them of enormous potential in the creation of biosensors since they are capable of increasing detection sensitivity by a great margin. Modern advancements have shown that CNT-modified electrodes can reduce the impact of surface fouling so that more authentic measures, like those utilized in clinical diagnostics where timely and accurate detection of disease-causing pathogens is critical, can be made.

The intrinsic diversity of carbon materials thus marks them as a barrier in the recent exploration of electrochemical sensor technologies and applications in biomedicine and other fields (Davidson et al., 2009). Besides carbon-based materials, 2D materials have been seen as possible rivals for electrochemical interfaces with properties that distinguish them from traditional materials. Graphene and hexagonal boron nitride (h-BN), for instance, have unique electronic properties due to their band structure and dimensionality. These materials enable deeper understanding of the electronic interactions within the semiconductor/liquid interface, which is crucial in achieving optimized sensor performance. Macroscale measurements and emergent techniques such as scanning electrochemical cell microscopy (SECCM) have provided useful information concerning the electrochemical performance of these interfaces. The voltage's sensitivity to changes in the electrochemical potential ensures that 2D materials can be synthesized with specific functionality by manipulating their interaction with liquid electrolytes. Such achievement is vital in the design of more sensitive and efficient electrochemical devices (Lewis et al., 2020). Synthetic diamond materials incorporated into another notable class of electrochemical interfaces exhibit excellent electrochemical characteristics and biocompatibility. The nanocrystalline-diamond thin films and nanowires synthesized have paved the way for new biosensing possibilities, combining chemical inertness with conductivity. The diamond interfaces not only allow efficient electron transfer but also provide an adaptable surface for grafting biomolecules, which makes them more useful to the biomedical community. Diamond nanowires, for example, have been found to be a reliable tool for monitoring biochemical interactions with unprecedented sensitivity according to recent studies. This versatility and capacity to create complex bio-inorganic interfaces highlight the promise of synthetic diamond to act as a lead material in prospective electrochemical biosensors, a quantum leap in the electrochemical interface technology (Davidson et al., 2009).

### 8.2.1. Solid-Liquid Interfaces

The study of solid-liquid interfaces is central to the understanding of electrochemical processes, and especially in batteries and fuel cells. In this case, the contact between liquid electrolytes and solid electrodes is of specific relevance in defining the overall efficiency and functionality of energy storage systems. The ion exchange and charge transfer kinetics at the solid-liquid interface control other effects, for example, electric double-layer formation that is central to electrochemical equilibrium. The interface possesses certain surface characteristics and mechanics of structure controlling nucleation, ion mobility, and growth. Because of the complexity of the interface, developing models that accurately depict these interactions is essential, particularly in batteries like lithium-ion in which solid-electrolyte interfaces can have a significant impact on cycle stability and energy density. Enhancements in understanding of solid-liquid interfaces are critical in order to achieve better electrochemical device performance and lifetime. To model such complex interactions, researchers have developed sophisticated mathematical frameworks.

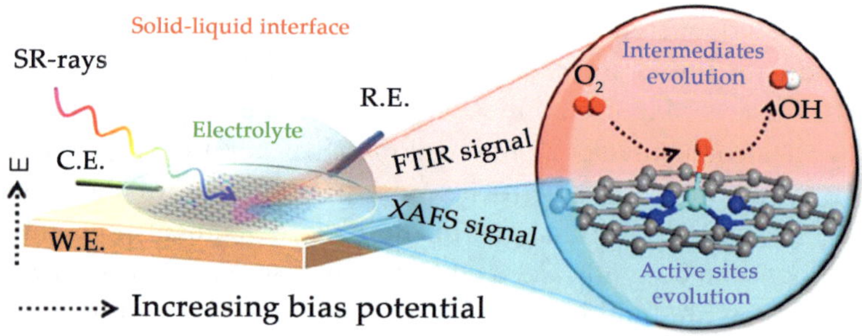


**Figure 8.5.** Solid-liquid interfaces.

*Source:* https://external-content.duckduckgo.com/iu/?u=https%3A%2F%2Fxpic.x-mol.com%2F20200716%252F10.1021_jacs.0c04231.gif&f=1&nofb=1&ipt=c627c40129e38659a8537b995d0437f46c1e95f71f770ce3ddddbd694e1acce8.

One such special method is the Electrochemical Mean-Field Kinetic Equations (EMFKE), which is derived from a microscopic lattice-gas model of charged species. It contains significant factors such as the Poisson equation, which deals with how electric fields influence ion migration at the solid-liquid interface. The equations provide insight into equilibrium conditions and kinetic behavior with charge/discharge cycles by taking into account the electrochemical growth and dissolution processes. Computer simulations of recent times have validated this model by reproducing charged double layers at plane electrodes successfully and thorough explanation of the one-dimensional electrochemical cell growth kinetics. Such comprehensive models enhance our understanding of solid-liquid interfaces by modeling correctly the conditions under which such important processes occur, thus paving the way for better battery design and utilization (Léger et al., 2003). The application of solid-liquid interfaces in the real world may particularly be observed in advancements in all-solid-state batteries, which involve high-level materials for efficiency. With Al-substituted $Li_7La_3Zr_2O_{12}$ (LLZO) as the solid electrolyte, the batteries have developed significantly with composite positive electrodes. The bilayer structure, formed by combining dense LLZO membranes and porous scaffolds, demonstrates an efficient method of driving ion diffusion. The one-way character of the pores ensures better penetration of cathode components with reduced path length that lithium ions have to cover, thereby maximizing the overall ionic conductivity. Furthermore, use of a soft ionically conductive solid ensures better contact among battery components at the solid-liquid interface, which is essential for maximizing electrochemical performance. These developments illustrate the key role that solid-liquid interfaces play in electrochemical devices in order for them to achieve practically meaningful, high-performance levels (Alvarado et al., 2020) (Figure 8.5).

### 8.2.2. Solid-Gas Interfaces

The study of solid-gas interfaces is a significant aspect of developing electrochemical interfaces, particularly in electrochemical devices such as fuel cells and batteries. The interfaces are characterized by interfacial interactions between the interface of a solid electrode material and a gaseous reactant. In solid oxide fuel cells (SOFCs), for instance, operation of the cell depends significantly on porous electrode design to ensure gas diffusion and catalytic operation. The reduction of operating temperatures from 1000°C to 800°C through the use of thick porous layers is a dominant technology trend that aims to enhance efficiency at the expense of less complex infrastructure requirements. These advances call for more understanding of the transport mechanisms governing gas flow, and the catalytic and electrochemical reaction taking place in these complicated structures (2nd Micro and Conference NF (MNF2009) et al., 2009). Moreover, solid-gas interface behavior supplies vital information on reaction kinetics, especially when interfaces of nanoscale sizes are concerned. Experiments on nanoscale behavior of solid-gas interfaces, such as that of the Pt|$CsHSO_4$ interface, reveal a two-component mechanism for oxygen reduction kinetics. Two separate processes were found to contribute to the overall reaction, demonstrating the significance of overpotential in managing such interactions. The ability to untangle such mechanisms through techniques such as conducting atomic force microscopy is one demonstration of the cutting-edge methodologies that are developing in this field. This level of detail in kinetic characterization not only determines the underlying processes at play but also suggests potential avenues for optimizing fuel cell performance through control of interfacial dynamics (Haile et al., 2010). Understanding of solid-gas interfaces is thus central to continuing advances in electrochemical devices, which rely on efficient mass transport and surface reaction dynamics. Recent research emphasizes the role of interface structure in determining electrochemical performance, with particular attention given to the role of nanostructuring in optimizing the interfaces. For example, microstructuring porous electrodes facilitates internal reforming reactions that allow gaseous fuels like methane to be converted into hydrogen and carbon monoxide, major components for optimal energy production. With advances in computational fluid dynamics (CFD) and modeling, the complete understanding of heat transfer, mass generation, and species interaction at solid-gas interfaces is increasingly feasible. This data will guide the design of next-generation electrochemical systems, eventually leading to enabling them for use in general commercial applications (2nd Micro and Conference NF (MNF2009) et al., 2009).

## 8.2.3. Liquid-Liquid Interfaces

Liquid-liquid interfaces play a vital role in many electrochemical applications, particularly in battery technology and energy storage devices.

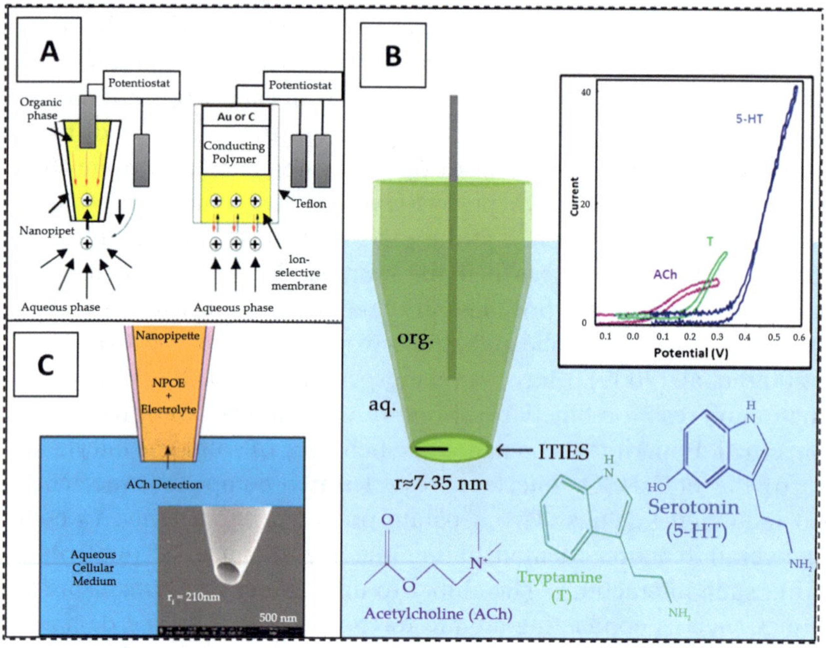


**Figure 8.6.** (A) The reaction mechanism at the liquid–liquid interface is depicted; 134 (B) nanopipette-based liquid–liquid interface probes for the electrochemical detection of ACh, tryptamine, and 5-HT via ionic transfer is represented; and 135 (C) The process of ACh detection at NPOE/water nanoITIES, with an SEM image representing a nanopipette with an approximate radius.

*Source:* https://www.researchgate.net/figure/A-The-reaction-mechanism-at-the-liquid-liquid-interface-is-depicted-134-B_fig5_381202002.

These interfaces, which are formed between immiscible liquids, play important roles in processes such as charge transfer, mass transport, and electrolyte stabilization. In battery research, data regarding the nature of liquid-

liquid interfaces can have a significant influence on the performance and lifespan of energy storage devices. By studying the interactions occurring at such interfaces, researchers are able to come up with better electrolyte formulations that enhance ion conduction and eliminate unwanted side reactions. Research becomes significant as next-generation battery development transcends traditional lithium-ion systems, where it takes on complex interfacial dynamics poorly understood (Crabtree et al., 2015). The interfacial structural occurrences and molecular interactions at liquid-liquid interfaces are intricate in character and thus comprise the central door to advancing the science of electrochemistry. Investigations utilizing advanced techniques such as Density Functional Theory (DFT) have proved the impact of imposed voltages on interfacial dipole moments and charge separation across these interfaces. For instance, the distinctive behavior of ions and organic solvents at the interface can lead to extreme contrasts in electrochemical performance, especially charge transfer rates and stability at the electrodes. These factors are significant when looking to advance more efficient and less degrading batteries as manipulation of such interfacial characteristics can aid enhanced electrode passivation and suppress parasitic reactions (Leenheer et al., 2015) (Figure 8.6).

In addition, the behavior of liquid-liquid interfaces in electrochemical systems has relevance to broader application, e.g., in fuel cells and sensors. The optimization of design of these interfaces can lead to maximum energy conversion efficiency and enable the fabrication of new electrochemical devices. Increasing understanding of the involved molecular and electronic effects enables scientists to design materials and structures to be optimized for particular applications. With the altering energy landscape, research in liquid-liquid interfaces is becoming increasingly important not only to address new technology needs but to irreversibly push forward the scientific knowledge base with regard to electrochemical interactions. The development in this area foretells a remarkable leap towards acquiring sustainable and efficient energy storage in the future (Crabtree et al., 2015).

## 8.3. CHARACTERIZATION TECHNIQUES

The study of electrochemical interfaces calls for characterization techniques that are thorough in detail and uncover material properties and behavior at the molecular level. A recent method of characterization is the use of dual-barrel theta conductance micropipets, with which researchers are able to examine dissolution kinetics of ionic crystals at unprecedented rates and with extreme specificity. Since this technique targets single crystal surfaces in a selective manner, it is an advantage that localized dissolution processes can be observed. This makes possible solid-liquid electron transfer mechanism precise studies. As an example, during a NaCl dissolution observation, researchers were able to estimate a first-order heterogeneous rate constant, highlighting controlled experimental conditions as a determining factor in electrochemical interface behavior (Alex et al., 2013). These developments in characterization techniques, collectively, expand the scope of electrochemical investigations to allow for thorough interrogation of interfacial processes. Another critical area in the characterization of electrochemical interfaces addresses the development of biosensors. Responding to growing needs for rapid and selective biomarker detection of disease, researchers are increasingly employing nanomaterials like graphene quantum dots (GQDs). These compounds maximize sensor performance in terms of improved selectivity and sensitivity and are both efficient fluorophores as well as electro-catalysts. Their stability, biocompatibility, and water solubility render GQDs highly appropriate for use in optical, electrochemical, and photoelectrochemical biosensing techniques. As defined in recent reviews, the combination of GQDs with advanced fabrication and sensor engineering is the trend of characterization tools shifting towards more efficient and practical application in the biomedical sector (Altintas et al., 2020). This union of electrochemistry with nanotechnology forms the crux of propelling advancements in the comprehension of electrochemical interfaces in complex biological systems. By way of conclusion, characterization techniques for exploring the subtlety of electrochemical interfaces disclose the multifaceted behaviors governing interactions at these interfaces. The application of the dual-barrel theta conductance micropipet technique allows for precision measurements that are well-suited for extensive kinetic examinations to enhance our theoretical foundation. Conversely, leveraging the strategic use of nanomaterials like GQDs in biosensors demonstrates the interdisciplinary approach required to treat the challenge of contemporary disease diagnosis and treatment. As a result of continual improvement and evolution of skincare regimes, scholars can add further to research and understanding of the electrochemical interfaces, ultimately resulting in innovation in fields as varied as materials science, biomedical engineering, and environmental monitoring. Such characterization advances easily bridge gaps between fundamental research and real-world applications, demonstrating the broader relevance of advances in research on electrochemical interfaces.

### 8.3.1. Electrochemical Impedance Spectroscopy (EIS)

Electrochemical impedance spectroscopy (EIS) is a powerful analytical method that provides information on electrochemical processes at interfaces. Employing a small perturbation potential and monitoring the response current, EIS measures the dynamic system response of an electrochemical system over a range of frequencies. This method is particularly convenient to use for characterizing the behavior of electrochemical interfaces since it allows scientists to examine various contributions to impedance, such as charge transfer resistance and double-layer capacitance. The versatility of EIS makes it flexible enough to examine various applications from energy storage devices like batteries and supercapacitors to biosensing and corrosion studies. Significantly, the fact that these impedance spectra can be simulated by equivalent circuits gives further insight into the mechanisms at work, and therefore how electrochemical systems may be designed and optimized. An interesting application of EIS is the evaluation of long-term functionality of implantable medical devices. In particular, the success of encapsulation techniques can be monitored through EIS by measurement of the change of impedance with time.

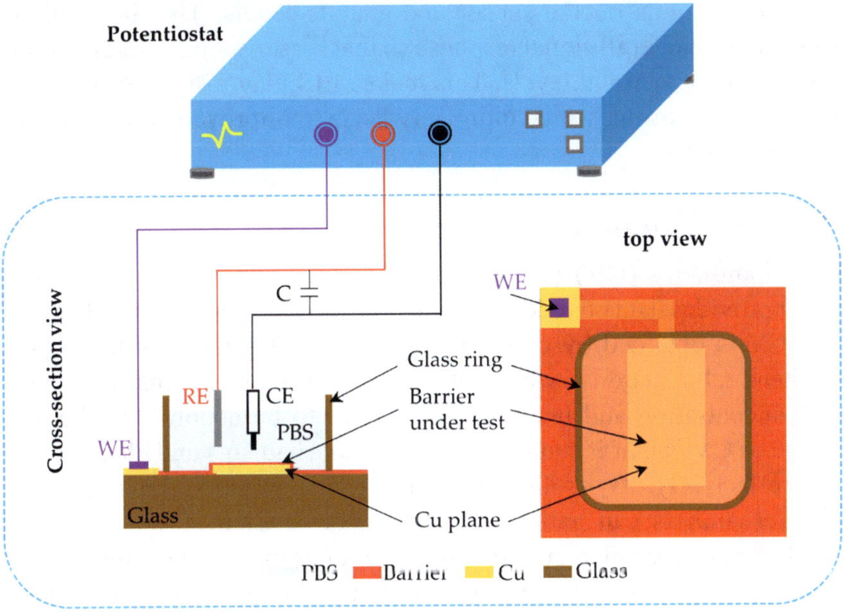


**Figure 8.7.** Schematic diagram of the electrochemical impedance spectroscopy (EIS) test setup.

*Source:* https://www.researchgate.net/figure/Schematic-diagram-of-the-electrochemical-impedance-spectroscopy-EIS-test-setup-WE_fig2_335786697.

For instance, one recent article shows how a polyimide-based process of hermetic encapsulation, augmented with atomic layer deposition of $HfO_2/Al_2O_3$, maintained outstanding electrochemical performance over 2.8 years at elevated temperatures (Cauwe et al., 2019). With the use of EIS, in situ observation of encapsulation stability was possible, providing critical information on the long-term survivability of the devices protective coatings. EIS, thus, not only serves as an analytical tool but also facilitates the development of more robust materials and structures for future generations of biomedical devices, which will further enhance device performance and patient outcome. EIS has also been shown to be a vital methodology in the production of biosensors, particularly those from nanomaterials like graphene quantum dots (GQDs). Such nanomaterials enhance the sensing performance of sensors in terms of sensitivity and selectivity toward specific biomarkers. For example, GQDs exhibit excellent electrochemical capabilities and are promising candidates for the fabrication of electrochemical biosensors for detecting ultralow target concentrations for disease detection (Altintas et al., 2020). With the application of EIS in testing such biosensors, researchers can explore the electrochemical interface interactions on the sensor and analyte levels. This interaction also contributes to the overall signal response so that biosensor designs are optimized to operate at their optimal levels. This makes EIS play a significant role in the evolution of biosensing, which ultimately leads to improved clinical diagnostic devices (Figure 8.7).

## 8.3.2. Cyclic Voltammetry (CV)

Cyclic voltammetry (CV) is a major technique of electrochemical interface study that provides data on redox behavior and kinetic properties of electroactive species. CV is obtained by cyclically changing potential in order to produce electrochemical reactions, which allows the current responses accountable for the concentration and presence of analytes to be quantified. The greatest advantage of CV is that it can report reaction mechanisms and electron transfer kinetics effectively. By following current responses at various potential sweeps, researchers can gather extensive data about reaction pathways and intermediates, hence enhancing the understanding of electrochemical systems. As a result, CV has become a wide-ranging technique in research involving various applications, from energy storage devices to biomolecular sensors, demonstrating its effectiveness and flexibility in exploring electrochemical processes. Moreover, the use of CV is also evident in electrochemical sensor high-performance material design, e.g., those fabricated from carbon materials. The current research has shown the incorporation of carbon nanotubes and other derivatives into the construction of sensors to enhance the electrochemical reactivity of biomolecules significantly. The unique properties of these carbon materials, combined with the ability to promote electron transfer reactions, allow for better sensitivity and specificity in the measurement of specific analytes. As implied, carbon nanotubes (CNT) not only have superior conductivity but also provide an electroactive surface for immobilization of biomolecules, which is a key need in diagnostic monitoring and biosensing. The combination of CV and these new materials also enhances the ability of electrochemical sensors, which become useful tools for environmental monitoring and medical diagnostics (Davidson et al., 2009) (Figure 8.8).

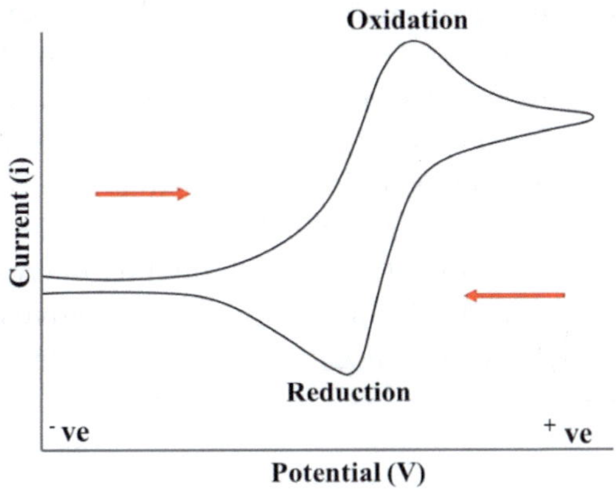


**Figure 8.8.** Representative of cyclic voltammetry (CV).

*Source:* https://mtxlabsglobal.com/wp-content/uploads/2024/09/Screenshot-363-1.png.

Besides, the integration of CV with novel nanostructured materials has revolutionized the notion and usage of electrochemical interfaces. For instance, certain advances in the fabrication of nanocrystalline diamond films have shown considerable ability to enhance sensor performance. The films have superior electrochemical properties and chemical inertness and hence represent an ideal option for interfacing with various biological systems. With the aid of CV along with these advanced materials, researchers have been able to prepare very sensitive sensors for real-time monitoring of key biological markers. This accomplishment is a remarkable shift in bioanalytical chemistry where electrochemical sensors attain record-breaking detection limits, therefore enabling prompt medical intervention in diseases such as chronic hepatitis (Cassayre et al., 2013). These developments underscore the primacy of cyclic voltammetry in advancing the science of electrochemical interfaces and sensor technology.

## 8.3.3. Scanning Tunneling Microscopy (STM)

The contribution of STM to the branch of electrochemical interfaces cannot be overemphasized. STM offers unparalleleled insight into the surface processes on molecular scales, and researchers can visualize the structure and function of individual atoms in electrochemical systems. Its ability to provide topographical maps of atomic resolution makes STM crucial in understanding the rudimentary processes at electrode surfaces. For example, STM can account for species self-assembly at electrode interfaces, and subsequent formation of ordered arrays with final, determinant effects on electrochemical reactivity. It is through understanding such ordering that theorists can build models relating macroscopic electrocatalytic performance and atomic-level phenomena, and thereby refine design of high-performance electrode materials for all possible applications. Once more, another field that contributes to STM's utility is its capacity for operation under electrochemical conditions, as highlighted by recent advances in in-situ electrochemical scanning tunneling microscopy (EC-STM). With EC-STM, researchers have been watching redox-active molecules such as tetra(N-methyl-4-pyridyl)-porphyrin cations self-assemble in experiments on gold-modified surfaces.

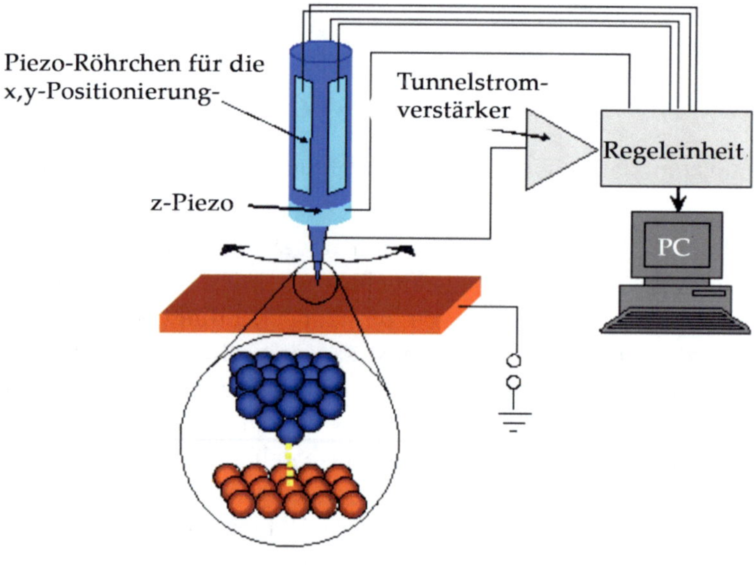


**Figure 8.9.** Scanning tunneling microscopy (STM).

*Source:* https://www.komet335.physik.uni-mainz.de/files/2017/03/svpo9poa.gif.

This technique offers fundamental information about the charge states and electrochemical properties of adsorbed species and hence allows for a full appreciation of the solid-liquid interface. The conclusions derived from potentio-dynamic EC-STM measurements demonstrate how specific interactions—ranging from van der Waals forces to electrostatic attractions—govern the organization of adsorbates under various potential conditions. These findings reveal the complexity of electrochemical processes and extend further the science of interfacial electrochemistry. Besides, coupling STM with other imaging techniques builds a strong ground to investigate electrochemical interfaces. With the application of correlative techniques that couple STM with spectroscopies, researchers can get a multi-spectra image of electrochemical activity in materials, particularly next-generation carbon nanomaterials. For instance, knowledge of the improved electrochemical performance of graphene and graphite can be developed with high-resolution STM imaging, which reveals information on surface reactivity previously hidden. This combination facilitates researchers probing structure-activity relations in electrochemistry more thoroughly, shedding light on determinants of catalytic efficiency. By gaining access to the relationship between material properties and electrochemical performance through advanced imaging techniques like STM, researchers are in a good position to lead new electrode materials with improved efficiency and functionality in electrochemical reactions, ultimately informing energy storage and conversion technology (Figure 8.9).

## 8.4. APPLICATIONS IN ENERGY STORAGE

The energy storage field has made immense progress in the past few years primarily based on advances in electrochemical interfaces. These interfaces are central to battery and supercapacitor performance and have a direct impact on the efficiency of ion transport and electron transfer. The structural morphology of such interfaces is the decisive factor for energy storage properties, with effects extending beyond performance to longevity and stability as well. Recent studies suggest that it is feasible to enhance the overall energy storage efficiency of devices through the manipulation of interfacial structures by controlled electrochemical reactions. This is achieved through applying non-equilibrium thermodynamics, in which operators can define the optimum conditions for the optimization of interface morphology and the creation of target charge transfer pathways (Bazant et al., 2020). The possibility of successfully tuning such characteristics is an exciting advance for many energy storage systems. The intersection of discovery science and engineering practice offers opportunities for new ideas in energy storage technologies. The Joint Center for Energy Storage Research (JCESR) is the prime example of this strategy because it is focused on developing next-generation battery systems with designs for high-performance and low-cost electricity storage. By promoting collaboration across more than one discipline—material science, electrochemistry, and engineering—JCESR is attempting to speed up the process of moving from theoretical ideas to practical use. This combined approach not only accelerates the innovation process but also allows for ongoing exploration of beyond-lithium-ion battery technologies, an untapped energy storage application frontier (Crabtree et al.,

2015). Such innovations hold the promise of future gains in energy efficiency and sustainability, complementing the next generation of energy grids. Next-generation electrochemical interfaces technologies also hold significant advantages for uses beyond traditional batteries. For instance, the principles from ion intercalation research are being employed to innovate next-generation supercapacitors that offer fast charging and long cycle life. Interface morphology breakthroughs, encouraged by thorough scientific investigation, have opened up paths for novel energy storage ideas, such as hybrid energy systems capable of effectively integrating properties from batteries and supercapacitors. Moreover, the density of states in electron donor materials can significantly be optimized to increase the efficiency of energy storage and harvesting (Bazant et al., 2020). With diversification and expansion of the electrochemical interface field, the potential of revolutionary application in energy storage becomes increasingly evident, leading to a sustainable future for energy.

## 8.4.1. Batteries and Supercapacitors

The development of supercapacitors and batteries is inherently related to electrochemical interface dynamics, in which the material of electrodes and interactions between electrolytes dictate overall performance.

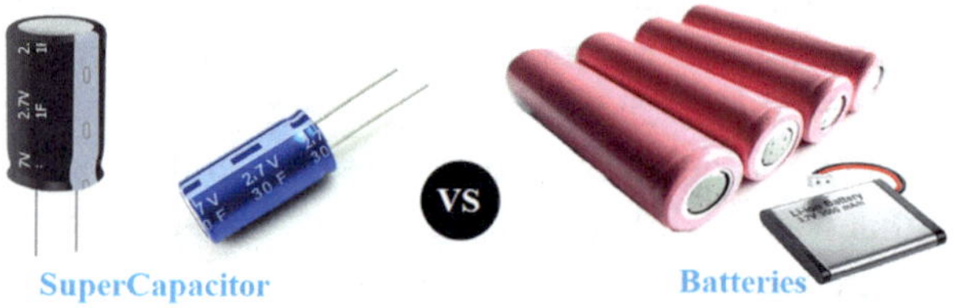


**Figure 8.10.** Batteries and supercapacitors.

*Source:* https://circuitdigest.com/sites/default/files/projectimage_tut/Battery-VS-Supercapacitor.png.

Electrochemical interfaces are key sites for charge transfer, and the nature of such interfaces contributes significantly to energy density, charge/discharge rates, and overall cycling stability of energy storage devices. For instance, the use of advanced materials, primarily carbon hybrids, has shown great potential in enhancing charge transfer efficiency. Functionalization of electrode surfaces may be crucial, with surface morphology and chemical composition tuned to enhance interactions at such surfaces, leading to improved performance parameters that characterize modern batteries and supercapacitors. Besides that, insights derived from ILs research have also transformed the conceptualization of interfacial phenomena in these devices. Studies on ILs' interactions with carbonaceous electrodes have demonstrated important mechanisms for electrochemical reactions. Utilizing techniques such as electrochemical scanning tunneling microscopy (EC-STM), scientists have illustrated how differences in alkyl chain lengths of imidazolium-type ILs affect cation intercalation and the structural stability of the electrode surface (Figure 8.10).

The findings indicate that longer alkyl chains increase van der Waals contacts with the graphite surface, influencing charge storage processes and, in the long run, the efficiency of capacitive energy storage. Such fundamental insight not only extends theoretical frameworks but also has design implications for more efficient rechargeable batteries and supercapacitors and indicates the multidisciplinary nature of electrochemical interfaces. In addition, the embedding of flexible, self-powered energy modules that integrate solar charging and energy storage holds promising prospects in wearable and mobile technologies. As one may observe in recent studies, innovative designs involving printed magnesium ion aqueous asymmetric supercapacitors reveal remarkable mechanical toughness and cycling stability. The optimality of such units serves to emphasize the significance of electrochemical interface optimization because effective communication between energy storage and conversion devices results in optimized performance indicators, including energy density and capacitance retention. Such enhancements serve to prove the viability of designing sustainable efficient power sources that address growing demands for portable energy solutions, further underscoring the significance of electrochemical interfaces in future battery and supercapacitor development.

### 8.4.2. Fuel Cells

Fuel cells provide the clearest indication of the importance of electrochemical interfaces in energy conversion technology. Fuel cells effectively convert chemical energy to electrical energy through electrochemical reactions, that is, hydrogen and oxygen. The efficiency of a fuel cell is largely determined by the capability of its electrolyte/electrode interfaces, where judicious reactions such as hydrogen oxidation and oxygen reduction occur. Overpotentials at such interfaces have significant impacts on deciding the overall efficiency of fuel cells. Scientists are able to measure such overpotentials in solid-oxide fuel cells using spatially-resolved X-ray photoemission spectroscopy, getting an insight into the electro-catalytic efficiencies of various electrode materials, such as nickel and platinum. These advances enhance our understanding of the reaction kinetics and eventually result in the development of more efficient fuel cells that are required for the next-generation clean energy technologies (Adler et al., 2010) (Figure 8.11).

**Figure 8.11.** General concept of a ($H_2$–$O_2$) fuel cell.

*Source:* https://www.researchgate.net/publication/318669013_Fuel_cells_for_electrochemical_energy_conversion.

The fuel cell interfaces are not only crucial to the electrochemical reaction but also very difficult to characterize. It is essential to disentangle the microscopic processes that take place in these interfaces using advanced techniques to probe the buried dynamics that dictate the performance. Traditional methods like electrochemical impedance spectroscopy offer valuable information on the charge transfer processes but possibly don't capture the full range of interfacial phenomena. New technologies, such as infrared spectroscopy, provide an avenue to bypass these issues by sensing molecular fingerprints at the electrochemical interfaces. Tailoring materials like graphene into these spectroscopic measurements increases sensitivity and selectivity, allowing scientists to monitor chemical species activity in real time.

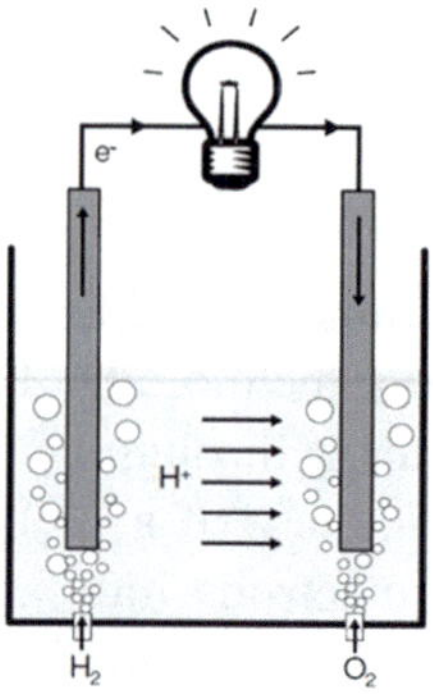


**Figure 8.12.** A simple fuel cell.

*Source:* https://www.researchgate.net/publication/318669013_Fuel_cells_for_electrochemical_energy_conversion.

This capacity to interrogate concealed electrochemical interfaces not only serves to optimize fuel cell design but improves the understanding of underlying electrochemical principles as well (Bie et al., 2015). Furthermore, recent innovations in fuel cell interface characterization are poised to find applicability in other fields of research, including electrocatalysis and energy storage. By a deep understanding of the reaction mechanisms at electrode surfaces, principles from research on fuel cells could be applied to maximize catalytic processes for various purposes. For instance, information gained through the observation of electrode materials' behavior within fuel cells can assist in creating more efficient catalysts for water splitting or carbon dioxide reduction. This interrelatedness highlights the significance of electrochemical interfaces in energy technology disciplines. As investigators continue to refine method and their knowledge base of these pivotal interfaces, opportunities for technological innovation toward greater fuel cell efficiency and longer-term applications of energy technology move nearer to realization. All endeavors toward an energy future powered by alternative technologies necessarily depend on continued advancement in electrochemical interface science (Adler et al., 2010; Bie et al., 2015) (Figure 8.12).

## 8.4.3. Solar Energy Conversion

The conversion of solar energy to useful power is an integral aspect of modern energy solutions, particularly via electrochemical interfaces. Through these interfaces, vital functions such as charge separation and transfer become possible, allowing technologies such as fuel cells and solar fuels to be more efficient. One of the most prevalent methods is via the electrochemical dissociation of water into hydrogen and oxygen, with the resultant recombination producing electricity while emitting only water as a by-product. However, the performance of these systems is hindered by kinetic constraints inherent in oxygen-based electrochemical reactions. The exploration of metal oxides as electrocatalysts has been prompted by their low cost, stability, and inherent capacity for storage of oxygen, promoting elevated reaction rates and, consequently, more efficient overall efforts for solar energy conversion, as unveiled in research on electrochemical devices (Bluhm et al., 2015). In solar energy conversion, dye-sensitized solar cells (DSSCs) are a groundbreaking approach which unites electrochemical interfaces with great effectiveness. Ever since ORegan and Grätzel first discovered DSSCs in 1991, the technology has manifested a promising future in comparison to traditional silicon-based photovoltaics, but just for low-power systems. Though DSSCs are less efficient than their silicon-based counterparts, compatibility with a less expensive and simpler fabrication process positions them well for commercial use. Significantly, the efficiency of these systems is very sensitive to nanostructures employed in anodes and cathodes. The past decades' progress of materials history, particularly in p-type and n-type semiconductors, stresses the potential for progress in efficiency along with the expanding horizon of DSSCs' energy conversion applications, referring to their pivotal role in sustainable energy technologies (Bonomo et al., 2017). For enhancing solar energy conversion through electrochemical principles, present research assigns great value to advanced characterization techniques at electrochemical interfaces. Investigation of the physical and chemical phenomena occurring in situ at such interfaces is at the heart of the breaking down of the complexity of energy conversion systems. The construction and use of new in situ characterization tools can provide dramatic insight into mechanisms associated with charge separation and transfer. These insights are not only critical to elucidating the underlying processes that govern the operating behavior of electrochemical reactors but also crucial for materials and design breakthroughs. By paying close attention to the processes at such interfaces, science can offer routes to high-performance systems with the promise of greater efficiency in converting solar energy, ultimately enabling a sustainable energy future (Bluhm et al., 2015) (Figure 8.13).

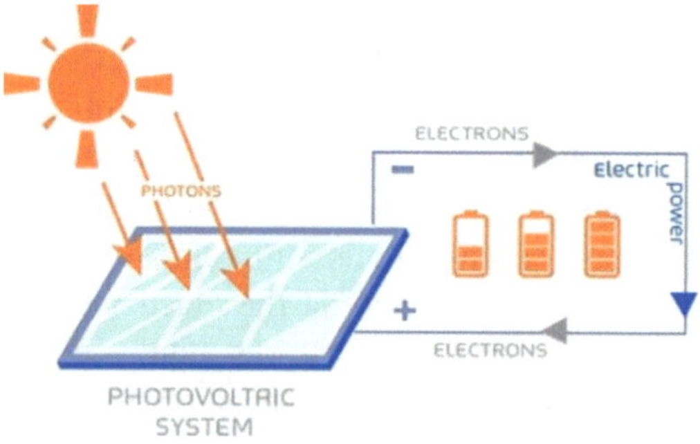


**Figure 8.13.** Solar energy conversion.

*Source:* https://cdn.prod.website-files.com/65b217df08a6f1f6aba3d4ce/681b30f43c936079d3ea0002_pasted-image-0--5--compressed--8-.jpeg.

## 8.5. ROLE IN CORROSION SCIENCE

Corrosion science is one of the important areas of study of the fundamental interactions involved at electrochemical interfaces. Corrosion represents the degradation of materials, primarily metals, by electrochemical reaction under conditions of exposure to environmental surroundings such as moisture and contaminants. The anodic and cathodic area distribution on a metal surface is a central feature. This is particularly observed under conditions where localized corrosion occurs and is a factor of the different potentials of the electrochemical reactions. Stochastic cellular automata (CA) modeling approach, which involves some diffusion principles as well as acid-base neutralization, assists in explaining how the reactions are spatially separated to induce corrosion rate fluctuations. Such modeling offers an insight into the conditions under which corrosion becomes localized or generalized, thus enhancing our understanding of the corrosion process as a dynamic interplay between environmental factors and electrochemical activity (Larbi et al., 2017). Furthermore, the morphology of materials significantly affects their corrosion behavior on electrochemical interfaces. Surface roughness and the existence of passive oxide layers are just a few of the factors that define to what degree a material is prone to corrosion. Passive layers, for instance, may work to create protective layers that alter the kinetics of electrochemical reactions, slowing down corrosion rates. However, failure of protective layers will trigger accelerated corrosion processes, resulting in massive material degradation. This combination of material structure and electrochemical dynamics makes clear the importance of surface engineering within corrosion science. By manipulating the surfaces of metals in such a manner as to optimize their passive behavior, scientists can produce more effective prevention strategies for corrosion within various environments, making clear the central role morphology plays in the broader scheme of electrochemical interfaces (Larbi et al., 2017). Modern advances in materials science have introduced new corrosion protection solutions, most significant of which are the applications of graphene and boron nitride (BN) nanosheets. Even though graphene is very impermeable to oxidation, it unintentionally creates galvanic cells that promote substrate corrosion. BN nanosheets, with their greater thermal and chemical stability and electrical insulation properties, however, are a more appropriate option to prevent corrosion. Experiments confirm that BN-coated metals show significantly reduced oxidation under extreme conditions compared to uncoated metal, highlighting performance benefits of the materials. The electrochemical tests not only verify enhanced open circuit potential for BN-coated surfaces but also suggest that protective efficacy relies on the nanosheets' quality. This information emphasizes the need for ongoing research to develop better protective coatings and achieve their purpose in preventing corrosion in electrochemical interfaces.

## 8.5.1. Mechanisms of Corrosion at Interfaces

Mechanisms of corrosion at interfaces is a complicated process of great significance in defining material degradation in electrochemical systems. Corrosion may assume diverse forms, but behind all lies the interaction between anodic and cathodic reaction at localized spots on a metal surface. In the course of these reactions, the resulting spatial separation of anodic and cathodic zones leads to micro-environments that significantly affect rates of corrosion. For instance, localized corrosion can be caused by the presence of disturbed passive layers, leading to enhanced dissolution at anodic sites with cathodic activity elsewhere. This dynamic response underscores the significance of studying and modeling such interactions since they are central to long-term material performance in many applications, from infrastructure to biomedical devices, where electrochemical interfaces are of crucial significance (Larbi et al., 2017). Moreover, the ionic and species transport mechanisms in electrochemical interfaces are the foundation of the corrosion process. Diffusion in the control of the movement of the reactants and products within the electrolyte is essential in the definition of how corrosion proceeds. Localized corrosion due to the weakening of passive oxide layers can, again, be induced by the accelerated local acidification, revealing how microenvironmental changes can influence corrosion kinetics. The control of such transport processes is essential to the optimization of efficient corrosion-resistance schemes.

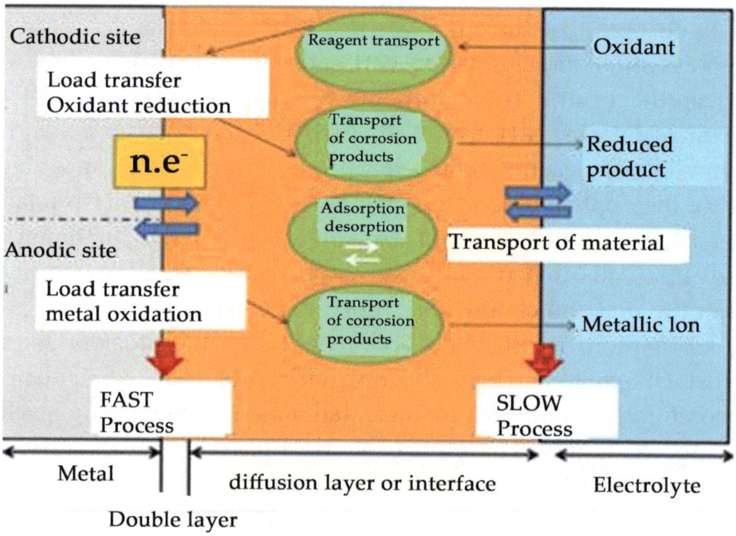


**Figure 8.14.** Different mechanisms taking place at the interface during the corrosion of a metal in a liquid medium.

*Source:* https://www.researchgate.net/figure/Different-mechanisms-taking-place-at-the-interface-during-the-corrosion-of-a-metal-in-a_fig4_344328152.

Understanding how material morphology influences such kinetic parameters, researchers can engineer better materials and coating systems to resist corrosion challenges, thereby ensuring extended lifespan and reliability of components exposed to hostile environments (Cauwe et al., 2019). In the context of solution creation for avoiding corrosion, the application of new materials like barrier layers has become increasingly significant. The barrier layers are used to cover metals and provide a second barrier against electrochemical attack.

For instance, work on atomic layer deposition (ALD) illustrates how specific barrier layers can significantly alter the functioning of implantable medical devices by providing long-term corrosion protection against corrosive agents. These advances not only enhance device reliability under physiologically relevant conditions but also open up inroads to optimized corrosion test practices. Through the presentation of the highest conceivable temperatures for accelerated tests, as seen in research on ALD-coated barriers, an increased insight into mechanisms of degradation can be achieved, which translates into more effective design strategies in corrosion-vulnerable applications (Cauwe et al., 2019) (Figure 8.14).

## 8.5.2. Protective Coatings and Their Effectiveness

The effectiveness of protective coatings to keep metals away from corrosion and oxidation is largely a function of their composition and structure. New advances have recognized materials that are more effective than usual coatings, including the like of boron nitride (BN) nanosheets. BN nanosheets, in addition to being very impermeable, therefore efficiently blocking corrosion, are also very thermally and chemically stable. Notably, unlike graphene, BN itself will not create galvanic cells with the substrate metal and therefore will not unintentionally lead to corrosion over time. Experiments show that copper foil covered with BN nanosheets shows much less oxidation upon heating in air for long periods of time. Such findings emphasize the importance of selecting proper coating material in order to achieve maximum corrosion protection and metal substrate longevity, especially in oxidative stress conditions (Chen et al., 2014). In infrastructure, and more precisely in oil and gas pipelines, the role of protective coatings is essential in order to guarantee pipeline integrity. Steel pipes, usually buried in the ground, are protected by a combination of coatings like above and cathodic protection (CP) systems. Deterioration of the external surfaces, to a great extent, by coating failure or lack of CP can lead to catastrophic failure and huge loss of money. There is a need, therefore, for a thorough understanding of the corrosion processes under multifaceted conditions. Various techniques, such as electrochemical impedance spectroscopy and polarization analysis, are applied to monitor coating performance and evaluate corrosion damage in real time. This integrated approach not only increases reliability in pipeline operation but also facilitates the development of predictive maintenance strategies critical for modern energy infrastructure (Forsyth et al., 2013). The effectiveness of protective coatings is also established by how they interact with electrochemical interfaces, necessary to understand corrosion behavior. The interfaces determine how the protective films react to conditions, particularly moisture and electrolytes. The performance of coatings therefore relies on their intrinsic material behavior in addition to their ability to form stable electrochemical interfaces that can withstand corrosive attack. For instance, while BN nanosheets create a robust barrier against oxidants, their protective time can be influenced by the electrochemical conditions under which they operate. Hence, ongoing research is directed at streamlining coatings in a way that they become electrochemically compatible with various substrates and environmental conditions. This entails studying the correlation of coating properties with electrochemical reactivity, which subsequently results in developing more effective protective systems that cater to the demands of different industrial applications (Chen et al., 2014) (Figure 8.15).

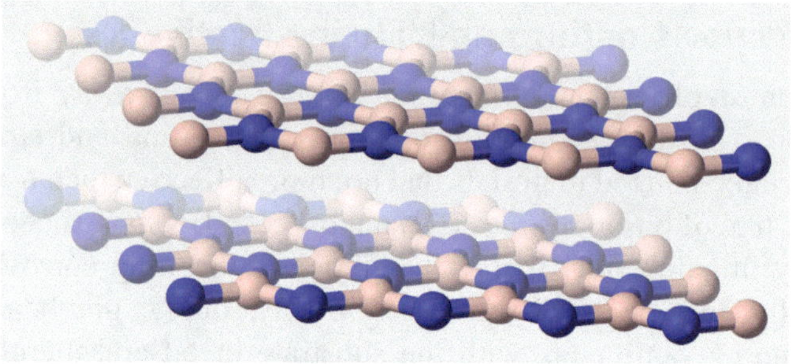

**Figure 8.15.** Boron nitride nanosheet.

*Source:* https://upload.wikimedia.org/wikipedia/commons/f/fa/Boron-nitride-%28hexagonal%29-side-3D-balls.png.

### 8.5.3. Strategies for Corrosion Prevention

Prevention of corrosion is crucial in providing the integrity and life of metallic structures, particularly in industries where environmental exposure accelerates degradation. The best approach among these involves the utilization of protective coatings. Protective coatings offer a covering between corrosive materials and the metal substrate, thus reducing oxidation as well as damage. Improved advances in sol-gel have gained attention for its potential to prepare hybrid coatings combining organic and inorganic elements, thereby achieving optimal mechanical, optical, and anticorrosive properties. Such coatings can be developed for specific application, for example, safeguarding major components of various industries, from construction to electronics. However, successful application of these novel materials is based on deep knowledge regarding their chemistry and the factors influencing their performance in field conditions, and thus research in this area is required for effective corrosion mitigation. In addition to coatings, another extremely promising method for avoiding corrosion is the use of cathodic protection systems. This method most often involves the flow of an electric current onto a metal surface, maintaining it in a lowered state and therefore less susceptible to oxidation. Cathodic protection can be divided into two general categories: impressed current and galvanic systems. The first system employs a power source external to the vulnerable metal that continuously feeds it electrons, whereas galvanic systems employ sacrificial anodes of more reactive metals. Both systems have been effective in protecting pipelines, storage tanks, and offshore structures. While their success, deployment of cathodic protection systems needs to be wisely employed relative to the electrochemical interfaces utilized and the specific environments in which these systems are deployed. Ongoing research for enhancing these systems is essential to optimizing the durability and safety of metal structures (Baj-Rossi et al., 2012). Lastly, incorporation of monitoring technologies into corrosion protection programs provides a fascinating new horizon in asset management. Real-time monitoring of corrosion indicators is enabled by continuous monitoring with the aid of microelectronics and biosensor-based technologies. This enables the development of totally integrated devices that monitor corrosion indicators in real time. The ability can enable better decision-making processes in terms of timing for maintenance and resource management, leading ultimately to cost savings through repair and replacements. Additionally, the application of such technologies becomes increasingly viable in sophisticated applications, especially in the healthcare and critical infrastructure sectors, where corrosion can have gravely damaging consequences. As the technologies improve, the need for effective electrochemical interfaces becomes all the more significant, which further calls for ongoing research and development in developing corrosion resistance over the long term (Baj-Rossi et al., 2012).

## SUMMARY

Electrochemical interfaces research reveals an interdisciplinary terrain where materials such as carbon nanostructures and synthetic diamond are paving the way for next-generation biosensing devices. The remarkable properties of carbon nanotubes, such as their high electrochemical reactivity as well as sensitivity to surface adsorbates, have made them the leader in the development of next-generation electrochemical sensors. These sensors have a broad range of applications from medical diagnosis to ecological analysis, as seen in their high sensitivity in detecting viral infections. Continued improvement in carbon-based materials not only optimizes these sensors but also provides the door to real-time monitoring, with potentially profound effects in the fields of biomedicine and nanotechnology (Davidson et al., 2009). In addition, recent advancements in the knowledge base of electrochemical activity and electrode structure underscore the necessity of catalyst optimization for high-efficiency energy utilization, particularly in solar technology. An investigation into the electrochemical characteristics of various crystallographic orientations reveals that various facets manifest various catalytic activities, which have dramatic effects on device performance, e.g., dye-sensitized solar cells. The application of methods such as scanning electrochemical cell microscopy enables researchers to detect these subtle differences in polycrystalline materials. Through the realization of electrochemical heterogeneity, further research can engage in optimizing electrode materials to enhance overall efficiency and, in the process, enhance the growing trend of renewable energy supplies (Aaronson et al., 2014). Finally, the progress in the field of electrochemical interfaces is a promising direction for both scientific investigation and technology implementation. The ability to develop and utilize materials that foster efficient electron transfer and signal transduction will remain to shape the world of electrochemical sensors and energy devices. With ongoing research devoted to optimizing these interfaces, the potential for sensor development, bioelectronics, and renewable energy is vast. Through advancements, these multi-disciplinary activities will be working fundamentally towards rectifying global issues like healthcare, the environment, and the use of energy and thereby bring to the forefront the unstoppable importance of electrochemical interfaces in future research.

Among the most significant advances in electrochemical interfaces is the application of the soft X-ray spectroscopy techniques, the mapping of resonant inelastic X-ray scattering (mRIXS) in particular. Using this technique, there is increased sensitivity for examining battery chemistry, an essential factor in the determination of interfacial reaction in energy storage systems. Soft X-ray absorption spectroscopy (sXAS) has been an important tool for exploring transition-metal states and surface chemistry for a long time but presents challenges, particularly when applied to oxygen redox reactions. Conclusions based on mRIXS can best fingerprint oxygen redox activity to rectify previous misconceptions and improve our understanding of electrode behavior in batteries. By shedding light upon such intricacies, scientists can develop plans for optimizing electrodes and enhancing the performance of batteries because the interfacial effects are instrumental in achieving effective energy solutions. This development is a significant step forward in battery science, pointing out the importance of advanced characterization techniques in unveiling the complexities of electrochemical interfaces (Devereaux et al., 2018). The other basic component in the analysis of electrochemical interfaces is examining electric double layers (EDL) over the electrode-surface film boundary. Density Functional Theory (DFT) comes through as a powerful tool here, showing how voltages applied create authoritative transformations at atomic length scales like charge splitting and interfacial dipole moments. Notably, electrodes such as $Li_3PO_4$, $Li_2CO_3$, and $Li_xMn_2O_4$ have complex interactions when subjected to voltages, under which adsorbed organic solvent molecules play an important role in computed voltages under vacuum conditions. Moreover, managing surface dipoles—a component scarcely investigated in research on batteries—is a possible scope for passivating electrodes. This subtle insight permeates the kinetics of lithium-ion insertion and parasitic electrolyte decomposition challenges, emphasizing a need for all-encompassing models capturing electrochemical behavior at the nanoscale level (Leenheer et al., 2015). Lastly, interfacial processes play an enormous role in electrochemistry since they dictate paramount events like $Li^+$ insertion dynamics and electrodeposition under different conditions. The distinction between voltage governing electron motion instantaneously and lithium cohesive energy voltage is a sophisticated concept for advancement in battery technology. The understanding of the self-consistency criterion's delayed response at interfaces versus more prompt electrochemical responses enables further sophistication in the comprehension of how batteries react under diverse conditions. This detailed comprehension is important in the overcoming of problems such as overpotentials and the misinterpretation of electrochemical activity. The exploration of such electrochemical interfaces not only highlights the importance of interfacial properties but also sparks innovations in battery design and efficiency, which ultimately pave the way for future realization of more efficient energy storage devices (Leenheer et al., 2015).

# REFERENCES


1. Aaronson, B. D. B., Lai, S. C. S., & Unwin, P. R. (2014). Spatially resolved electrochemistry in ionic liquids: Surface structure effects on triiodide reduction at platinum electrodes. https://core.ac.uk/download/29190112.pdf (accessed on 3rd October 2025).
2. Adler, A., Ahn, J., Andújar, M., McDaniel, A. H., Appleby, A. J., Bessler, W. G., Bockris, J. O., et al. (2010). Measuring individual overpotentials in an operating solid-oxide electrochemical cell. http://arxiv.org/abs/1003.0035 (accessed on 3rd October 2025).
3. Altintas, Z., & Mansuriya, B. D. (2020). Applications of graphene quantum dots in biomedical sensors. https://core.ac.uk/download/322792122.pdf (accessed on 3rd October 2025).
4. Alvarado, J., Chen, G., Doeff, M. M., et al. (2020). All-solid-state batteries using rationally designed garnet electrolyte frameworks. https://core.ac.uk/download/286728971.pdf (accessed on 3rd October 2025).
5. Andersen, T. R., Cottam, S., Elkington, D. C., et al. (2020). Manipulating nanoscale structure to control functionality in printed organic photovoltaic, transistor and bioelectronic devices. https://core.ac.uk/download/288432664.pdf (accessed on 3rd October 2025).
6. Aranda, P., Dico, G. L., Lazzara, G., Lisuzzo, L., Ruiz-Hitzky, E., & Wicklein, B. (2019). Multicomponent bionanocomposites based on clay nanoarchitectures for electrochemical devices. https://core.ac.uk/download/286528415.pdf (accessed on 3rd October 2025).
7. Baj-Rossi, C., Burleson, W., Carrara, S., et al. (2012). Fully integrated biochip platforms for advanced healthcare. https://core.ac.uk/download/286729684.pdf (accessed on 3rd October 2025).
8. Bazant, M. Z., & Fraggedakis, D. (2020). Tuning the stability of electrochemical interfaces by electron transfer reactions. http://arxiv.org/abs/2003.03057 (accessed on 3rd October 2025).
9. Bazant, M. Z., & Smith, R. B. (2017). Multiphase porous electrode theory. http://arxiv.org/abs/1702.08432 (accessed on 3rd October 2025).
10. Ben Cheikh Larbi, F., Chaussé, A., De Lamare, J., et al. (2017). Contribution of cellular automata to the understanding of corrosion phenomena. http://arxiv.org/abs/1710.01545 (accessed on 3rd October 2025).
11. Bie, Y. Q., Horng, J., Ju, L., et al. (2015). Vibrational spectroscopy at electrolyte/electrode interfaces with graphene gratings. https://core.ac.uk/download/288432924.pdf (accessed on 3rd October 2025).

12. Blanco, M., Granozzi, G., Kosmala, T., et al. (2018). Potential driven non-reactive phase transitions of ordered porphyrin molecules on iodine-modified Au(100): An electrochemical scanning tunneling microscopy (EC-STM) study. https://core.ac.uk/download/168406406.pdf (accessed on 3rd October 2025).
13. Bluhm, H., Crumlin, E. J., Hong, W. T., et al. (2015). Insights into electrochemical reactions from ambient pressure photoelectron spectroscopy. https://core.ac.uk/download/83235318.pdf (accessed on 3rd October 2025).
14. Bonomo, M., Cavallo, C., Di Pascasio, F., et al. (2017). Nanostructured semiconductor materials for dye-sensitized solar cells. https://core.ac.uk/download/54538567.pdf (accessed on 3rd October 2025).
15. Cassayre, L., Chamelot, P., Gibilaro, M., et al. (2013). Investigations of Zr(IV) in LiF–$CaF_2$: Stability with oxide ions and electroreduction pathway on inert and reactive electrodes. https://core.ac.uk/download/12044130.pdf (accessed on 3rd October 2025).
16. Cauwe, M., Li, C., Mader, L., et al. (2019). Accelerated hermeticity testing of biocompatible moisture barriers used for the encapsulation of implantable medical devices. https://core.ac.uk/download/286086386.pdf (accessed on 3rd October 2025).
17. Cauwe, M., Li, C., Mader, L., et al. (2019). Ultra-long-term reliable encapsulation using an atomic layer deposited $HfO_2$/$Al_2O_3$/$HfO_2$ triple-interlayer for biomedical implants. https://core.ac.uk/download/237010489.pdf (accessed on 3rd October 2025).
18. Cha, H., Cho, J., Kim, J., et al. (2020). Improvements to the overpotential of all-solid-state lithium-ion batteries during the past ten years. https://core.ac.uk/download/323384107.pdf (accessed on 3rd October 2025).
19. Chen, Y., Jones, R., Li, L. H., et al. (2014). Boron nitride nanosheets for metal protection. http://arxiv.org/abs/1503.01295 (accessed on 3rd October 2025).
20. Colburn, A. W., Bode, A. A. C., Bode, A. A. C., Burgess, J., Charola, E., Chen, C. C., Chen, C. C., et al. (2013). Dual-barrel conductance micropipette as a new approach to the study of ionic crystal dissolution kinetics. https://core.ac.uk/download/19553440.pdf (accessed on 3rd October 2025).
21. Crabtree, G. (2015). The Joint Center for Energy Storage Research: A new paradigm for battery research and development. http://arxiv.org/abs/1411.7042 (accessed on 3rd October 2025).

22. Davidson, J. L., Gurbuz, Y., Gürbüz, Y., et al. (2009). Review on carbon-derived, solid-state, micro and nano sensors for electrochemical sensing applications. https://core.ac.uk/download/pdf/11741398.pdf (accessed on 3rd October 2025).

23. Dayeh, S. A., Liu, Y., Vishniakou, S., et al. (2015). Engineering metamaterials to control lithium-ion transport pathways. https://core.ac.uk/download/323064259.pdf (accessed on 3rd October 2025).

24. Devereaux, T. P., & Yang, W. (2018). Anionic and cationic redox and interfaces in batteries: Advances from soft X-ray absorption spectroscopy to resonant inelastic scattering. https://core.ac.uk/download/287622621.pdf (accessed on 3rd October 2025).

25. Eren, B., Fang, H. T., Feng, B., et al. (2019). Operando STM study of the interaction of imidazolium-based ionic liquid with graphite. https://core.ac.uk/download/287621982.pdf (accessed on 3rd October 2025).

26. Feinauer, J., Hein, S., Iliev, O., et al. (2018). MULTIBAT: Unified workflow for fast electrochemical 3D simulations of lithium-ion cells combining virtual stochastic microstructures, electrochemical degradation models and model order reduction. http://arxiv.org/abs/1704.04139 (accessed on 3rd October 2025).

27. Figueira, R. M. B. M., Fontinha, I. R., Pereira, E. V., et al. (2016). Hybrid sol-gel coatings: Smart and green materials for corrosion mitigation. https://core.ac.uk/download/143977354.pdf (accessed on 3rd October 2025).

28. Forsyth, M., Gupta, R. K., Hinton, B. R. W., et al. (2013). An overview of methods for simulating and evaluating pipeline corrosion. http://hdl.handle.net/10536/DRO/DU:30069099 (accessed on 3rd October 2025).

29. Güell, A. G., Zhang, G., Unwin, P. R., Patten, H. V., & Schmickler, W. (2016). Nanoscale electrochemistry of $sp^2$ carbon materials: From graphite and graphene to carbon nanotubes. https://core.ac.uk/download/293881802.pdf (accessed on 3rd October 2025).

30. Haile, S. M., Hightower, A., & Louie, M. W. (2010). Nanoscale electrodes by conducting atomic force microscopy: Oxygen reduction kinetics at the Pt|$CsHSO_4$ interface. https://core.ac.uk/download/4884969.pdf (accessed on 3rd October 2025).

31. Jirka, S., Pearlman, J. S., & del Río Fernandez, J. (2016). Oceans of Tomorrow sensor interoperability for *in-situ* ocean monitoring. https://core.ac.uk/download/81579673.pdf (accessed on 3rd October 2025).

32. Kaner, R. B., Liu, Z., Shao, Y., et al. (2019). Printable magnesium ion quasi-solid-state asymmetric supercapacitors for flexible solar-charging

integrated units. https://core.ac.uk/download/323077575.pdf (accessed on 3rd October 2025).

33. Leenheer, A., & Leung, K. (2015). How voltage drops are manifested by lithium-ion configurations at interfaces and in thin films on battery electrodes. http://arxiv.org/abs/1507.02334 (accessed on 3rd October 2025).
34. Léger, A., Léger, C., Léger, C., Barkey, D., Grier, D., Kessler, D. A., Grier, D. G., et al. (2003). A mean-field kinetic lattice gas model of electrochemical cells. http://arxiv.org/abs/cond-mat/0303072 (accessed on 3rd October 2025).
35. Lewis, N. S., Simpson, B. H., & Thompson, A. C. (2020). Macroscale and nanoscale photoelectrochemical behavior of p-type Si (111) covered by a single layer of graphene or hexagonal boron nitride. https://core.ac.uk/download/287633393.pdf (accessed on 3rd October 2025).
36. Shin, J., Lee, J., Park, Y., & Choi, J. W. (2020). Aqueous zinc ion batteries: Focus on zinc metal anodes. Chemical Science, 11, 2028–2044. https://doi.org/10.1039/d0sc00022a (accessed on 3rd October 2025).
37. Zheng, Y., Vasileff, A., Zhou, X., Jiao, Y., Jaroniec, M., & Qiao, S. Z. (2019). Understanding the roadmap for electrochemical reduction of $CO_2$ to multi-carbon oxygenates and hydrocarbons on copper-based catalysts. Journal of the American Chemical Society, 141, 7646–7659. https://doi.org/10.1021/jacs.9b02124 (accessed on 3rd October 2025).

# CHAPTER 9

# CHARACTERIZATION TECHNIQUES FOR SURFACES

## Contents



## INTRODUCTION

The study of surface characterization techniques is a basic field of inquiry in a number of scientific fields, from materials science through chemistry to engineering. Understanding the properties and characteristics of surfaces is crucial because it makes it possible to tailor the performance and functionality of materials. Surfaces are critical in this respect because they directly engage with the environment, and they determine phenomena like adhesion, corrosion, and catalysis. Characterization techniques are excellent tools to probe surface properties at the nanoscale and microscale, providing a better vision of material behavior. As the demand for improved materials increases, particularly in areas like nanotechnology and renewable energy, the demand for effective surface characterization techniques increases, which emphasizes the need for continued improvement and innovation in these techniques. There are various methods of characterizing surfaces, each appropriate to meet various aspects of surface property, like topography, composition, and electronic structure. Some of the most widely used techniques for high-resolution imaging of surface morphology are scanning electron microscopy (SEM) and atomic force microscopy (AFM), which provide information on surface roughness and texture. In addition, spectroscopic methods such as Raman spectroscopy and X-ray photoelectron spectroscopy (XPS) yield valuable information regarding chemical composition and molecular interactions at surfaces. Of specific interest is the use of soft x-ray and electron spectroscopy since it enables the investigation of chemical and electronic properties at surfaces and interfaces, hence constituting a comprehensive understanding of surface properties and interactions. With the scientific community working towards creating newer characterization techniques, the window of opportunity for the discovery of new materials and applications widens. The merging of surface characterization with technological advancements has created much improved device performance, especially in devices like photovoltaics. Thin-film solar cells, for instance, require careful characterization techniques to make them work and be effective. This is achieved through methods such as capacitance spectroscopy and photoluminescence analysis, which quantify electrical properties and material characteristics essential to the effectiveness of energy conversion. Furthermore, developments in data modeling and simulation provide greater possibility to predict material response based on surface characteristics observed. As suggested in recent research where polymeric microspheres with surface-modified properties exhibited significant ion exchange characteristics, the inclusion of characterization techniques leads to the developed materials that effectively address environmental challenges. Ultimately, continuous development and application of these techniques remain essential to the success of scientific progress and technological innovation in surface characterization.

## *Definition of Surface Characterization*

Surface characterization is a significant material science that provides intrinsic data regarding the physical and chemical properties of surfaces. Surface characterization comprises various techniques charged with shedding light on surface topography, composition, and material properties. By observing the surface in the regimes of microscales and nanoscales, researchers are capable of establishing essential features such as texture, roughness, and chemical reactivity, which significantly contribute to a functionality of materials in applications. Surface characterization techniques are not merely essential in material development but also play a pivotal role in quality control and production process optimization for manufacturing, such that surfaces meet specified performance criteria for target applications, including coatings, electronics, and biomaterials. Multiomic analysis, which incorporates genomic and proteomic data for the determination of surface marker profiles of cells, is one of the most persuasive surface characterization techniques. For instance, this method has been successfully utilized in research on chronic myeloid leukemia (CML) stem cells and has made important findings regarding cell surface characteristics differentiating leukemic cells from normal counterparts. Identification of aberrant surface markers like TIM3 via comprehensive multiomic analysis underscores the intricacy of cellular surfaces and their relevance in disease pathology. Such detailed analysis not only enables better understanding of the physical properties of the cells but also leaves space for targeted therapies, demonstrating the relevance of surface characterization techniques in the development of medical science and therapeutic approaches (Nilsson et al., 2023).

In addition to biological applications, surface characterization is highly crucial in engineering fields, such as civil and materials engineering. For instance, aggregates' morphology and texture in pavement materials play a critical role in their long-term performance. In the past, the determination of these attributes was carried out using subjective, manual methods with inefficiency and variability. Advances in digital image processing and smartphones brought about solutions for rapid and standardized morphological description of aggregates, which solved the limitations of traditional methods. By accurately differentiating different aggregate shapes and textures using computational techniques, new criteria can be established to improve the performance of materials in transport infrastructures (Heidelberg et al., 2020). This illustrates how surface characterization is not only important for the characterization of material properties but also for real-world applications across broad disciplines.

## *Importance in Material Science and Engineering*

In materials science and engineering, an understanding of the properties and response of materials at the molecular and macroscopic level is essential to innovation and use. Recent advances in characterization techniques enable scientists to probe material surfaces with unprecedented resolution, providing valuable information on their structural, chemical, and physical response. Some of these innovations are confocal laser scanning microscopy (CLSM), which has been very important in allowing non-invasive imaging that is capable of recording reaction processes and surface structures in real time. Such a degree of scrutiny is crucial when developing new materials or enhancing existing ones, for it becomes possible to detect defects as well as probe surface interactions influencing performance and lifetime. Thus, effective characterization has direct implications for the optimization of material properties, rendering engineered materials suitable for a specific application, and minimizing failures in service, which may be economically and socially burdensome. Furthermore, the integration of new characterization technologies into materials science has far-reaching implications for environmental sustainability and recycling practices. The global increase in plastic waste, for example, emphasizes the urgent need for efficient biodegradation technologies. Biocatalytic processes involved in plastic degradation can benefit immensely from the development of characterization methods to identify effective enzymatic processes for degrading permanent materials.

Through precise methods of measuring the effectiveness of enzymes and their compatibility with plastics, researchers can enhance biocatalysts and address the emerging environmental problem with plastic waste. By advancements in characterization in this field, researchers not only become capable of optimizing material properties but also contribute to propelling sustainable approaches through novel recycling technologies, which create important fields for minimizing environmental impacts, facilitating waste valorization, and defining a circular economy. In addition, the intended deployment of characterization tools will be a key driver of interdisciplinary collaboration between material science and engineering.

The integration of chemistry, physics, and biology wisdom by employing sophisticated characterization methodologies can lead to new materials having tailored properties for specific applications. For example, CLSM not only yields surface characterization but also allows the visualization of dynamic processes that occur at interfaces, such as chemical reactions or biological interactions. Such information can trigger innovation in applications from biomedical engineering to electronic materials. With researchers increasingly using techniques like those detailed in the referenced works, they create an environment of work in which competence and knowledge can cross time lines to make promising innovations in engineering materials of tomorrow to meet advanced technology and sustainability issues. This ultimately maximizes the validity and efficiency of material creation within academic and commercial circles.

# 9.1. OPTICAL MICROSCOPY

Optical microscopy is a simple tool in surface characterization that gives a wide range of insight into material morphology at various scales. As it uses visible light to light samples, optical microscopy gives researchers the option to observe features from submicron to micrometer scales. This versatility is extremely useful in materials science and biology where accurate visualization of surface features is paramount. Moreover, progress in optical microscopy techniques, such as brightfield, fluorescence, and differential interference contrast (DIC) microscopy, has broadened its use from mere imaging to dynamic examination of surface interaction and chemical character. These methods allow real-time monitoring of biological processes, structural dynamics, and the assessment of surface topographies, confirming optical microscopys focal place in contemporary research practice. While powerful, traditional optical microscopy is resolution-limited by the diffraction limit, which restricts visualization of objects less than approximately 200 nm in size.

This has led to the development of advanced methodologies such as near-field scanning optical microscopy (NSOM) and infrared scattering scanning near-field optical microscopy (IR-sSNOM). Such methods take advantage of the dynamics in evanescent wave interactions and nonlinear optical effects to achieve diffraction-limited resolutions. For instance, IR-sSNOM facilitates simultaneous chemical and topographic nanoscale characterization. Such an approach resolves complexities in analyzing polymer materials, as seen in Matěj et al. (2024), to provide close-up information on chemical composition and microstructural differences. Visualization of heterogeneities at the surface is a significant step towards understanding and enhancing the properties of materials and thereby the capabilities of optical microscopy for surface characterization. Apart from that, compatibility with other supporting analytical techniques further enhances its efficiency in surface characterization. For instance, the combination of optical microscopy with spectroscopic methods like Raman microscopy and Fourier transform infrared (FTIR) microscopy facilitates global investigations on the chemical and physical properties of surfaces. The hybrid methods not only provide morphology but also determine chemical compositions, phase distribution, and surface functional groups, as reported in Matěj et al. (2024).

This multi-parametric strategy is particularly useful in the characterization of modified polymer membranes or engineered materials, in which a high-resolution understanding of surface properties is required for optimizing performance under novel applications, including biomedicine and nanotechnology. Thus, the development of optical microscopy as a broad-spectrum tool, supplemented by other methods of characterization, highlights its importance in improving our understanding of complex surface interactions at micro and nanoscale levels (Figure 9.1).

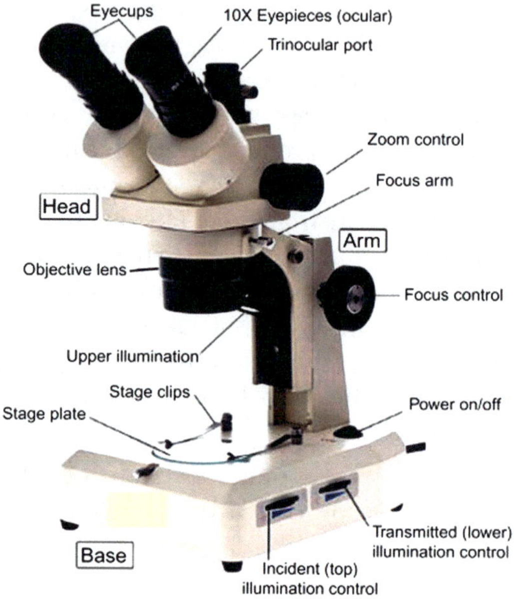


**Figure 9.1.** The basic of optical microscopy.

*Source:* https://ars.els-cdn.com/content/image/3-s2.0-B9780323264358000046-f04-03-9780323264358.jpg.

### 9.1.1. Principles of Optical Microscopy

Optical microscopy is a simple technique applied in surface characterization with details on microstructural properties of materials. Its principles are founded on the manipulation of light with the use of lenses to enlarge specimens, allowing observations as tiny as at the micro and nano levels. The sample is being illuminated by the source of light, typically a lamp or a laser, and varying configurations of objectives form the image with higher resolution and contrast. The modularity of this method is particularly significant in the study of biological samples and materials science since it enables the scientists to view surface properties and behavior that are relevant to many applications. Along with the progressive advancements in optical microscopy, recent methods such as phase contrast and fluorescence microscopy have been developed that have increased the potency of this versatile tool for surface characterization (Díaz et al., 2012). It should be noted that modern optical microscopy techniques have adapted to the needs of detailed sample analysis, particularly in biomaterials and nanotechnology. The advent of super-resolution microscopy shattered the diffraction limit, allowing scientists to resolve among slender details of structures that were once thought of as indistinguishable. This improved ability is crucial in probing surface behaviors, such as cell adhesion on biomimetic scaffolds, which is extremely relevant in tissue engineering applications. Measurement of adhesion strength accurately forms the core of fundamental biological research and biomedical design. Optical microscopy can be used to analyze cell-surface interactions, with information on the responses of cells towards any material interfaces, thus establishing the design and functionality of new biomaterials (Gerecsei et al., 2019). The technological uses of optical microscopy to profile surfaces are extensive, contributing to beneficial advancements in all spheres. Optical microscopy is an essential tool for quality control and failure analysis in materials engineering, helping evaluate surface defects, roughness, and coatings. Furthermore, by integrating imaging with examination, e.g., spectroscopy, researchers can be given complementary information regarding surface chemistry and morphology. Ongoing development of optical techniques is anticipated to enhance sensitivity, speed, and ability to examine dynamic processes in real time. Even as scientists pursue the application of these developments, the principles of light microscopy remain at the heart of providing innovative solutions to scientific inquiry and real-world surface characterization applications (Díaz et al., 2012).

## 9.1.2. Types of Optical Microscopy Techniques

Optical microscopy techniques have significant application in the meticulous examination of surfaces in a large number of scientific disciplines. Bright Field Microscopy is one of the most standard forms of optical microscopy in which specimens are illuminated under visible light. The primary advantage of the technique is simplicity, through which one can observe contrast differences as a result of various thicknesses and refractive indices of the sample. This mode is better in cell structure and tissue pattern analysis and is thus crucial in biological and medical sciences. Its resolvable limit is, nonetheless, restricted by the wavelength of light, making analysis of finer surface details complicated. Thus, while Bright Field Microscopy is a baseline method, researchers prefer to investigate more advanced techniques in high-precision surface characterization, as indicated by the recently occurring trends in Optical Microscopy applied in Materials, with continuous advancement of such techniques in scientific studies (Díaz et al., 2012) (Figure 9.2).

In contrast, Phase Contrast Microscopy shows more about transparent samples by converting phase variations in light that passes through the sample to amplitude variations, producing improved contrast without the use of stains. This makes it particularly well-suited for live-cell imaging, where live functional dynamics are observable in real time. Furthermore, this technique allows specimens usually hard to study with conventional microscopy methods to be observed, making the reach of the field extendable to fine biological structures. Nonetheless, the optimal results rely on the right calibration as the technique is very sensitive to variations in the optical pathway.

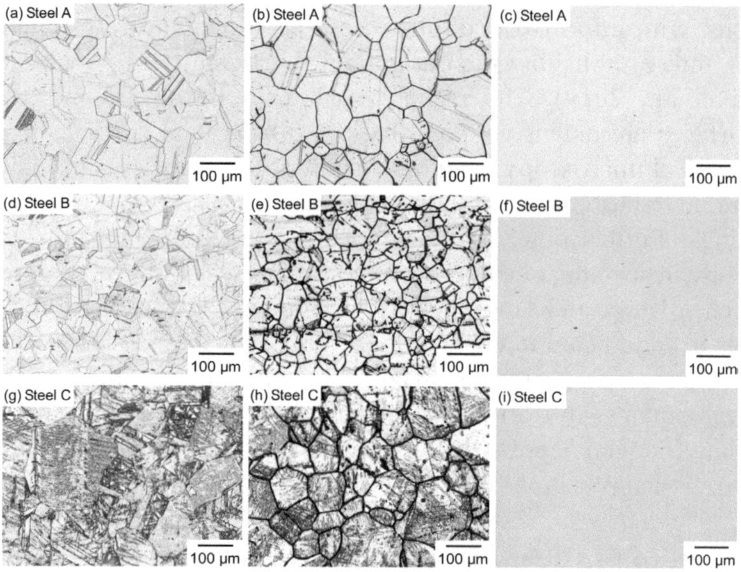


**Figure 9.2.** Optical microscope images of specimen surfaces: (a–c) steel (A); (d–f) steel (B); and (g–i) steel (C).

*Source:* https://www.researchgate.net/figure/Optical-microscope-images-of-specimen-surfaces-a-c-steel-A-d-f-steel-B-and-g-i_fig2_338807821.

Researchers acknowledge the capability of Phase Contrast Microscopy in providing information on live samples without interfering with their natural environments, thus establishing its value as a complex surface characterization method (Díaz et al., 2012). Differential Interference Contrast (DIC) Microscopy is yet another significant optical microscopy technique whose contrast is maximized by two beams of polarized light interfering with each other while traveling through a sample. This technique is especially useful for imaging thin and transparent samples because it generates a pseudo-three-dimensional impression which enhances topographical contrast making differences more evident. The characteristic contrast produced by DIC enhances visualization of cell membranes, organelles, and other subcellular structures, thus providing a more detailed understanding of their functional context. These capabilities demonstrate the integration of advanced optics in microscopy as the technique has been utilized in a variety of fields, including materials science and biophysics, to facilitate extensive surface analysis. Ongoing developments, including those with technologies like spontaneous parametric down-conversion, demonstrate ongoing optimization and adaptation of optical microscopy techniques for high-quality surface characterization.

## 9.1.3. Applications in Surface Characterization

In surface characterization, numerous applications are relevant, primarily in optimizing material properties for specialized industrial needs. Surface tailoring features is the centerpiece of maximizing the functionality of materials across industries. For instance, techniques such as Direct Laser Interference Patterning (DLIP) allow for the creation of patterned surface topologies, which can improve functionality dramatically, ranging from friction reduction to enhanced wettability and enhanced antimicrobial functionality. These biomimetic applications allow engineers to mimic the complex structures of biological systems, which possess improved properties compared to their original forms. Following the discovery of these engineered surfaces leads to tremendous advancements in areas such as biomedical devices and microelectronics, underlining the importance of effective surface characterization (Delfino et al., 2024). One of the main aspects of surface characterization is the application of contact angle measurements to study surface modifications. Sessile drop is a fundamental technique for measuring the hydrophilicity or hydrophobicity of surfaces and plays a significant role in applications like coating technologies and adhesion studies. But the interpretation of contact angle measurements can sometimes be tricky, particularly when making the evaluation of complex surface topographies with mixed chemistries. In order to bypass such complications, such as High-Precision Drop Shape Analysis (HPDSA), has been created to enable more reproducible, precise measurements of the contact angle. By translating images captured from sessile drop experiments, HPDSA provides a concrete model for calculating contact angles that can be utilized to assist in designing and optimizing coatings and other surface treatments (Heib et al., 2018). Apart from this, understanding and improving the surface properties of materials is required in order to advance technologies in material science and nanotechnology. Surface roughness and morphology are significant in governing properties including adhesion, wear resistance, and corrosion performance. Knowledge in these areas not only advances our basic understanding of surfaces but also drives product design creativity and process development. In fact, a combination of DLIP with post-processing techniques such as electropolishing has the potential to generate smoother surfaces while maintaining valuable topography. These synergetic methods exemplify the enormous potential of surface engineering, enabling applications from pharmaceuticals through to microelectronics. Continued development of these techniques indicates the dynamic nature of surface characterization and its instrumental role in technology and industry (Delfino et al., 2024).

## 9.2. SCANNING ELECTRON MICROSCOPY (SEM)

SEM is an essential surface characterization instrument that obtains high-resolution, detailed images required for the nanoscale examination of materials. SEM uses a focused beam of electrons to interact with the surface of a specimen and generate multiple signals that convey information about its topography, composition, and physical characteristics.

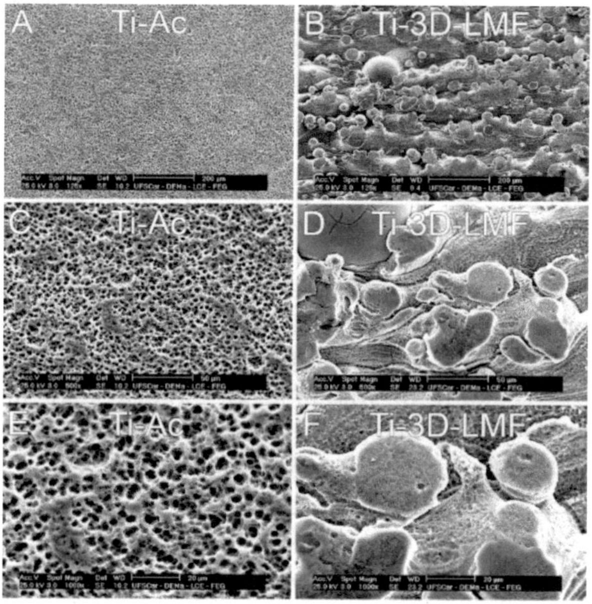


**Figure 9.3.** Surface characterization by scanning electron microscope (SEM). The columns represent the groups of microtextured Ti discs produced by acid etching (Ti-Ac, A, C, E) and microtextured Ti discs produced by laser metal fusion 3D printing (Ti-3D-LMF, B, D, F). The lines represent the different SEM magnifications: 125x (A, B), 500x (C, D), and 1,000x (E, F).

*Source: https://www.researchgate.net/figure/Surface-characterization-by-Scanning-Electron-Microscope-SEM-The-columns-represent-the_fig1_378675327.*

SEMs ability to achieve magnifications of up to a million times allows researchers to visualize surface structures and defects that cannot be observed with normal optical microscopy. SEM integration with other techniques is an important step in advancing the understanding of surface phenomena and making it a ready tool for material scientists conducting surface characterization. One of the main advantages of SEM is its capacity to provide not only surface morphology but also compositional information through techniques like Energy Dispersive X-ray Spectroscopy (EDS). Through the combination of SEM and EDS, researchers can study elemental composition at the micro and nanolevel. This is particularly useful in applications where extensive study of surface treatments and modifications needs to be conducted. For instance, in a study of various substrates treated differently, SEM was combined with atomic force microscopy (AFM) and contact angle determination to investigate the effects of the treatments on surface properties of flexible materials such as polyethylene terephthalate (PET) and polycarbonate (PC) (Chmela et al., 2015) (Figure 9.3).

This multi-disciplinary view not only explains structural changes but also helps to develop effective surface engineering solutions. Besides, the applicability of SEM is perceived through the examination of particle interaction in suspensions, such as in research investigating the electrokinetic properties of nanosuspensions. For example, recent research into zirconia ($ZrO_2$) nanosuspensions used SEM to verify the interaction between nanoparticles and dispersants and found the role of the surface coatings in stabilizing suspensions to be paramount. The SEM imaging confirmed the presence of attached polyelectrolyte layers on the particles and showed how surface treatments affect properties such as viscosity and zeta potential (Anvari et al., 2015). Such an ability to see under the SEM such interactions provides vital information that can inform the creation of high-performance materials and their use in industries, and it emphasizes the technique's central role in characterizing surfaces.

### 9.2.1. Working Principle of SEM

SEM operation is fundamentally a phenomenon of interaction of a focused electron with a sample surface. In SEM, a very focused electron beam is scanned over the surface, which removes secondary and backscattered electrons from the topmost few nanometers of the sample. This emitted electron information is used to send back information about the surface topography and composition, and it is converted into a high-resolution image. The electron beam's essential interaction with the specimen not only reveals topographical data but also provides information about material properties, including crystallography and elemental composition. It is an involved process that is vital to the characterization of nanostructured materials that exhibit different properties compared to their bulk forms, as noted in the most recent studies on the importance of accurate characterization tools in understanding the action and use of nanoparticles in various fields, including medicine and electronics (Joudeh et al., 2022). In addition to topography imaging, SEM works through a variety of detection modes that expand its diagnostic capacity. The ability to utilize various detectors gives the researchers the ability to analyze various properties of the sample. For instance, energy-dispersive X-ray spectroscopy (EDS or EDX) is able to be combined with the SEM system, yielding elemental analysis of the sample. This opportunity is particularly useful in the analysis of nanomaterials, which have particular chemical and physical properties that are significantly different from particles larger in size. In addition, the method aids in the characterization of biogenic nanoparticles, which is noted in studies aimed at ecologically friendly synthesis techniques (N/A, 2021) (Figure 9.4).

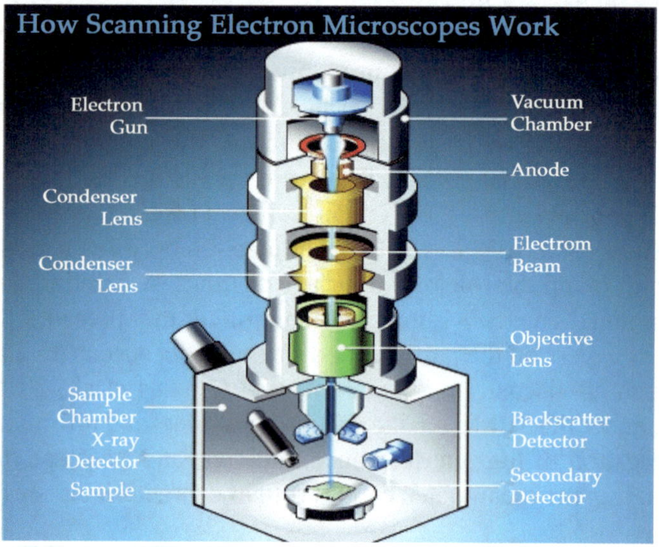


**Figure 9.4.** Working principle of SEM.

*Source:* https://www.researchgate.net/publication/330951958/figure/fig2/AS:723874438205440@1549596523382/Working-principle-of-SEM-source-7.ppm.

This cross-disciplinary approach enhances how scientists discover material properties at the nanoscale, revealing vital information that drives innovation in many industries, from catalysis to pharmaceuticals. Understanding the functioning of SEM greatly contributes to advancing surface characterization techniques in material science. Techniques high-resolution imaging provides scientists with the ability to see details generally out of reach with conventional optical microscopy. As technology in SEM continues to improve, developments in electron detectors and imaging computer software will continue to enhance characterization techniques. With these developments, research can be carried out on surface behavior in detail, particularly for novel nanomaterials and applications. As has been noted in many investigations into nanotechnology, the value of precise surface characterization cannot be overemphasized, given that it directly contributes towards the synthesis of materials with specified properties for specific applications (Joudeh et al., 2022). As we continue to build further knowledge in this field, SEM will prove to be a valuable tool for surface characterization in many scientific disciplines.

## 9.2.2. Advantages of Using SEM for Surface Analysis

SEM use in surface analysis is numerous advantages that enhance our understanding of material properties. Among its advantages is the high-resolution imaging capability, with the ability to see surface details at the nanometer scale. This level of resolution is critical to understand the microstructure of materials since it has the capability of revealing intricate topographical variations that can be related to mechanical behaviors or properties.

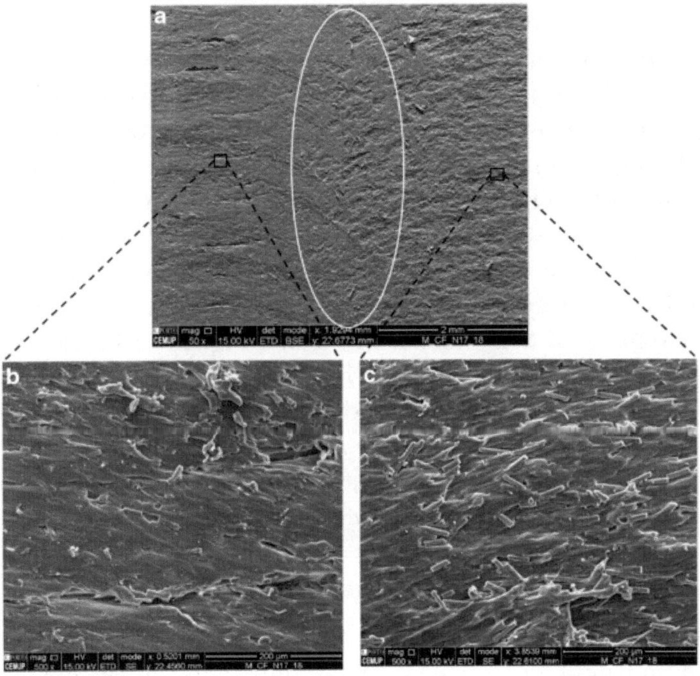


**Figure 9.5.** Surface analysis via SEM.

*Source:* https://www.researchgate.net/figure/Surface-analysis-via-SEM-a-Parameter-sets-1-and-2-b-Details-of-the-regions-machined_fig5_339394254.

Additionally, SEM can be used to produce three-dimensional-like pictures through various imaging modes, a quality of which is beneficial for accurate surface morphology interpretation. For instance, in biomaterial use, the surface characterization can benefit in the consideration of how topography of the surface influences cell adhesion and proliferation, an important concern in tissue engineering applications, according to the advancements on bioactive glass ceramics (Paramita et al., 2021). Another major advantage of SEM is that it can provide compositional information through techniques such as Energy Dispersive X-ray Spectroscopy (EDX) in conjunction with SEM imaging. This has the potential to enable scientists to not only observe surface topography, but also ascertain the elemental composition of materials in real time. By combining structural and compositional information, researchers can have a greater understanding of material properties and performance. For example, recently produced materials like MXenes have been in the limelight for various applications varying from energy storage to medicine, precisely due to their surface chemistry tunability (Alhabeb et al., 2017) (Figure 9.5).

The capability to directly observe surface composition versus overall performance interactions with SEM significantly enhances the process of materials development, allowing for property tailoring to a particular application through optimization. In addition to its capacity to provide high-resolution imaging and compositional analysis, SEM is versatile and highly versatile across a range of different disciplines and is a highly valuable tool in numerous fields such as materials science, biology, and nanotechnology. Its versatility allows examination of a wide range of material types, from metals, ceramics, and polymers to biological specimens. Furthermore, advances in SEM technology, such as environmental SEM (ESEM), allow samples to be examined under natural conditions, useful for the study of hydrophilic or biological samples that would be destroyed under conventional vacuum conditions. This versatility renders SEM an indispensable surface analysis technique such that scientists can comprehend complex interactions at surfaces, whether in the synthesis of new biomaterials or in the study of new electronic materials like MXenes.

## 9.2.3. Limitations and Challenges of SEM

The application of SEM in surface characterization has several inherent deficiencies and limitations, which may significantly affect the accuracy of resulting data. Sample preparation is one of the characteristic challenges that usually require desiccation and/or conductive coating. Such processes can modify the natural structure of delicate biological or organic materials and lead to inaccurate conclusions regarding surface properties. Moreover, the electron beam with high energy used in SEM also possesses the ability to induce radiation damage, particularly with samples composed of low atomic number materials. Radiation sensitivity may lead to distortion of image data, which complicates analysis of organic compounds or soft material where native structures need to be retained. It is therefore necessary for scientists who want to interpret SEM data correctly in relation to material characterizing techniques applied to surfaces to be aware of these limitations. Furthermore, the resolving power and imaging ability of SEM are principal challenges in extensive surface characterization. While SEM is capable of producing high-resolution images, finer details can become particularly hard to achieve for features at the nanoscale or low-contrast material. Techniques such as wet SEM were developed to enhance imaging in liquid environments; however, they have their own set of problems such as poor adhesion of samples to imaging windows and low intrinsic material contrast. Following recent developments, surface treatments to improve sample fixation and use of contrast improvers have been promising in overcoming the shortfalls, but their use within routine practice remains complex and sample-dependent (Borum et al., 2025). This, therefore, drives the need for increased resolution and improved imaging in SEM to continue demanding better methodological development in sample preparation and imaging techniques. In addition to instrumental limitations of SEM, interpretive data analysis challenges that are intrinsic to the process are also a significant bottleneck for researchers. Electron-beam interactions with materials may be so intricate that they produce secondary and backscattered electrons that improve image construction but could obscure inherent material properties. An example of this scenario is analysis of heterogeneous or diverse surfaces, where variations in topography could result in misinterpretations about surface composition and morphology. The intricate interdependence of particle characteristics and image resolution complicates the characterization process even further. As evident in studies utilizing electron microscopy for nanomedicine, immense work has to be done in deconvoluting the interactions of nanoparticles within cells, i.e., while SEM remains a crucial tool, the pitfalls result in judicious interpretation (Akhtar et al., 2025). This constant struggle with the limitations of SEM says a lot about the constant requirement for novel approaches in surface characterization.

## 9.3. ATOMIC FORCE MICROSCOPY (AFM)

AFM is an advance in surface characterization techniques that enables researchers to study materials at the nanoscale with unprecedented precision. The method employs a sharp probe to traverse the surface of a sample, measuring interactions between the probe and the surface to produce high-resolution topographical maps. Through overcoming many of the limitations of traditional microscopy techniques, AFM allows probing of surface roughness, texture, and morphology in high resolutions, which play a critical role to understand in comprehending many of the material properties. This capability renders AFM highly beneficial in material science, biology, and nanotechnology, where surface characteristics can significantly influence the functionality as well as the performance of materials or biological organisms. The technique of AFM is transferred to quantitative study of microstructure and wettability characteristics of intricate materials, such as shales used in oil recovery processes. Traditional methods for the quantification of wettability are usually compromised in that they are plagued by defects that emanate from surface roughness and mixed wettability behavior. AFM provides a robust alternative through the ability to conduct nanoscale characterization of surface interactions that control wettability, which enables researchers to better understand oil/water interactions under reservoir conditions.

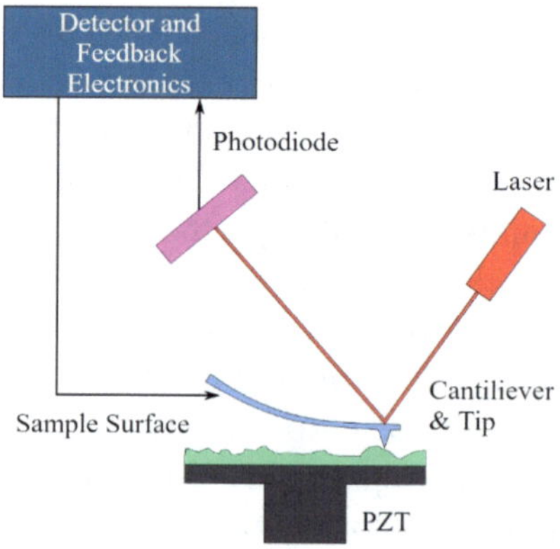


**Figure 9.6.** Atomic force microscopy.

*Source:* https://microbenotes.com/wp-content/uploads/2020/03/Atomic-Force-Microscope-AFM.jpeg.

This is one topic that has come to the forefront of recent studies, demonstrating how AFM is able to detect the intricate interplay of mineral composition and wettability behavior and make seminal contributions towards oil recovery strategy development (Huo et al., 2023). Furthermore, AFM-guided nanomechanical characterization techniques shed light on the mechanical properties of the material, further taking the level of understanding of surface properties to new levels. By direct visualization of the micromechanical properties, AFM facilitates the study of stress distributions and deformation behaviors in materials with specific interest to complex polymeric composites. The technique's high spatial resolution allows localized mechanical variations developed through microstructural features to be observed and hence its broader applications in engineering and materials design fields. By adding AFM with extensive mechanical characterization, one acquires extremely informative data that is most critical to maximize performance of materials in different applications and enable the researcher to position the properties optimally and creatively (Liang, 2023) (Figure 9.6).

## 9.3.1. Fundamentals of AFM Technology

Atomic Force Microscopy (AFM) is currently a standard method for nanoscale surface measurement. Unlike traditional microscopy methods utilizing light, AFM employs a nanoscale cantilever with a sharp tip to scan and interact with the sample surface. When the tip approaches the surface, there are forces such as van der Waals forces and electrostatic interactions that cause deflection of the cantilever.

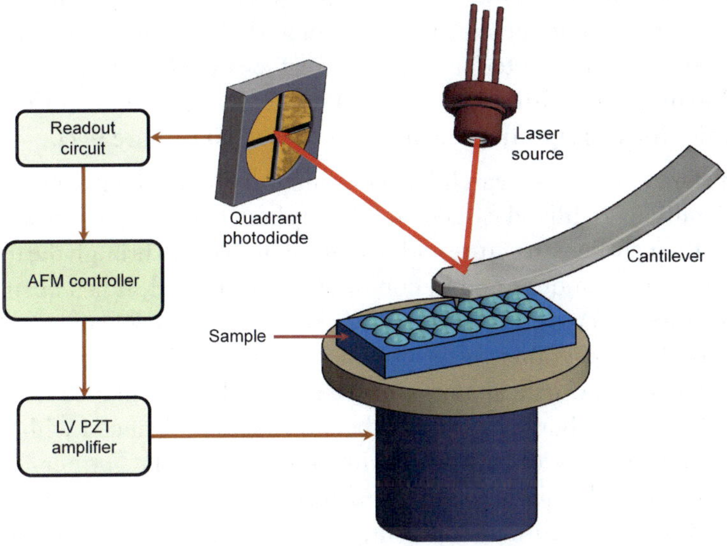


**Figure 9.7.** General schematic of the atomic force microscope (AFM).

*Source:* https://www.creative-biostructure.com/atomic-force-microscope-afm-for-exosome-characterization.htm?srsltid=AfmBOoqJqQVMen8lPrdH0BJOdgMthymuzAfNywAMkt4XtqPietYh7o-L.

This deflection is precisely quantified, enabling one to reconstruct the sample topography at the atomic scale. The multimode character of AFM makes it helpful in a variety of fields such as materials science, biology, and nanotechnology, allowing one to gather information on surface roughness, morphology, and mechanics. With its ability to create three-dimensional topographic maps, AFM is a fundamental tool in advancing material properties and surface interactions knowledge, and eventually to advances in surface fabrication and engineering. In efforts towards optimizing material performance, AFM technology facilitates the study of phenomena at the nanoscale, which influence macroscopic properties. Techniques such as Force Distance (FD) measurements enable scientists to quantify the surface mechanical properties, such as adhesion and elasticity, through the measurement of the tip-sample interaction forces. It can also be used in conjunction with other techniques such as Scanning Tunneling Microscopy (STM) or X-ray Photoelectron Spectroscopy (XPS), which increases the information gathered from the research. For example, surface chemistry alteration at the nanolevel can be comprehended extremely effectively by combining AFM with spectroscopic techniques. Multidimensional study of this nature is necessary to develop new materials because it will facilitate the connection between surface properties and overall performance parameters. These characteristics highlight the importance of AFM not only for characterization at the surface but also as a means to design and develop materials. Utilization of AFM in the assessment of fresh composite materials signifies its significance in current research (Figure 9.7).

For instance, recent research has combined AFM with new materials such as eco-friendly epoxidized soybean oil acrylate nanocomposite coatings to investigate their surface structure and structural integrity. Through the use of new characterization techniques and in conjunction with AFM, it has been possible to demonstrate improved corrosion resistance and mechanical properties by nanoparticle dispersion within polymer matrices. Such studies exemplify the influence of nanoscale features on the overall performance of advanced materials. Also, as technology in AFM becomes more advanced with improved resolution and automation, its usage in laboratory work further establishes its place in materials studies. Scientific research involving AFM provides a foundation for continued development, advancing surface characterization as depicted in the development of high-performance nanocomposite structures and devices like CuO-based memristors, indicating the broad spectrum of uses of this generalized tool in the field of surface science.

### 9.3.2. Types of Measurements Obtained from AFM

Atomic Force Microscopy (AFM) provides a generic way to obtain a range of measurements required to characterize nanoscale surfaces. Topographic imaging is a significant one, where the AFM tip scans across the sample surface to create high-resolution three-dimensional topography maps. This is applied to precisely measure surface roughness and morphology with informative details about material properties such as adhesion and friction. Second, AFM may be utilized to measure the mechanical properties of materials by techniques like force spectroscopy, where interaction tip-sample force is quantified in units of distance. The ability to obtain quantitative data on the elastic modulus and other mechanics from it further explains material behavior under various conditions. Aside from plain surface imaging, AFM extends this by enabling the measurement of nanoscale electrical and magnetic properties. It does so through probes tailored to the method of interest, e.g., those employed in magnetic-force microscopy (MFM), to measure localized magnetic fields. Researchers can hence image magnetic domains and examine the associated interactions within materials (Abelmann et al., 2002). The techniques provide critical data for use in applications like spintronics and material science. Similarly, electrostatic force microscopy (EFM) can be employed to image surface charge topographies, essential for the determination of the electrical properties of semiconductor materials and nanostructures. This generalized procedure is a testament to AFMs versatility in investigating a variety of scientific questions through many measurement modalities. In addition to mechanical and topological measurements, AFM can be employed to investigate chemical properties through techniques such as chemical force microscopy (CFM). This method involves functionalizing the AFM tip using specific chemical groups that interact with the surface of the sample and enabling one to measure chemical bonds and molecular interactions at the nanoscale. The special advantage of CFM is its ability to correlate chemical properties and topographical information, thereby augmenting the process of characterization and providing a higher level of understanding of material composition. Moreover, the emerging AFM tip designs, such as carbon fiber tips, exhibit increased sensitivity and durability to enable high-quality readings even in hostile environments (Castellanos-Gomez et al., 2010). The integration of these varied measurement techniques places AFM as a key tool in the complicated surface characterization world.

### 9.3.3. Comparison with Other Microscopy Techniques

Among the techniques of surface characterization, microscopy is a mainstay for the understanding of material properties at the micro and nanoscale. The Atomic Force Microscopy (AFM) and Scanning Electron Microscopy (SEM) are two of the most common methods which possess unique strengths and limitations compared to other microscopy techniques. SEM provides high-resolution surface morphology images due to its ability to resolve in the nanometer scale, making it of critical importance to analyze materials like titanium implants where osseointegration is directly related to surface topography. It is the capability to imagine stacks of titanium oxide nanotubes and alterations under doping with ions like zinc and strontium, as indicated by research showing improved cellular adhesion and growth on modified surfaces (Bhattacharjee et al., 2024). Though its forte, SEM typically comprises sample preparation that influences the surface properties, and thereby is compared with AFM, whose strength lies in the fact that it can investigate surface forces and nanomechanical properties without significant sample alteration. Both SEM and AFM are distinguished from TEM in that Transmission Electron Microscopy enables unparalleled information on internal material structure at atomic resolution and hence forms a surface characterization complementary technique. While SEM is focused on surface morphology, TEM is key to understanding crystallinity and phase distribution in materials, particularly weldments, as illustrated in the characterization of AA-2219-T6 alloy friction stir welding (Hussain et al., 2024). Bringing these techniques into focus allows for widespread understanding of microstructural elegance and mechanical properties as far as surface morphology is concerned. For instance, the rise in hardness with water-cooled samples as attested by TEM imaging indicates how internal microstructural features such as particle dispersions impact heavily on the surface behavior of wear and corrosion. Thus, the use of several microscopy techniques improves the information gained from surface characterization activities so that the material performance becomes appreciated within an overall context. Moreover, the preferred microscopic method is often decided based on the type of analytical goals and the material being studied. For biomedical applications, engineered surfaces designed for enhanced biocompatibility require careful investigation to establish parameters influencing cell-material interaction. Techniques such as X-ray Photoelectron Spectroscopy (XPS) may be coupled with microscopy techniques to provide chemical state information on the surface changes. The synergy of XPS with SEM analysis can be particularly effective in the characterization of doped titanium surfaces by giving the needed morphological and compositional data to optimize osseointegration (Bhattacharjee et al., 2024). On the other hand, in welded material analysis, applications of SEM with microhardness and X-ray diffraction illustrate the role of cooling techniques on surface and bulk properties to guide the design of optimized processes toward further enhancing mechanical strength (Hussain et al., 2024). Finally, the proper use of varied microscopy techniques is important in enhancing surface characterization methodologies and, by association, material performance across various fields.

## 9.4. X-RAY PHOTOELECTRON SPECTROSCOPY (XPS)

XPS is a key technique of surface property analysis, which allows researchers to probe the elemental composition and chemical surroundings of materials with an incredible spatial resolution of below 10 nanometers. This high spatial resolution is particularly useful in the characterization of thin films and complex surface topology in materials science and nanotechnology as well as catalysis. The technique is based on the process of bombarding a substance with X-rays, leading to the ejection of core electrons from surface layer atoms. By measuring kinetic energy of the emitted electrons, the inference of binding energy, elemental character, and oxidation states becomes feasible, thus enabling one to determine a complete understanding of the surface chemistry and physical properties of the target substance (Krishna et al., 2022). Applications of XPS extend beyond elemental analysis to include chemical bonding and oxidation states data, thus enabling advanced interpretation of surface activity. The technique provides high-resolution valence band information, which are crucial to understanding electronic activity and material reactivity.

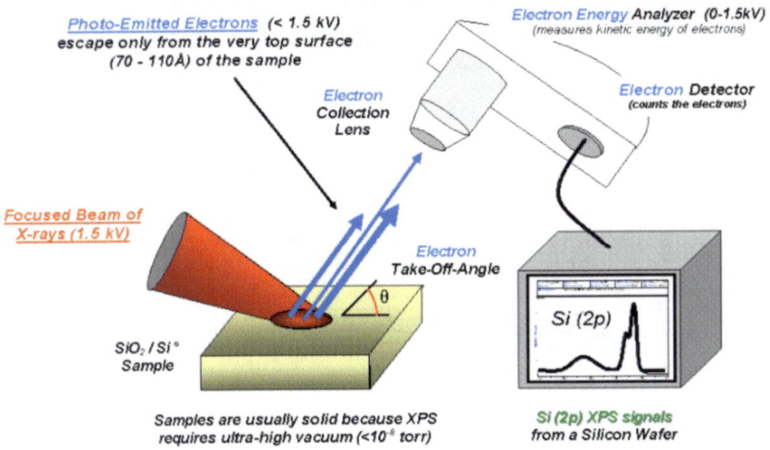


**Figure 9.8.** Basic components of a monochromatic XPS system.

*Source:* https://en.wikipedia.org/wiki/X-ray_photoelectron_spectroscopy#/media/File:System2.gif.

The latest advancement in XPS has brought to the forefront its importance in finding the phase composition at the nanoscale level, particularly for new materials such as nanostructured semiconductors and catalysts. For example, using the technique of XPS, researchers have been able to differentiate the intricate surface characteristics of copper oxide nanopetals and uncover their optical and catalytic properties driven by their structural morphology (Khan et al., 2020). This versatility in surface characterization exhibits the utility of XPS in material development across a range of scientific fields. In addition, while advantageous, XPS has its challenges, specifically with regard to surface complexity and spectral interpretation. Sample charging and overlapping peaks can inhibit proper analysis, requiring a sound grasp of the technical underlying principles. Ongoing improvement in XPS techniques aims to address such challenges, thereby improving data consistency and interpretation accuracy. As researchers push deeper into novel applications of XPS, it remains an essential tool for surface characterization in the majority of applications, from corrosion science, electrochemistry, and nanomaterials, to become the foundation in progressing materials discovery and development (Krishna et al., 2022) (Figure 9.8).

## 9.4.1. Basic Principles of XPS

XPS spectral analysis is a surface-sensitive method vital to the characterization of numerous materials. Using monochromatic X-ray radiation, XPS induces the photoelectric effect in which photons knock core-level electrons out of surface atoms.

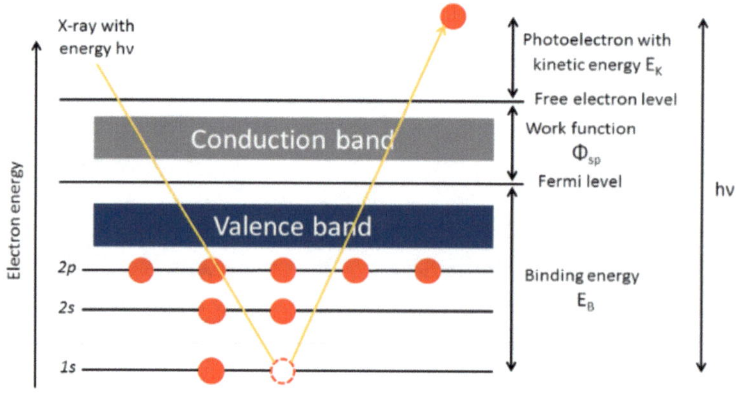


**Figure 9.9.** Principle of XPS.

*Source:* https://www.researchgate.net/figure/Principle-of-XPS_fig50_312493700.

The photoelectrons released from the surface are examined to provide elemental composition, chemical state, and electronic state data of the material surface. The core underlying principles of XPS are founded on the fact that there is a correlation between the emitted electrons' binding energy and respective elements present. This binding energy depends on the local chemical environment, thus making XPS exceptionally good at being able to resolve surface chemistry differences. Aside from elemental analysis, XPS can also give data on the oxidation state and bonding chemistry of elements, which makes it a very useful tool in surface chemistry, materials science, and nanotechnology studies (Magdy, 2023). The application of XPS is governed by different instrumental attributes that make it a more effective analytical tool. The technique is generally carried out in ultra-high vacuum to prevent contamination and accurately measure emitted electrons. The XPS system includes an electron energy analyzer, detector, and source of X-ray all combined to detect and quantify the photoelectrons that are emitted. The electron energy analyzer is capable of operating in various modes including hemispherical and time-of-flight setups, allowing the researcher to achieve high energy resolution and surface composition sensitivity. Decoding the spectra through characteristic peaks enables the identification of the elements and their respective chemical states, crucial to applications ranging from semiconductor fabrication to analysis of surface evolution in chemical catalysts (Magdy, 2023) (Figure 9.9).

Though it has its own strengths, XPS is not without its weaknesses that must be dealt with by users. The technique probes only the topmost several nanometers of a surface, and this may not represent bulk band properties of a material. In addition, processing of complex materials or multilayer systems can lead to difficulties in the interpretation of overlapped spectral features. The errors in data analysis, particularly in surface characterization, indicate the necessity to learn XPS basics and techniques in depth. As discussed, misinterpretation risk exists for those with less training on these advanced techniques. Thus, with the exclusive dependence on initial information without full comprehension of principles engaged might worsen existing issues in scientific literature, triggering a joint initiative among stakeholders to encourage right and well-informed applications of XPS in the scientific community (George et al., 2023).

## 9.4.2. Surface Chemical Composition Analysis

Analysis of surface chemical composition is a foundation in material characterization, as it yields fundamental information on the molecular interactions and reactivity of surfaces. Knowledge of surface chemistry is crucial in applications from catalysis to semiconductor production, where surface properties determine performance and efficiency. Techniques such as X-ray photoelectron spectroscopy (XPS) and energy-dispersive X-ray spectroscopy (EDX) are important in identifying the elemental makeup and oxidation state of surface species. Scientists can determine, through such analyses, how reactivity and stability are impacted by surface changes, particularly when materials undergo exposure to harsh environments. The comprehensive understanding gained through these techniques enables material design with desired properties, ultimately optimizing their real-world applications across a variety of industrial markets. In addition, surface chemical analysis advances have also been followed by sophisticated model techniques that describe the structure-function relationship. For instance, X-ray absorption near-edge structure measurements examine the evolution of surface and bulk composition upon chemical treatments, for instance, lithiation process of SiO anodes. Outcome of such studies indicates that heat treatment significantly contributes towards modifying surface composition and structural integrity and hence affecting electrochemical performance. The effects observed suggest a complex interaction of factors, such as the formation of dielectric layers that are inhibitive to ion transport. Understanding these effects is vital in the optimization of battery materials, where a precise composition and structure directly correlate to efficiency of operation and longevity, as in evidence in the literature in the investigation of microstructure variation induced by processing conditions (Xu et al., 2022). The surface functional groups of materials are also important in determining their chemical properties, particularly in the areas of environmental and bioengineering applications. For example, the characterization of pyrolyzed biochars at different temperatures demonstrates the degree to which the elemental composition and specific surface area can influence infrared absorption properties. Analysis of 92 biochars determined that the presence of some functional groups is related closely with the process parameters governing their synthesis, and hence the necessity for exercising strict control over preparation conditions. Techniques such as Diffusive Reflection Fourier Transform Infrared Spectroscopy (DRIFTS) in combination with statistical analysis can assist in establishing these relationships, offering insight into how adsorption properties and chemical reactivity are affected by surface chemistry. Such data turns out to be valuable from the perspective of designing biochars for specific applications, for instance, land improvement or pollutant adsorption, thus defining surface chemical makeup and real-world implications in environmental management (Janu et al., 2021).

## 9.4.3. Applications in Thin Film Characterization

The characterization of thin films is of primary importance to many scientific and technological procedures from semiconductor technology to advanced coatings. Among the most prominent of the techniques employed for such applications is the X-ray photoelectron spectroscopy (XPS), which allows for the investigation of surface chemistry at below 10 nm depths. This highly useful tool provides data on chemical bonding and oxidation states, of utmost importance in establishing the composition and integrity of thin films. The ability to probe core-level peaks and valence band structures further expands the insight into the surface properties and phase composition, which is most important for maximizing material performance in applications such as catalysis, photovoltaics, and electronics (Krishna et al., 2022). The constant improvements in XPS highlight its pivotal role in the discipline of material science due to its multiplicity of utility in various fields of study. To achieve comprehensive thin film characterization, support techniques will normally provide invaluable contribution. One such case is the pairing of XPS with attenuated total reflection Fourier infrared spectroscopy (ATR-FTIR), wherein chemical structure analysis with varying sensitivity depths is enabled. This blending is particularly useful when examining heterogeneous thin films, where the quantity of material undergoing examination can have a huge impact on the interpretation of results. In a such study, PTFE thin films produced by pulsed laser ablation were examined using these techniques, which yielded critical information regarding the chemical structure and the surface morphology (Piwowarczyk et al., 2019). The integration of different characterization techniques, such as scanning electron microscopy and optical microscopy, further adds to the depth of knowledge of physical properties and deposition mechanisms, highlighting the importance of a multi-disciplinary approach to surface analysis. The applications of thin film analysis extend beyond simple characterizations to industry and quality control. The resilience of techniques like XPS and ATR-FTIR not only ensures greater understanding of material properties but also makes process optimization feasible in areas like electronics and coatings. Accurate characterization directly influences product reliability as well as performance and is thus mandatory for manufacturers to place appropriate testing procedures in place for thin films. The knowledge gained from the surface characterization techniques not only contributes to design creativity and production but also results in nanotechnology and materials engineering breakthroughs. The applications of the thin film characterization techniques continue to contribute crucially to furthering progress in numerous areas such that materials are designed for maximal utilization in their respective uses.

## 9.5. SECONDARY ION MASS SPECTROMETRY (SIMS)

SIMS is a very sophisticated analytical technique applied widely for the surface examination of materials. Its primary use is bombardment of the surface of a solid with a collimated beam of ions, resulting in the ejection of secondary ions from the material. The secondary ions are then mass-analyzed, providing accurate information about the elemental and molecular composition of the surface. The advantage of SIMS is its high spatial resolution and sensitivity, with trace elements and small molecular species being detectable, making it crucial for use in materials science, semiconductor science, and biomaterials research. Aside from that, SIMS can also generate two-dimensional maps and three-dimensional depth profiles, which reflect species distribution and may be significant to describe the layered architecture of complicated materials, especially in polymer systems, where identical constituents separation is most critical (Mei et al., 2021). The Time-of-Flight Secondary Ion Mass Spectrometry (ToF-SIMS) variant has subsequently further enhanced the advantages of standard SIMS. ToF-SIMS offers high mass resolution and detection over a very broad mass range, making it of incalculable worth when identifying polymers that may have similar elemental compositions. The method eliminates chemical labeling techniques, which might alter material properties and complicate things into being hard to interpret. It provides complete mass spectra for each pixel in an image and enables a more accurate analysis of the surface chemistry.

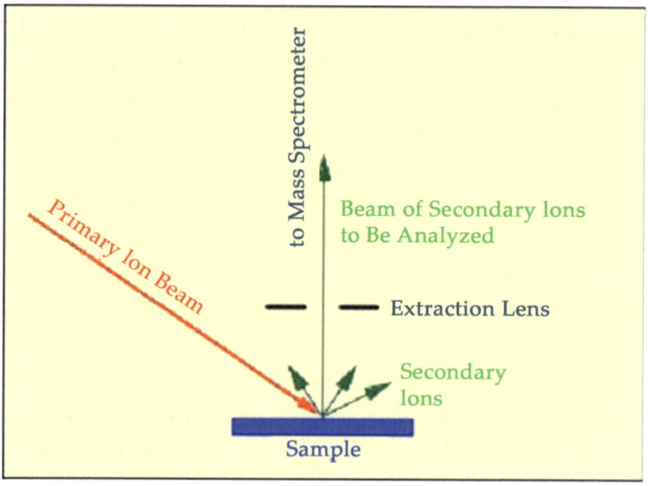


**Figure 9.10.** Schematic depiction of SIMS.

*Source:* https://cdn.serc.carleton.edu/images/research_education/geochemsheets/techniques/SIMS4.JPG.v2.jpg.

This multi-dimensional capability enables researchers to conduct spectral analysis, imaging, and depth profiling concurrently, and thereby attain an integrated understanding of the surface characteristics (Mei et al., 2021). The versatility of ToF-SIMS enables it to be applied in various domains ranging from polymer blends to composites and electronic materials, where information on surface and near-surface chemistries is critical for the optimization of performance. Advances in data visualization and analytics technology have significantly improved the interpretive power of SIMS results. Toroidal self-organizing map (SOM) algorithm optimization has been the subject of recent efforts to better visualize hyperspectral data obtained from ToF-SIMS imaging. With the SOM, scientists are able to generate multicolor similarity maps that retain mass spectral similarities at image pixels, creating a dense map of the chemical landscape on the sample surface. These are supplemented by methods like Relational Perspective Mapping (RPM), which give a two-dimensional representation of high-dimensional data and help explain intricate relationships within the spectral data. Together, these top-level analytical tools enable a more detailed examination of surface chemistries, thereby enabling greater insight into material properties and potential applications in a wide variety of scientific disciplines (Figure 9.10).

## 9.5.1. Overview of SIMS Technique

The SIMS approach offers a new surface characterization method allowing scientists to obtain high-resolution chemical compositions of materials in the nanoscale. The working mechanism of SIMS involves the bombardment of an ion-solid sample by a primary beam of ions. This interaction results in the emission of secondary ions from the sample surface. These secondary ions generated are then measured and analyzed to derive compositional information about the subject material. One of the primary advantages of SIMS is its capacity to acquire qualitative and quantitative information with high depth resolution, which is critical in the analysis of layered materials and coatings. The ability of the methods to analyze very small areas also makes it easier to investigate multicomponent and heterogeneous materials, such as solid electrolyte interfaces (SEI) in battery electrodes, to acquire information on their performance and longevity (Lombardo et al., 2023). The versatility of the SIMS technique is not confined to simple compositional analysis; it also encompasses imaging, which is pivotal in visualizing the elemental or compound spatial distribution on a surface. A case in point is Time-of-Flight Secondary Ion Mass Spectrometry (ToF-SIMS), one of the variants of SIMS, which enhances spatial resolution and facilitates the generation of high-resolution images. This technique is particularly valuable in applications such as semiconductor manufacturing and materials science, where spatial distribution of materials is paramount. Secondary ion emission imaging from the sample allows ToF-SIMS to reveal trace elements and contaminants, which are key contributors to device function and reliability (Chan et al., 2016). Such sophisticated surface examination not only speeds up material development but also enhances manufacturing processes by detecting potential defects prior to failure. Coupling machine learning with SIMS data analysis is the next big thing in the field, opening up possibilities for richer interpretation of intricate datasets. Recent breakthroughs have used generative adversarial networks to reconstruct three-dimensional models from two-dimensional ToF-SIMS images, allowing for a better understanding of microstructural aspects. Such combination not only augments the conventional ability of SIMS but also allows study of sophisticated surface composition versus material property relationships, e.g., electrochemical performance in battery materials. As scientists continually improve the SIMS technique and hybrids with advanced computation methods, the potential for breakthrough discovery across many disciplines, such as energy storage and sensor technologies, increases exponentially. These innovations make SIMS an unequivocally pivotal method for surface characterization (Lombardo et al., 2023) (Figure 9.11).

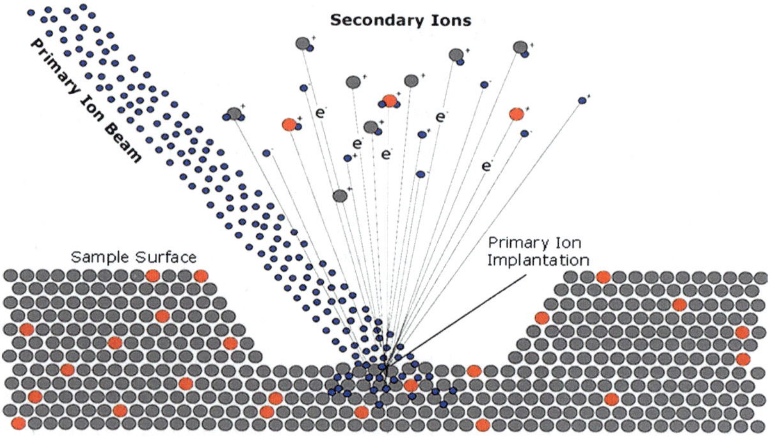


**Figure 9.11.** Secondary ion mass spectrometry (SIMS) technique.

*Source:* https://www.cameca.com/products/sims/technique.

## 9.5.2. Depth Profiling Capabilities

State-of-the-art characterization equipment's depth profiling capability is at the heart of understanding surface structures, particularly in the high-resolution context. Techniques such as hard x-ray ptychographic reflectometry imaging illustrate the excellent synergy between high-resolution imaging and depth profiling. The method is a synthesis between the two-dimensional imaging capabilities of ptychography and the high-resolution depth profiling capacity of x-ray reflectivity. By accessing both amplitude and phase data, researchers can image not only the surface topography but also critical data on interfacial roughness and local structural features. Such a holistic picture is crucial for materials, particularly those in nanodevices and quantum technologies, where functional behavior is tightly entangled with surface characteristics, providing unmatched insight into the function–structure connection (Myint et al., 2024). Depth profiling also makes the examination of sophisticated architectures and heterogeneous materials, which are typical of cutting-edge material systems, possible. The use of increasingly sophisticated techniques, such as high-definition reservoir mapping during drilling (HD-RMWD) technology, demonstrates the manner in which these uses enrich our understanding of rock formation subsurface layers. The HD-RMWD utilizes ultradeep azimuthal resistivity measurements to construct precise resistivity profiles around boreholes, facilitating real-time geosteering and improved reservoir characterization. This approach minimizes pilot well requirements to a very large extent and allows effective handling of geological uncertainties while drilling. With the help of advanced depth profiling methods, geologists are able to make informed well placement decisions and optimize hydrocarbon recovery operations (Vieira et al., 2024). Depth profiling capabilities are of extreme value not only in geological applications but also in a wide range of surface characterization problems. In semiconductor processing and sensor technologies, high-resolution surface metrology provided by advanced depth profiling techniques can identify faults and direct design optimization. By identifying the depth of various layers and material compositions, scientists have gained enhanced understanding of the way interfaces speak to each other. This is what ultimately directs the tuning of process parameters in manufacturing electronic devices and nanostructured materials. These capabilities are necessary across all sectors, enabling product development innovations and also efficiency of manufacturing processes. With more research advances, the depth profiling techniques will further become more integrated and influential within the analyses of complicated surface and subsurface structures (Myint et al., 2024; Vieira et al., 2024).

### 9.5.3. Advantages and Limitations of SIMS

SIMS possesses singular strengths in quantitative surface composition analysis at very high spatial resolution. Its high sensitivity, for instance, allows SIMS to readily detect trace elements and isotopes and renders it highly valuable for materials science, geochemistry, and forensic science applications. The technique is superior to the resolution of obtaining high compositional maps of complex surfaces, which provide information on material interfaces, coatings, and thin films. SIMS can further analyze a wide range of solid samples, for instance, metals, ceramics, and polymers. The ability to perform depth profiling also contributes to its application by enabling compositional variations along layers of a material to be investigated, as is relevant in microelectronics and coating technology. This capability makes SIMS a potent tool for material property and behavior analysis. Apart from its numerous advantages, SIMS is also subject to a number of significant limitations that researchers would have to consider. One of such primary concerns is the surface alteration which takes place during the SIMS process, as the bombardment of the main ions can lead to severe damage or modification of the analyzed surface. This may provide a misleading depiction of the initial composition of the samples. In addition, even though SIMS is extremely sensitive, it is not necessarily suited for any material; some samples, such as soft or delicate material, may not withstand the high-energy ion beam bombardment. The technique also requires extensive calibration and matrix effect understanding, thus quantitative analysis becomes complicated. Its long sample preparation and requirement for special equipment may also deter its application in most research environments. These limitations make in-depth understanding of SIMS critical to realize its maximum utility in material characterization. The use of SIMS with other techniques can overcome some of the limitations, enhancing its overall utility in material characterization. The integration of SIMS with techniques such as X-ray photoelectron spectroscopy (XPS) or Fourier-transform infrared spectroscopy (FTIR), for example, can provide a better overall view of the surface composition and chemical state of materials. Such complementary methods can offer additional validation of SIMS data, resulting in more effective analysis of intricate surfaces. On top of that, ongoing SIMS technology advancement is optimizing spatial resolution and limiting surface damage, further optimizing its application across a range of fields. With scientists pushing the boundaries of SIMS techniques and exploring new applications, the synergistic benefit of using two or more characterization approaches could potentially create more stable and meaningful analysis, ultimately contributing to improved understanding of material properties and behaviors across a variety of scientific disciplines, such as cultural heritage object restoration and conservation, as discussed in John et al. (2019) and António et al. (2019).

## 9.6. SURFACE PROFILOMETRY

Surface profilometry is a significant material surface characterization technique that provides valuable information on the topographical properties of material surfaces at micro and nano scales. The technique is particularly adapted to understand the texture, roughness, and variation in height of the surface, which are significant factors controlling the functional behavior of material. Quantitative measurement of surface features is possible through methods such as contact profilometry, optical profilometry, and atomic force microscopy. This quantitative information can then be utilized in making decisions regarding surface finish specification in a variety of applications, from manufacturing to materials science. The benefit of surface profiling is that it can be employed to correlate surface properties with performance parameters like adhesion, friction, and wear, thus enhancing the understanding of how alterations in surface topology can impact overall performance. One of the fascinating uses of surface profilometry is in studying traditional Asian lacquer surfaces, where the technique helps to characterize and classify different types of lacquers based on surface topography. Lacquers such as laccol, thitsi, and urushi possess certain topographical features that can be quantified by non-contact, non-destructive methods.

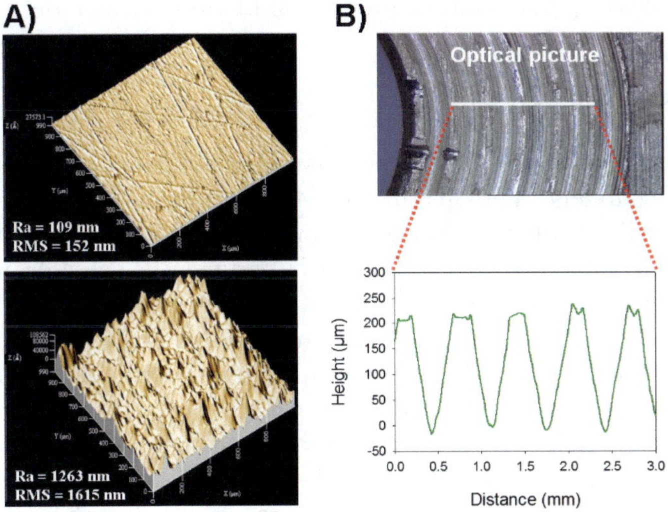


**Figure 9.12.** (A) Surface morphological images showing parts of a metal surface having different wear (roughness); and (B) a profile showing a thread having a pitch of ~0.7 mm and a height of 0.2 mm.

*Source:* https://www.surfacesciencewestern.com/analytical-services/surface-profilometry/.

The union of confocal microscopy and advanced data analysis techniques, including convolutional neural networks (CNN), allows for a powerful study of such materials. Not only does such advanced analysis provide a better characterization of the cultural heritage objects, but it also provides a better understanding of the aging of these lacquers. Through the application of extensive surface profiling techniques, researchers can determine the unique properties of such art surfaces and support preservation that honors their historical significance (Ravines et al., 2025). On an industrial front, surface profilometry is essential in the examination of specialized coatings like diode laser cladding (DLC) surfaces that have found interest due to their improved behavior under severe conditions like nuclear applications. Compared to traditionally machined surfaces, whose topographical properties are well known, DLC surfaces present more intricate ground with wave-like shapes and random asperity distributions. Their characterization would only be precise through very advanced profilometric techniques since linear measurements will not be able to capture the intricacy present. Using laser confocal microscopy in investigating control parameters, researchers can accurately measure the quality and structure of DLC coatings. These holistic evaluations are the foundation of standard quality metrics in industrial settings, ultimately contributing to surface engineering development and the production of stronger materials (Mandal et al., 2024) (Figure 9.12).

## 9.6.1. Techniques Used in Surface Profilometry

Surface profilometry employs several varied techniques to quantify surface properties accurately. Perhaps the most important of these is confocal microscopy, which offers surface imaging in high-resolution, non-contact terms. The advantage of this technique is particularly valuable in sensitive sample situations, such as those experienced in art conservation work, where preservation of material integrity is paramount above all else. Confocal microscopy facilitates the acquisition of quantitative surface texture data, and hence it is feasible to differentiate between various surface types and conditions. For instance, in Asian lacquer surface research, confocal microscopy was employed to analyze types of lacquer—laccol, thitsi, and urushi—through an examination of their topographical differences without disturbing their surface integrity. This technique not only verifies the validity of surface profiling but also enhances the quality of identification and categorization of art and cultural heritage objects with efficiency, suggesting its role in the preservation of historical artifacts under scientific study (Ravines et al., 2025). Additionally, issues arising from unique surface structures necessitate integrating advanced analytical techniques. The complexity of the characterization of diode laser cladding (DLC) surfaces illustrates this, with these surfaces having wave-like characteristics with metal tracks and randomly located asperities that make the surfaces difficult to assess. Conventional profilometry methods, where measurements are based on straightforward linear measurements, fail to meet the mark when used for these complex patterned surfaces. To that response, scientists have also come up with state-of-the-art methods, e.g., laser confocal microscopy, to capture both the areal and profile measurement of DLC surfaces such that their topography and quality can be analyzed in finer detail. By their exploration of how variations in manufacturing parameters impact surface properties, such methods not only promote uniform quality measurement but also expand the information about the mechanical performance of DLC coatings, as pertinent to situations where the surface condition must be highly stringent, such as nuclear usage (Mandal et al., 2024). The integration of data science techniques into surface profilometry is a significant advancement for the characterization process. Techniques such as feature engineering and convolutional neural networks (CNN) allow robust analysis of the numerical surface texture data acquired by profilometry. The technique allows for automatic discrimination of surface classes and measurement of temporal changes, a critical aspect in studies monitoring aging processes. With these advanced analytical techniques combined with conventional profilometry approaches, it is possible to develop a better understanding of the surface characteristics of various materials. For example, the aging effect on the surface of lacquer can be made apparent by combining quantitative data with machine learning algorithms, where valuable information will be gained in the fields of art conservation and material science. Hence, these advancements underscore the paramount significance of technology in the evolution of surface profiling techniques in the broad context of material characterization (Ravines et al., 2025).

## 9.6.2. Importance of Surface Roughness Measurement

The appreciation of the importance of surface roughness measurement is crucial in a wide range of fields, but perhaps more so in materials science and engineering. Surface roughness is defined as a measure of surface texture in terms of irregularities and is of primary importance to the mechanical, optical, and even chemical properties of materials. In coating and adhesive bonding, for instance, roughness can be a determining factor in adhesion strength and wear resistance. Accurate measurement allows scientists and engineers to refine processes such as machining and surface finishing, ultimately resulting in improved product durability and performance. In addition, detailed information regarding surface profiles facilitates the tailoring of materials for specific uses so that they are not only appropriate for structure but for appearance, in industries ranging from consumer products to aerospace. Techniques for measuring surface roughness have also become more mature, incorporating multiple methods, each of which has its strengths and weaknesses. Techniques such as Atomic Force Microscopy (AFM) and Scanning Electron Microscopy (SEM) provide high-resolution images that capture the fine detail of surface topography. Such newer methods are critical to measuring materials in the advanced research of nanotechnology and semiconductor production. Through developing a complete picture of surface roughness, scientists are able to make correlations between specific surface features and how the material will behave. This comprehension drives innovation, leading to the development of high-performance materials that can withstand harsh environments or facilitate higher-order function, like catalysis in electrochemical devices, as investigated in studies of copper oxides (Permyakova et al., 2019). In addition, the surface roughness application also extends to 6G wireless communication systems, where surface interaction at terahertz (THz) frequencies is becoming increasingly important. As development and research progress, it is a key piece of information to know how the roughness of surfaces affects signal transmission and antenna efficiency. Surface characterization techniques that are efficient can be used in designs keeping energy loss low and transmission quality high and hence enable next-generation wireless technologies. Surface roughness measurements and findings that are derived lead to conclusions that help researchers and scientists counter performance enhancement and reliability of systems with high frequency-based applications (Theodore et al., 2019). All these developments culminate in the necessity for the incorporation of surface roughness characterization into the broader surface science and engineering context.

### 9.6.3. Applications in Various Industries

The versatility of surface characterization techniques has made possible revolutionary applications in numerous areas. In nanotechnology, the ability to precisely characterize surface properties such as topography, roughness, and chemical composition allows researchers to design nanoparticles with tailored functionalities. For instance, the unique optical, electronic, and magnetic characteristics from novel surface chemistries facilitate innovation in drug delivery systems and biosensors. By using these techniques, industries can optimize the performance of nanomaterials, and thus, the effectiveness and efficiency of products. With applications extending from environmental pollution remediation to catalysis, sophisticated characterization tools are absolutely important in developing sustainable solutions for numerous pressing global problems, illustrating the role of societal contributions of nanotechnology studies (Kumari et al., 2023). Another area of emphasis affected by surface characterization techniques is the development of green materials, that is, natural fiber-reinforced polymer composites (NFRPC). The composites, which are based on natural fibers and polymers, exhibit unique surface properties that may be tailored with advanced characterization techniques, creating new possibilities in the construction and automotive industries. Successful incorporation of these composites hinges on overcoming fiber/matrix adhesion issues and processing techniques. By the application of surface characterization techniques, researchers are able to improve the matrix-fiber interfacial properties that consequently contribute to enhanced mechanical properties in NFRPCs. Such agreement in properties opens up possibilities for lighter and stronger materials that facilitate green manufacturing practices, thereby increasing industrial sustainability (Kamarudin et al., 2022). Surface characterization techniques have also revolutionized the manufacturing and quality control operations in industries ranging from electronics to coatings. The comprehensive surface characterization assists in the development of advanced coatings with enhanced durability, corrosion resistance, and aesthetics. As an example, in the aerospace industry, precise surface analysis ensures coatings meet rigorous regulating standards while delivering best performance within aggressive environments. Additionally, the combination of automated characterization methods has enhanced production productivity and accuracy while reducing waste. As industries become increasingly concerned with sustainability and performance, the role of surface characterization is only going to become more important in enabling innovation and product reliability within different applications (Kamarudin et al., 2022).

## SUMMARY

In summarizing research into surface characterization techniques, it is important to underscore the pivotal place such techniques have in advancing scientific and industrial inquiry. Using various techniques, researchers can obtain true and precise information about surface characteristics, which provide the very foundation of most disciplines, ranging from materials science to biology. The very critical study of surface-active agents, as debated in numerous research studies, is especially crucial for knowing their possible applications in industrial processes. However, it has been noted that inaccuracies in such studies undermine their validity and create significant impediments to advancements within the field, referencing the significance of robust methodological designs and standard operating procedures that yield reliability and reproducibility in experimental outcomes, as argued by leading researchers within the field (Matthew et al., 2020). The impact of surface characterization extends beyond academic studies; it directs useful applications that have implications in technology development and innovation. The definition of effective characterization approaches empowers societal and industrial players to capitalize on the properties of the surface in functional enhancement. The connection between accurate characterization and technological advancement is critical, particularly in the context of new materials and mixtures whose promise in alternative applications, like medicine and the environment, is eagerly awaited. For instance, knowledge on the biomechanical behavior of materials through characterization can enhance biocompatible implant design or advance pollution control technology. Thus, a commitment to enhancing surface characterization methods can result in further applications supporting worldwide sustainability and health goals (Colavita et al., 2020). Ultimately, surface characterization research is reliant on continuous refinement and modification of methods to respond to evolving issues. As the field progresses, interdisciplinary collaboration will become crucial for cross-pollinating different approaches and disciplines so that more intricate and in-depth analyses can be performed. Interdisciplinary collaboration may lead to hybrid methods that combine standard characterization methods with new technologies such as nanotechnology and artificial intelligence. With new solutions discovered and academic standards maintained, the potential for taking huge strides in surface science can be achieved, thereby facilitating the continued push for the optimization of materials for various applications, which consequently will benefit society as a whole (Matthew et al., 2020).

Methods of characterization play a fundamental role in the investigation of surface properties and behaviors, particularly in materials science and engineering. The techniques vary from microscopic to spectroscopic and provide different types of information on surface features including topography, composition, and electronics. Scanning Electron Microscopy (SEM) is particularly renowned for the ability to yield high-resolution images of the surface structure that facilitate extensive morphological investigations. Moreover, techniques like Atomic Force Microscopy (AFM) provide three-dimensional nanoscale surface topography, revealing fine textures that have a significant influence on material properties. By integrating these techniques, researchers can obtain a general characterization of surfaces, which is essential in designing materials to meet different applications across industries from electronics to biomaterials. Another critical surface characterization method is X-ray Photoelectron Spectroscopy (XPS), through which surface chemistry can be studied by measuring elemental composition and states of oxidation. It is particularly valuable in the determination of functional groups and chemical states on surfaces and provides insights essential to catalysis and corrosion science applications. Furthermore, Fourier Transform Infrared Spectroscopy (FTIR) is incorporated in order to improve the analysis since it identifies some molecular vibrations within specific functional groups, allowing for greater understanding of the organic and inorganic surface interactions. An integration of XPS and FTIR allows for rigorous analysis of chemical properties resulting in a conclusive conclusion about the creation of surface treatments and modifications for improved performance and durability. Lastly, one cannot stress enough the importance of the use of quantitative methods such as contact angle measurement, since it provides direct information on surface wettability and energy. This is a critical parameter for cases of adhesion, coatings, and fluid interface. Surface wettability can significantly affect the application of the surface, controlling such processes as self-cleaning, antifogging, and anti-corrosion. Techniques such as the Sessile Drop method of contact angle measurement can help elucidate the contact of surfaces with liquids, in order to provide a measurable platform to determine surface properties. By combining macroscopic measurements with atomic-level characterization techniques, researchers are able to achieve an integrative picture of surface response that guides the development of future materials and optimizes their functionality in real-world use (Kaltenbach et al., 2020; Th Céry et al., 2018).

# REFERENCES


1. Abelmann, L., Heskamp, I. R., Heskamp, I., et al. (2002). The CantiClever: A dedicated probe for magnetic force microscopy. https://core.ac.uk/download/pdf/11460492.pdf (accessed on 3rd October 2025).
2. Abou-Ras, T., Kirchartz, U., & Rau, U. (2016). Advanced Characterization Techniques for Thin Film Solar Cells. https://www.semanticscholar.org/paper/4942c0ae5859ee6c435da1f13a4f2d8dff01c94b (accessed on 3rd October 2025).
3. Akhtar, S., & Zuhair, F. (2025). Advancing nanomedicine through electron microscopy: Insights into nanoparticle cellular interactions and biomedical applications. International Journal of Nanomedicine, 20, 2847–2878. https://www.semanticscholar.org/paper/34a4f683bc86095b7e58bdbc38272be19087c21f (accessed on 3rd October 2025).
4. Alhabeb, M., Maleski, K., Anasori, B., Lelyukh, P., Clark, L., Sin, S., & Gogotsi, Y. (2017). Guidelines for synthesis and processing of two-dimensional titanium carbide ($Ti_3C_2T_x$ MXene). Chemistry of Materials, 29(18), 7633–7644. https://doi.org/10.1021/acs.chemmater.7b02847 ((accessed on 3rd October 2024).)
5. Anvari, B., Bartovska, L., Havrda, J., et al. (2015). Direct probing of dispersion quality of $ZrO_2$ nanoparticles coated by polyelectrolyte at different concentrated suspensions. https://core.ac.uk/download/323067243.pdf (accessed on 3rd October 2025).
6. Attia, A., Mohamed, M. B. I., Hegazy, M. A., Ghobashy, M., El-Wahab, H. A., & Abdelhai, F. (2025). Enhancing corrosion resistance with chemically modified aluminum oxide in UV-curable coatings applied to steel surfaces. Scientific Reports, 15. https://www.semanticscholar.org/paper/7e97a0cde34a6f29ce72060d21e07931cce9e94c (accessed on 3rd October 2024).
7. Bhattacharjee, A., Pereira, B. L., Soares, P., & Popat, K. (2024). Titania ($TiO_2$) nanotube surfaces doped with zinc and strontium for improved cell compatibility. Nanoscale, 16, 12510–12522. https://www.semanticscholar.org/paper/5338862cc73ee038c60492b64b0d8d4dddaebad0 (accessed on 3rd October 2025).
8. Borum, R., Reichert, P., Li, Z., Mohan, A. E., Bothe, J. R., Chang, I. Y. T., Su, Y., et al. (2025). Enabling liquid scanning electron microscopy for therapeutic suspensions using vacuum-compatible liquid capsules. Analytical Chemistry. https://www.semanticscholar.org/paper/4541b7d40f80cced548b68a4f794447d0c7723d8 (accessed on 3rd October 2025).

9. Candeias, A., & Madariaga, J. M. (2019). Applications of Raman spectroscopy in art and archaeology. Journal of Raman Spectroscopy, 50, 137–142. https://doi.org/10.1002/jrs.5571 (accessed on 3rd October 2025).

10. Chan, A., Xie, W., Weng, L., Ng, K., & Chan, C. (2016). Characterization of HOPG, sputtered HPOG and graphene by ToF-SIMS and XPS. RAN. https://www.semanticscholar.org/paper/8918164fbb17fdd36f8684d78b63f3d6f19e1781 (accessed on 3rd October 2025).

11. Chmela, O. (2015). Different imaging techniques for investigation of treatment effects on various substrate surfaces. https://core.ac.uk/download/30277444.pdf (accessed on 3rd October 2025).

12. Colavita, F., Lapa, D., Carletti, F., Lalle, E., Bordi, L., Marsella, P., Nicastri, E., et al. (2020). SARS-CoV-2 isolation from ocular secretions of a patient with COVID-19 in Italy with prolonged viral RNA detection. Annals of Internal Medicine, 173(3), 242–243. https://doi.org/10.7326/m20-1176 (accessed on 3rd October 2024).

13. Cryan, J. F., O'Riordan, K. J., Cowan, C. S. M., Sandhu, K. V., Bastiaanssen, T. F. S., Boehme, M., Codagnone, M. G., et al. (2019). The microbiota-gut-brain axis. Physiological Reviews, 99(4), 1877–2013. https://doi.org/10.1152/physrev.00018.2018 (accessed on 3rd October 2024).

14. Delfino, P. M., Leonhard-Trautmann, P., Mücklich, F., Ott, et al. (2024). Fabrication of smooth, periodic surface structures: Combining direct laser interference patterning and electropolishing. https://core.ac.uk/download/651994621.pdf (accessed on 3rd October 2024).

15. Díaz Marcos, J., Oncins Marco, G., & Pinna, F. (2012). Advanced optical microscopies for materials: New trends. https://core.ac.uk/download/16207437.pdf (accessed on 3rd October 2024).

16. Fatima, S., Imran, M., Kanwal, F., Javaid, A., Latif, S., & Boczkaj, G. (2023). Design and preparation of magnetically-oriented poly(styr-co-MMA)-3MPS capped Fe (ZnO) hybrid microspheres for ion exchange removal of toxic pollutants from wastewater. Water. https://www.semanticscholar.org/paper/f7fa928d3d365131bff89df64eda9453634fe583 (accessed on 3rd October 2025).

17. Gardner, W., Maliki, R., Cutts, S. M., Muir, B. W., Ballabio, D., Winkler, D. A., & Pigram, P. J. (2020). Self-organizing map and relational perspective mapping for the accurate visualization of high-dimensional hyperspectral data. Analytical Chemistry, 92(15), 10450–10459. https://doi.org/10.1021/acs.analchem.0c00986 (accessed on 3rd October 2025).

18. Gerecsei, T., Horvath, R., Orgovan, N., et al. (2019). A practical review

on the measurement tools for cellular adhesion force. http://arxiv.org/abs/1905.12329 (accessed on 3rd October 2024).

19. Heib, F., & Schmitt, M. (2018). Statistical contact angle analyses with the high-precision drop shape analysis (HPDSA) approach: Basic principles and applications. https://core.ac.uk/download/287035103.pdf (accessed on 3rd October 2024).

20. Heidelberg, A., Lanotte, M., & Kutay, M. E. (2020). Three-dimensional reconstruction and morphological characterization of aggregates using imaging and spherical harmonics. Journal of Testing and Evaluation. https://www.semanticscholar.org/paper/ae3365f9d9f5be7d780c0518e9a8478675ef67cd (accessed on 3rd October 2025).

21. Huo, X., Sun, L., Yang, Z., Li, J., Feng, C., Zhang, Z., Pan, X., et al. (2023). Mechanism and quantitative characterization of wettability on shale surfaces: An experimental study based on atomic force microscopy (AFM). Energies. https://www.semanticscholar.org/paper/011a9b7869ae159329b984bc625999f0e5ed46e1 (accessed on 3rd October 2025).

22. Hussain, G., Shehbaz, T., & Alkahtani, M. (2024). Investigating impact of in-process cooling mediums on microstructural and mechanical properties of FSWed AA2219 T6 joints. Scientific Reports, 14. https://www.semanticscholar.org/paper/13e3a4ceea356a6cd704101d88d2ca637fd0252c (accessed on 3rd October 2024).

23. Jampani, S. C., Arige, S., Nagendra, S. L., Ramachandra Rao, M. S., & Dixit, T. (2024). Characterization of CuO thin films based memristive devices for resistive random-access memory. 2024 IEEE 9th International Conference for Convergence in Technology (I2CT). https://www.semanticscholar.org/paper/49980257df49dd010923fad2f340c241cd04764d (accessed on 3rd October 2024).

24. Janu, R., Mrlik, V., Ribitsch, D., Hofman, J., Sedláček, P., Bielská, L., & Soja, G. (2021). Biochar surface functional groups as affected by biomass feedstock, biochar composition and pyrolysis temperature. Carbon Resources Conversion, 4, 36–46. https://doi.org/10.1016/j.crcon.2021.01.003 (accessed on 3rd October 2025).

25. Joudeh, N., & Linke, D. (2022). Nanoparticle classification, physicochemical properties, characterization, and applications: A comprehensive review for biologists. Journal of Nanobiotechnology, 20, Article 100. https://doi.org/10.1186/s12951-022-01477-8 (accessed on 3rd October 2024).

26. Kaltenbach, T., Anderson, J. C., Burke, C. A., Dominitz, J. A., Gupta, S.,

Lieberman, D. A., Robertson, D. J., et al. (2020). Endoscopic removal of colorectal lesions: Recommendations by the US Multi-Society Task Force on Colorectal Cancer. The American Journal of Gastroenterology, 115(3), 435–464. https://doi.org/10.14309/ajg.0000000000000555 (accessed on 3rd October 2025).

27. Kamarudin, S. H., Mohd Basri, M. S., Rayung, M., Abu, F., Ahmad, S., Norizan, M. N., Osman, S., et al. (2022). A review on natural fiber reinforced polymer composites (NFRPC) for sustainable industrial applications. Polymers, 14(17), 3698. https://doi.org/10.3390/polym14173698 (accessed on 3rd October 2025).

28. Khan, M. A., Nayan, N., Shadiullah, S., Ahmad, M. K., & Soon, C. F. (2020). Surface study of CuO nanometals by advanced nano characterization techniques with enhanced optical and catalytic properties. Nanomaterials, 10(7), 1298. https://doi.org/10.3390/nano10071298 (accessed on 3rd October 2024).

29. Kmetík, M., Kopal, I., Král, M., & Dendisová, M. (2024). Characterization of modified PVDF membranes using Fourier transform infrared and Raman microscopy and infrared nanoimaging: Challenges and advantages of individual methods. ACS Omega, 9(29), 24685–24694. https://www.semanticscholar.org/paper/5b5d9d63141493a195409646bc79322816e031d6 (accessed on 3rd October 2024).

30. Kumari, S., Raturi, S., Kulshrestha, S., Chauhan, K., Dhingra, S., András, K., Thu, K., et al. (2023). A comprehensive review on various techniques used for synthesizing nanoparticles. Journal of Materials Research and Technology, 27, 1739–1763. https://doi.org/10.1016/j.jmrt.2023.09.291 (accessed on 3rd October 2025).

31. Liang, X. (2023). Visualization of nanomechanical properties of polymer composites using atomic force microscopy. Polymer Journal, 55(9), 913–920. https://www.semanticscholar.org/paper/e457ac4e1ea4767dc1e928067d2f9af81ca4d198 (accessed on 3rd October 2025).

32. Lombardo, T., Kern, C., Rohnke, M., Sann, J., & Janek, J. (2023). Machine learning-assisted battery electrode microstructure characterization through secondary-ions imaging. ECS Meeting Abstracts. https://www.semanticscholar.org/paper/b4e1c4c9a32b3e1ccd8124e98e03a51e6acdc1c0 (accessed on 3rd October 2025).

33. Magdy, M. (2023). X-ray techniques dedicated to materials characterization in cultural heritage. Chemistry Select. https://www.semanticscholar.org/

paper/2985d8c963a423d64f594da3efb0cc0f991d4f90 ((accessed on 3rd October 2024).)

34. Major, G. H., Pinder, J. W., Austin, D., Baer, D., Castle, S., Čechal, J., Clark, B. M., et al. (2023). Perspective on improving the quality of surface and material data analysis in the scientific literature with a focus on x-ray photoelectron spectroscopy (XPS). Journal of Vacuum Science & Technology A. https://www.semanticscholar.org/paper/a9cf0f0104c94170fbca0f35a3cd0b123a32fa78 (accessed on 3rd October 2024).
35. Mandal, R., Cavill, S., Bloomer, S., Carpenter, C., & Baufeld, B. (2024). Surface characterization of laser clad surfaces using profile and areal techniques. Advances in Materials Technology for Power Plants. https://www.semanticscholar.org/paper/f6783439adbf4a7703e9bbbe2205a95ba95cd2a0 (accessed on 3rd October 2025).
36. Mei, H., Laws, T. S., Terlier, T., Verduzco, R., & Stein, G. E. (2021). Characterization of polymeric surfaces and interfaces using time-of-flight secondary ion mass spectrometry. Journal of Polymer Science, 60(8), 1174–1198. https://doi.org/10.1002/pol.20210282 (accessed on 3rd October 2024).
37. Myint, A., Tripathi, A., Wojcik, M. J., Deng, J., Cherukara, M. J., Schwarz, N., Narayanan, S., et al. (2024). Three-dimensional hard X-ray ptychographic reflectometry imaging on extended mesoscopic surface structures. APL Photonics. https://www.semanticscholar.org/paper/8e8fe2f0c5183057626bbae072b69674ad4c6341 (accessed on 3rd October 2025).
38. Nanda Gopala Krishna, A., & Philip, J. (2022). Review on surface-characterization applications of X-ray photoelectron spectroscopy (XPS): Recent developments and challenges. Applied Surface Science Advances, 12, 100332. https://doi.org/10.1016/j.apsadv.2022.100332 (accessed on 3rd October 2024).
39. Nilsson, M., Komic, H., Gustafsson, J., Sheybani, Z., Paul, S., Rolfson, O., Hellstrand, K., et al. (2023). Multiomic single-cell analysis identifies von Willebrand factor and TIM3-expressing BCR-ABL1+ chronic myeloid leukemia stem cells. Blood. https://www.semanticscholar.org/paper/c8455bcb5f6d706bc81a241543be9fd066ab371e (accessed on 3rd October 2024).
40. Paramita, P., Murugesan, R., Narashiman, S., Selvamurugan, N., Sukumar, D. K., Chung, T. W., & Moorthi, A. (2021). Sol–gel-based synthesis and

biological properties of zinc integrated nano bioglass ceramics for bone tissue regeneration. Journal of Materials Science: Materials in Medicine, 32. https://doi.org/10.1007/s10856-020-06478-3 (accessed on 3rd October 2025).

41. Penin, A. N., Aleksandrovskii, A. A., Golenishev-Kutuzov, A. V., Kang, C. H., Hubert, C., Sibilia, C., Jundt, D., et al. (2005). Mapping of periodically poled crystals via spontaneous parametric down-conversion. http://arxiv.org/abs/physics/0505215 (accessed on 3rd October 2025).
42. Permyakova, A., Herranz, J., El Kazzi, M., Diercks, J. S., Povia, M., Mangani, L. R., Horisberger, M., et al. (2019). On the oxidation state of $Cu_2O$ upon electrochemical $CO_2$ reduction: An XPS study. ChemPhysChem, 20, 3120–3127. https://doi.org/10.1002/cphc.201900468 (accessed on 3rd October 2025).
43. Piwowarczyk, J., Jędrzejewski, R., Moszyński, D., Kwiatkowski, K., Niemczyk, A., & Baranowska, J. (2019). XPS and FTIR studies of polytetrafluoroethylene thin films obtained by physical methods. Polymers, 11(10), 1629. https://doi.org/10.3390/polym11101629 (accessed on 3rd October 2024).
44. Rappaport, T. S., Xing, Y., Kanhere, O., Ju, S., Madanayake, A., Mandal, S., Alkhateeb, A., et al. (2019). Wireless communications and applications above 100 GHz: Opportunities and challenges for 6G and beyond. IEEE Access, 7, 78729–78757. https://doi.org/10.1109/access.2019.2921522 (accessed on 3rd October 2025).
45. Ravines, P., Sheets, D., Webb, M., Mazurek, J., Schilling, M. R., & Khanjian, H. (2025). Surface characterization of Asian lacquers using surface metrology and data science: Introducing the roughness spectrum. Surface Topography: Metrology and Properties, 13. https://www.semanticscholar.org/paper/02f6c851a57f2ab92813d9d176c8adbbc2904087 (accessed on 3rd October 2025).
46. Teng, X., Li, F., & Lu, C. (2020). Visualization of materials using the confocal laser scanning microscopy technique. Chemical Society Reviews, 49(9), 2408–2425. https://doi.org/10.1039/c8cs00061a (accessed on 3rd October 2025).
47. Théry, C., Witwer, K. W., Aïkawa, E., Alcaraz, M. J., Anderson, J. D., Andriantsitohaina, R., Antoniou, A., et al. (2018). Minimal information for studies of extracellular vesicles 2018 (MISEV2018): A position statement of the International Society for Extracellular Vesicles and update of the MISEV2014 guidelines. Journal of Extracellular Vesicles, 7. https://doi.org/10.1080/20013078.2018.1535750 (accessed on 3rd October 2025).

48. Twigg, M. S., Baccile, N., Banat, İ. M., Déziel, É., Marchant, R., Roelants, S., & Van Bogaert, I. (2020). Microbial biosurfactant research: Time to improve the rigor in the reporting of synthesis, functional characterization and process development. Microbial Biotechnology, 14(1), 147–170. https://doi.org/10.1111/1751-7915.13704 (accessed on 3rd October 2024).

49. Vieira da Cunha, A. M., Koosah, M. P., Xavier, J. A. P., Coutinho, C. E. B., Pessoa, E. C., Mezzomo, E. H., da Hermida, A. S., et al. (2024). Enhanced reservoir characterization and horizontal well placement with the use of high-definition and three-dimensional reservoir mapping-while-drilling systems in Campos Basin, offshore Brazil. SPWLA 65th Annual Symposium Transactions. https://www.semanticscholar.org/paper/6d42d67caab9d6a5a03844ed480b7f4d59934ac2 (accessed on 3rd October 2025).

50. Xu, S., Hou, X., Wang, D., Zuin, L., Zhou, J., Hou, Y., & Mann, M. D. (2022). Insights into the effect of heat treatment and carbon coating on the electrochemical behaviors of SiO anodes for Li-ion batteries. Advanced Energy Materials, 12. https://doi.org/10.1002/aenm.202200127 (accessed on 3rd October 2025).

51. Zhu, B., Wang, D., & Wei, N. (2021). Enzyme discovery and engineering for sustainable plastic recycling. Trends in Biotechnology, 40, 22–37. https://doi.org/10.1016/j.tibtech.2021.02.008 (accessed on 3rd October 2025).

# CHAPTER 10

# SURFACE CHEMISTRY IN NANOTECHNOLOGY

## Contents



## INTRODUCTION

The field of nanotechnology is at the intersection of materials science and chemistry, studying phenomena occurring at the nanoscale, as typically 1 to 100 nanometers. At such a scale, properties of materials can radically deviate from their bulk counterparts, creating new performances and applications. Surface chemistry then becomes a deciding factor in these behaviors since nanoparticles possess a very high surface-to-volume ratio, making them reactive and more capable of interacting with the surroundings. The contrast makes nanoparticles valuable to use in a variety of industries, including pharmaceuticals, catalysis, and environmental remediation. The surface chemistry plays a particularly critical role in the domains related to nanoparticle efficiency, where modification on their surface can significantly alter their functionalities and properties, making way for innovative material science solutions. Surface chemistry is also important for considering the viability of nanoparticles in medicine. For example, iron oxide nanoparticles have found extensive applications based on their superparamagnetism and biocompatibility to make them viable candidates for targeted drug delivery and magnetic resonance imaging. Surface functionalization is solely responsible for the efficacy of these nanoparticles in biomedical applications that can enhance stability, enhance bioavailability, and decrease toxicity. As mentioned above, "keeping in mind the fact that surface chemistry of MNPs plays an important role in the practical application of in vitro and in vivo applications," it is evident that extensive research on the technique of surface modification is necessary to maximize the use of such nanomaterials for various applications, resulting in the development of medical science (Ganapathe et al., 2020). Furthermore, the surface chemistry is relevant in environmental science, particularly wastewater treatment. Nanomaterials such as graphene and graphene derivatives have been of great promise in the removal of heavy metal ions because of the tunability of their surface chemistry and high specific surface area. The nanomaterials can adsorb impurities with efficient performance beyond traditional approaches such as ion exchange and chemical precipitation. Notably, adsorption, given its effectiveness and low cost, has become a critical water purification method. Understanding the complex correlation between surface properties and impurities not only clarifies adsorption processes but also directs the synthesis of even more effective materials (Khaligh and Johan, 2020). With the continued advancements in surface chemistry, the number of new applications and remedies to pressing environmental issues keep increasing, highlighting the significance of the field in nanotechnology as well as materials science.

## *Importance of Surface Chemistry in Nanotechnology*

The subtle role of surface chemistry in nanotechnology cannot be overstated, considering its vital impact on nanoscale materials' properties and nanotechnology functions. Surface chemistry refers to the chemical reactions and processes occurring at the interface between materials and have the potential to greatly alter the way materials behave upon nanoscale miniaturization. Here, a significant proportion of the atoms in a material reside on its surface, and surface modification—physical or chemical—may be a dominant factor affecting such characteristics as reactivity, stability, and compatibility with biological entities. For instance, nanoparticles can show contrasting behavior compared to their bulk phase analogs due to differences in surface area-to-volume ratios and higher surface energy, hence the necessity of an entire knowledge of the surface chemistry to harness their potential to the fullest. Further, the application of biotemplates, for instance, plant virus capsids, illustrates the usefulness of surface chemistry in nanotechnology. These biological structures present a unique opportunity for the synthesis of new materials, as they are of high-precision, three-dimensional structures that are best suited for nanoscale applications. The nature of these biomolecules allows them to be synthesized into highly monodisperse amounts of material, and their failure to replicate themselves within mammalian cells means they are perfectly suited for safe use in biomedical applications. By examining the nanoscale characteristics of such biotemplates, researchers employ surface chemistry to design and synthesize high-performance materials with built-in functionality not accessible otherwise, demonstrating the role of surface interactions in nanotechnology (Aljabali et al., 2015). In addition, innovation of novel surface modification technologies, including plasma processing, demonstrates the intersection of surface chemistry with emerging manufacturing technologies. Plasma processes are used to modify the surface characteristics of various materials, including textiles and plastics, for the production of advanced surfaces to specified performance levels. This not only enhances adhesion, wettability, and resistance but also aligns with sustainable manufacturing goals by minimizing waste and maximizing process efficiency. As has been discussed in recent literature, emphasis on surface design through non-equilibrium plasma technologies has the potential to revolutionize materials science by propagating the growth of cost-conscious, eco-friendly, and high-speed processing across various industries (Arias et al., 2019). Hence, it is still important to learn about surface chemistry for further advancement in nanotechnology and the creation of improved materials.

## 10.1. ROLE OF SURFACE CHEMISTRY IN NANOMATERIALS

The role played by surface chemistry in nanomaterials is pivotal due to the unique properties that are bestowed by their nanoscale dimensions. At this scale, most of the atoms will reside at or on the surface and, thus, influence the general chemical and physical response of the materials profoundly. Surface treatments can alter many such properties such as reactivity, catalytic activity, and adsorption capacity. For example, the surface chemistry of nanomaterials can be customized to permit increased adsorption properties, which is particularly important for processes such as wastewater treatment. Adsorption methods, especially for the elimination of heavy metals and organic contaminants, exemplify the efficiency of nanomaterials in environmental cleaning. This is supported by evidence indicating that not only does nanomaterial possess a large specific surface area but also tunable surface properties which enable it to perform more effectively in pollutant removal processes (Khaligh and Johan, 2020). Further, interfacial interactions between nanomaterials and the environment dictate their performance. Surface chemistry dictates how nanomaterials interact with substrates and other chemical species. Surface modification to enhance stability over a range of pH or ionic strengths will significantly improve their utility in drug delivery and biosensing. The freedom in modulating the charge, hydrophobicity, and functional groups leads to optimal interaction with biological systems such as cells and biomolecules. These advancements emphasize the importance of designing nanomaterials with proper surface chemistries in a carefully manner to achieve desirable performance and behavior in a specific application. The use of engineered nanomaterials for heavy metal targeting further emphasizes their potential as effective adsorbents due to highly optimized surface characteristics (Wahab et al., 2012). In addition to environmental applications, surface chemistry is also significant in the area of nanomedicine, where the design of nanocarriers is essential for drug delivery. Surface characteristics of nanoparticles control biodistribution, cellular uptake, and ultimately therapeutic efficacy. Examples involve methods such as surface functionalization with targeting ligands or coatings, which can manage how nanoparticles interact with specific cell types or tissues, enhancing their therapeutic impact with fewer side effects. Surface chemistry interaction with biological constituents is a force driving drug formulation and disease treatment technology innovation, including for cancers. With constant research going on in this field, the design of nanomaterials with controlled surface chemistries is the key to potential advances in health science and medical applications, supporting the move towards increasingly sophisticated and capable therapeutic methods (Khaligh and Johan, 2020).

### 10.1.1. Synthesis of Nanoparticles

The synthesis of nanoparticles has been of significant interest as they have unique properties and will potentially be applied to various fields, including medicine, electronics, and environmental science. Common synthesis of nanoparticles is typically physical or chemical, which may be energetically demanding and may leak dangerous compounds into the environment. Due to the necessity of developing more efficient and cleaner processes, biogenic synthesis has been one possible solution. This process employs biological systems, such as plants and microbes, to produce nanoparticles under mild conditions, which in turn tends to reduce production costs as well as environmental disturbance. By employing nature's processes, scientists are discovering new means of synthesizing nanoparticles that not only preserve ecological balance but also enhance the material's function in various applications. Biological routes for nanoparticle synthesis, particularly with the assistance of plant virus capsids, provide a fascinating interface between biology and nanotechnology. Plant viruses serve as biotemplates in constructing precisely defined, three-dimensional nanoparticle architectures due to their natural tendency to self-assemble with high accuracy. These nanoscale viruses have highly useful qualities such as monodispersity, as well as the ability to be synthesized on a scale-up basis, making them highly attractive for use in many fields of materials science. The safety aspect arising from plant viruses' lack of ability to replicate within mammalian cells also contributes to their attractiveness in biotechnological advancement. Through studies on the synthesis pathways of nanoparticles from biological templates, researchers have the potential to disclose novel functional materials that are not only efficient but also environmentally friendly, paving the way for new breakthroughs in the field in the future. In efforts to simplify available synthesis protocols, research into metal nanoparticles has revealed that biosynthetic approaches outperform conventional approaches (Figure 10.1).

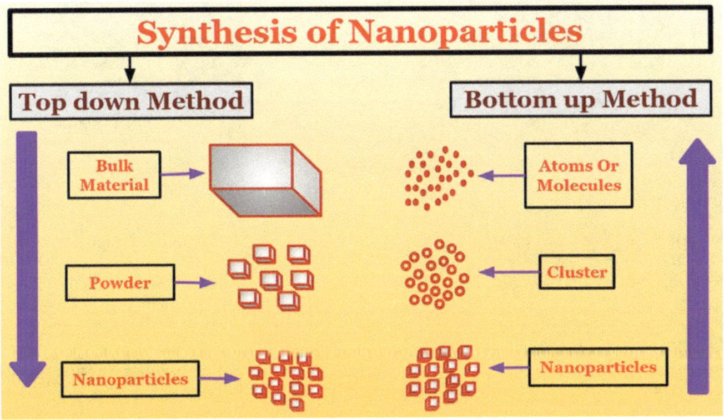


**Figure 10.1.** The synthesis of nanoparticles.

*Source: https://blogger.googleusercontent.com/img/b/R29vZ2xl/AVvXsEjCPIsRB-sZfiToJGExxlYAf6ZFHxsEIdcQe5qeh34PYUdGL9V40pk1k57YclV9o8C4PANn6K-uVNuLsl2jl-A6ow_7u5TSFHmSl3Bk0RFByye0UCySikvL1L1jyQwu7Z0kEzsuDu-og-cspdTUqlM5fXwGnGKAHZgl_tk2n7bl5hU9fQHTgfOm3mptQ-sgMY/s1267/Synthesis%20of%20Nanoparticles.jpg.*

The IRCSET-EMPOWER project, for instance, underscores the importance of designing better methodologies for metal nanoparticle synthesis, particularly in view of the environmental challenges posed by conventional approaches. Employing biogenic systems, e.g., fungi, researchers are not only surpassing certain of the drawbacks of existing approaches but also gaining insight into the biochemical processes involved in nanoparticle nucleation and growth. This controlled biosynthesis can potentially provide an improved way of synthesizing metal nanoparticles and elucidating the important factors influencing particle formation. Advancing knowledge in these processes will enable researchers to create low-cost routes for making nanoparticles, achieving a full exploitation of nanoparticle utilization in surface chemistry and materials science, potentially revolutionizing a number of industries (Aljabali et al., 2015; Das et al., 2009).

## 10.1.2. Functionalization of Nanomaterials

Functionalization of nanomaterials is one of the most important aspects of the evolution of nanotechnology, as it enhances the versatility and applicability of materials at the nanoscale. Functionalization is the process by which surface chemistry of nanomaterials is altered to confer desired characteristics for certain applications.

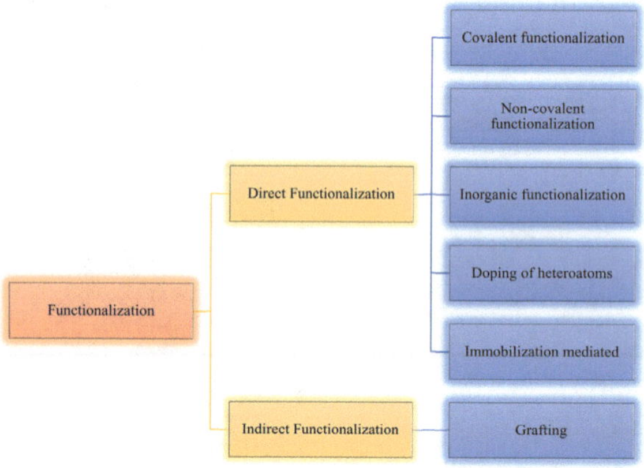


**Figure 10.2.** Different methods of functionalization of nanomaterials.

*Source:* https://www.researchgate.net/figure/Different-methods-of-functionalization-of-nanomaterials_fig2_354191150.

For instance, by chemical coating or grafting, the surface chemistry of nanomaterials can be tailored to increase their solubility, biocompatibility, or catalytic activity. These alterations create exemplary properties entirely different from their bulk counterparts, which are of extreme importance in fields such as medicine, where tailored-made nanomaterials can improve drug delivery systems or create targeted therapies. The dynamic involvement of chemical interactions on the surface itself is all-important in order to deliver the required functional properties, thus demonstrating the central position of functionalization in maximizing the application of nanomaterials in various fields. The most significant contribution of functionalized nanomaterials is in the area of nanomedicine. In this area, the precise control over the surface of nanomaterials allows for the development of complex drug delivery systems, which can target specific tissues or cells selectively.

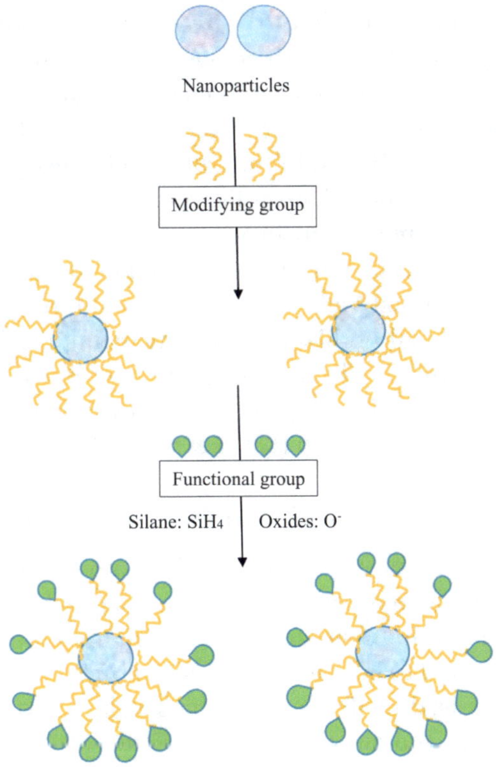


**Figure 10.3.** An illustration of direct functionalization of nanomaterials.

*Source:* https://www.researchgate.net/figure/An-illustration-of-direct-functionalization-of-nanomaterials_fig3_354191150.

Recent advancements in nanocarrier development have stressed the efficacy of the utilization of functionalized nanoparticles in loading therapeutic drugs to enhance their biocompatibility with fewer side effects. Furthermore, the bio-functionalization of nanomaterials with targeted ligands or antibodies enables targeted drug delivery, ensuring optimal therapeutic effects while reducing off-target effects. The therapeutic potential of such nanomaterials in the delivery of chemotherapeutic drugs and biologics has been hailed as a revolution in the treatment of chronic diseases, illustrating their revolutionary impact on healthcare. These innovative applications reflect the profound effect of functionalization on the treatment of patients and therapeutic strategies in modern medicine (Patra et al., 2018) (Figure 10.2).

Outside the area of medicine, functionalized nanomaterials are more and more well known in terms of application in environmental clean-up and sustainability. Surface modification to enhance the reactivity of nanoparticles can optimize their potential to adsorb contaminants or to catalyze degradation reactions. Functionalization of nanomaterials like metal oxides and mesoporous silica has been the focus of recent research focused on environmental performance improvement. By modifying surface chemistry, such nanoparticles can be designed to effectively pick up pollutants such as heavy metals or organic molecules and thus become useful in wastewater treatment and air filtration systems. This application is part of a broader trend in materials science, wherein the sustainability of nanomaterials is being explored through functionalization to address important global problems. With advancing discipline, the creation of functionalized nanomaterials that integrate improved properties with an environmental conscience will take center stage to technological development and ecologically sound stewardship (Chen et al., 2021) (Figure 10.3).

## 10.1.3. Stability and Reactivity of Nanoparticles

Nanoparticle stability and reactivity are largely governed by surface chemistry, which dictates their interaction with other materials and environments. Nanoparticles possess a high surface area-to-volume ratio that renders them highly reactive compared to bulk materials. This property is the very essence of applications in catalysis, drug delivery, and biosensing. Surface features such as charge, hydrophobicity, and functional groups dictate how nanoparticles interact with biological systems and chemical milieus. For instance, nanobiosensor design has been a central drive, taking advantage of such surface characteristics in order to achieve high sensitivity and specificity. This is done mainly through the integration of nanomaterials that enhance transduction processes and, as such, overcome problems associated with detecting biomolecules at low concentrations. This advancement highlights the pivotal role of surface chemistry in determining the stability and reactivity of nanoparticles in health care and environmental monitoring functions pivotal to well-being (Naresh et al., 2021). Metal-based nanoparticles' reactivity is unique in their non-specific antibacterial nature that is capable of addressing a wide array of pathogens. Experiments have shown that the particles, such as silver, gold, and copper nanoparticles, are able to change bacterial membranes and disrupt metabolic processes, thereby reducing the likelihood of developing resistance that is typical with conventional antibiotics. Physicochemical properties of the nanoparticles, such as size and shape, also influence their reactivity and bioavailability. Research emphasizes understanding how nanoparticles interact with microbes in order to achieve maximum efficiency as antimicrobial substances. The broader implications of this research have applications in biomedical application, where metal nanoparticles are used not just for their reactivity but even for their stability so that they are able to sustain their functional capacity under different conditions (Elena et al., 2020) (Figure 10.4).

**Figure 10.4.** Stability and reactivity.

*Source:* https://pubs.acs.org/doi/10.1021/acs.chemrev.7b00208.

Besides, the stability of nanoparticles during synthesis and application is also important for their maintenance of functional properties. Agglomeration, oxidative degradation, and environmental conditions are some of the influences that can degrade the stability of nanoparticles and affect their surface chemistry and reactivity. For instance, encapsulation of nanoparticles in protective matrices is a common strategy employed to stabilize them and maintain their effectiveness and sustained release in drug delivery systems. In the same way, functionalization of target ligands on nanoparticle surfaces improves their stability by shielding them from degrading agents and improving their interaction with target cells. Constant development of new materials and processes is aimed at creating nanoparticles that retain their stability and reactivity across various conditions, ultimately contributing to nanotechnology and material science progress. Hence, the understanding of the dynamic interaction between reactivity and stability remains relevant to the discovery of new applications (Naresh et al., 2021; Elena et al., 2020).

## 10.2. SURFACE CHEMISTRY IN CATALYSIS

The role of surface chemistry in catalysis cannot be overstressed, as it inherently fixes the efficiency and selectivity of catalytic processes. Catalysts operate by providing a pathway of lower activation energy, mostly by surface interactions. The nature of this surface, such as composition, morphology, and electronic structure, is the deciding factor in the way molecules adsorb, react, and desorb. Metal and metal oxide nanoparticles are particularly important in this context, as their high surface area-to-volume ratio enhances catalytic activity. Their ability to be exactly controlled in their surface characteristics allows scientists to tailor catalysis for many chemical reactions with further enhancement of performance in industrial applications. These advances demonstrate the critical interdependency of surface chemistry and catalysis, confirming that there is a necessity to explore and harness these materials for accelerating chemical reactions in an efficient manner. Engineered inorganic nanocrystals from nanotechnology exhibit exceptional catalytic potential due to their unique surface characteristics. By manipulating particle dimensions and surface functionality, scientists have engineered nanohybrids to efficiently catalyze different chemical reactions. These multi-functional materials not only enhance the rate of reactions but also offer targeted specificity in areas such as environmental remediation and biomedical investigation. For instance, nanohybrids may be engineered to enhance their surface chemistry for optimal interaction to help enhance their use in catalysis in key processes involving the degradation of pollutants or drug therapy. The variety of these nanostructures offers an auspicious avenue towards the creation of catalysis, pushing the boundaries of traditional methods, and providing new avenues for efficient industrial processes and environmentally benign solutions. Surface chemistry is paired with catalysts also in employing functional nucleic acids (FNAs) for therapeutic and bioanalytical uses. Recent advances in DNA nanotechnology have unveiled the possibility of designing and synthesizing self-assembled DNA nanostructures with FNAs, making use of their unique properties for targeted delivery. Such FNAs can control specific recognition, therapeutic efficacy, and even catalytic reactions at the molecular level. By functionalization of such nanostructures with FNAs, scientists can achieve higher specificity in cellular systems, which corresponds to better bioimaging and theranostic applications. As highlighted in the current paradigm of research, the synergy between FNAs and surface chemistry is a field of frontier catalysis. Continuation with such synergies will develop the foundation for revolutionized applications transversal to nanotechnology, materials science, and biomedical innovation and affecting various fields of study (Huang et al., 2020; Mujahid et al., 2022).

### 10.2.1. Heterogeneous Catalysis Mechanisms

Mechanistic research in heterogeneous catalysis is at the forefront of surface chemistry advances in nanotechnology and materials science. Heterogeneous catalysts, typically solids, come into contact with gaseous or liquid reagents, enabling chemical reactions to take place within the interface. Such interactions are crucial in accelerating reaction rates with the aim of saving energy.

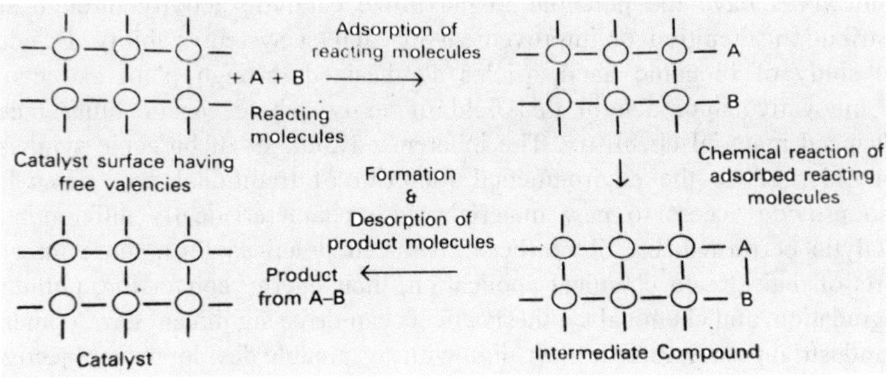


**Figure 10.5.** The mechanism of heterogeneous catalysts.

*Source:* https://d10lpgp6xz60nq.cloudfront.net/physics_images/VIK_CHE_QB_C04_E03_007_S01.png.

The activity of these catalysts is based on the characteristics of their surface, beginning with the surface area, porosity, and the presence of active sites. Identifying the nuance in these features allows scientists to fashion nanomaterials for specific catalytic purposes. Various nanoparticle systems, i.e., metal and metal oxide nanoparticles, possess intrinsic physicochemical properties that enhance their catalytic efficiency, rendering them good prospects for environmentally friendly processes in the operations of chemical industries, which is the overall vision of green chemistry elucidated in recent research (Joudeh et al., 2022). Furthermore, the working mechanisms of the heterogeneous catalysts are complex at the atomic scale. Such processes are usually categorized into several stages like adsorption, reaction, and desorption. The reactants are initially adsorbed on the surface of the catalysts, in which they must interact with the active sites in the right manner. Certain reactions then lead to the formation of the products, which are released or desorbed from the catalyst. The specifics of these steps are not uniform and can be a function of a range of variables, including temperature, pressure, and even the nature of the catalyst material itself. Recent findings underscore the mechanisms through which the nanoscale dimensions of catalysts alter such surface interactions, leading to enhanced reactivity and selectivity, critical in the design of new reactions and catalytic processes, like those being researched in green synthesis strategies highlighted in modern nanotechnology (Shafey, 2020) (Figure 10.5).

As researchers investigate deeper into mechanisms of heterogeneous catalysis, relevance of surface modification techniques is of prime importance. Techniques such as support material selection or the use of metal–organic frameworks have the potential to maximize catalytic activity through site distribution alteration or improvement in catalyst system stability. Besides, the study of biogenic nanoparticles synthesized through plant extracts is an innovative approach in this field of study, where sustainability meets advanced material chemistry. The inherent advantages of biogenic synthesis not only reduce the environmental footprint of traditional procedures but also provide access to new materials with characteristically differentiated catalytic performances. Since these advanced practices form more integral parts of mainstream chemical application, their energy conversion, pollutant degradation, and chemical synthesis effects can drive significant developments in industrial processes as well as align with sustainable development objectives (Joudeh et al., 2022).

## 10.2.2. Role of Nanostructured Catalysts

The use of nanostructured catalysts has come into prominence with an enhanced role in establishing green energy alternatives, particularly for hydrogen generation. As clean energy demands continue to grow continuously, the electrocatalytic hydrogen evolution reaction (HER) has been the subject of intense research attention as it is a fundamental process in water electrolysis. Recent research emphasizes the potential of low-cost and high-performance nanostructured catalysts made of noble and non-noble metals. These catalysts possess higher intrinsic activity and a higher density of active sites, which enables improved efficiency of hydrogen generation. With the innovative utilization of the unique surface properties and morphologies of nanostructured materials, researchers aim to optimize catalytic activity and thus facilitate the transition toward a hydrogen economy. Therefore, systematic research on reactivity as it relates to catalyst development is crucial for constructing effective solutions in sustainable hydrogen production (Zhu et al., 2019). Other than their application in the manufacture of hydrogen, nanostructured catalysts also play a leading role in the optimization of chemical reactions in various industries, with increased selectivity and reactivity. The modifiable physical and chemical properties of nanomaterials allow one to design catalysts that best drive desired reactions and suppress undesired by-products. The designing of these catalysts relies on the comprehension of the intrinsic properties of nanostructured materials, which can vary based on their size, shape, and composition. Nanoscale catalyst synthesis allows researchers to amplify surface area and optimize the number of accessible active sites for chemical reactions (Figure 10.6).

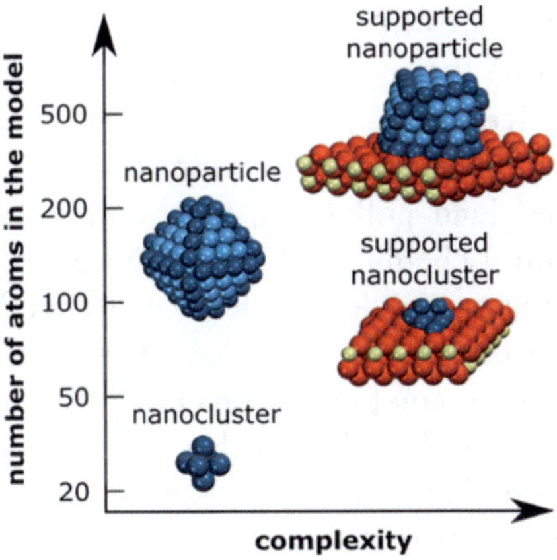


**Figure 10.6.** Typical models of nanostructured catalysts. Reliable electronic structure simulations of metal-oxide interactions require highly complex models composed of up to 500 atoms.

*Source:* https://www.researchgate.net/figure/Typical-models-of-nanostructured-catalysts-Reliable-electronic-structure-simulations-of_fig1_362023822.

These developments are vital in addressing the demand for efficient catalysis in energy conversion and environmental cleaning procedures. Thus, the introduction of advanced nanostructured catalysts is vital in guaranteeing innovation in materials science (Jeevanandam et al., 2018). Moreover, the production and use of nanostructured catalysts also have a factor to consider regarding potential impacts on the environment, that is, their toxicity. As the production and use of nanoparticles (NPs) increase, there is a need to examine their safety dynamics within the various ecosystems. The nanomaterial characteristics are known to yield unforeseen toxicological influences, and there are fears as to how they interact with the biological system. Hence, scientists are at a challenge in establishing the distinction between engineered and naturally occurring nanoparticles to understand their respective nanoscale features and assess associated risks. The nanotechnology regulatory landscape is in a state of continuous change, with countries implementing measures to minimize possible dangers of nanoparticles in the environment. The integrated approach ensures that while boosting catalysis and materials science, the stability of ecosystems and human well-being is not undermined (Jeevanandam et al., 2018).

### 10.2.3. Applications in Environmental Remediation

The increasing rate of environmental pollution justifies innovative methods of controlling contamination, and nanotechnology has the potential in solving such contamination. Because of their elevated surface area-to-volume ratio and capacity for adjusting the surface chemistry to their needs, nanomaterials are exceptionally effective for the clean-up of the environment. Nanomaterials have, in the recent past, been applied to various analytical applications in the fight against water and land pollutants, for example, for the adsorption of heavy metals and harmful chemicals. Further, these new materials can affect environmental pollutants at a molecular level, and this further enhances their capture and conversion processes. As indicated by recent studies, the use of nanotechnology in environmental operations not only aids in decontamination but also can affect the overall quality of damaged ecosystems of agricultural chemicals and thereby provides a twofold benefit of remediation and ecosystem safeguard (Bacchini et al., 2013). Among the significant uses of nanotechnology to environmental remediation is the development of targeted nanoparticles that are capable of targeting and correcting targeted contaminants. A good example is poly(acrylic acid)-coated iron oxide nanoparticles that have been shown to be very effective at adsorbing targeted contaminants such as methylene blue, a common model for dye adsorption study (Berret et al., 2012). Characterization experiments have revealed that the adsorption process of such nanoparticles facilitates the efficient complexation of the pollutants with extremely high adsorptive capacity for dyes. Not only is this native competence of nanomaterials a source of pollutant removal, but it also renders remediation processes more environmentally friendly. With particle design optimization and surface properties, researchers are able to further widen the application ranges of such materials in different environmental conditions, describing the role of surface chemistry in guaranteeing effective remediation processes. The application of nanotechnology in remediation techniques also eliminates shortcomings associated with standard practices, which are often hindered by inefficiencies and non-selective processes in contaminated environments. Traditional remediation processes are usually cumbersome and generate secondary waste, making the cleanup process more complex. Nanotechnology, on the other hand, enables specific treatments with little ancillary waste and lower operating expenses. The application of nanomaterials in agricultural uses, for example, not only cleans up chemical runoff-polluted soils but also boosts agricultural productivity through healthier soil. This merge of remediation and agricultural practice indicates the flexibility of nanotechnology and its potential to create synergies which, in the long term, contribute to cleaner environments. Thus, the ongoing R&D in this field promise to provide innovative solutions which can address the complex problems environmental pollution presents (Bacchini et al., 2013).

## 10.3. SURFACE CHEMISTRY IN BIOMATERIALS

An understanding of surface chemistry in biomaterials is vital for the design of medical applications, particularly for implantable devices, drug delivery systems, and biosensing technologies. Surface features like roughness, chemical content, and functionalization are determinant of cell adhesion, proliferation, and differentiation.

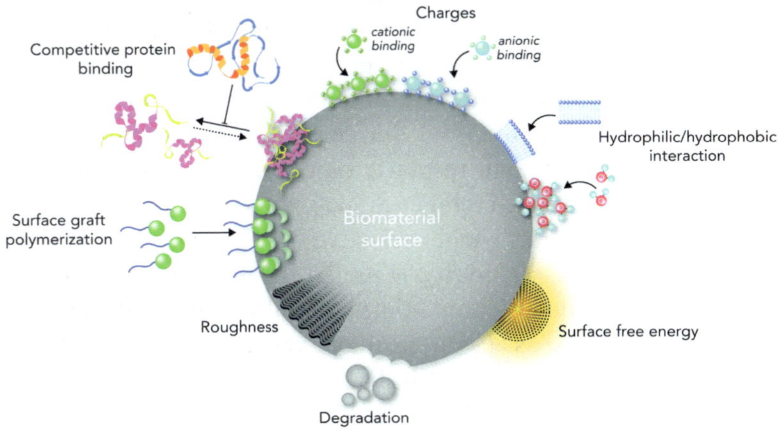


**Figure 10.7.** An illustration of the key surface physicochemical properties in directing biological responses to biomaterials.

*Source:* https://pubs.rsc.org/en/content/articlehtml/2020/cs/d0cs00103a.

The biological response to biomaterials is governed by these surface properties, determining biocompatibility and bioactivity. Recent progress in nanotechnology allows for the engineering of biomaterial surface structure at the nanoscale, maximizing their interaction with biological molecules and cellular architecture. Coatings and modifications have sought to enhance integration of such materials with biological environments, unveiling the immense potential for manipulation of surface chemistry to facilitate tissue regeneration and integration at the molecular scale (Liu, 2021). Therefore, future biomaterials development will increasingly be reliant on surface chemistry knowledge in order to solve clinical challenges and improve patient care. Nanotechnology has revolutionized biomaterials science by enabling the control of materials at the nanoscale with accuracy, thus greatly influencing surface chemistry. The control makes it possible for scientists to create new nanostructured surfaces with unique characteristics, such as enhanced mechanical strength, porosity, and surface area. Such changes are particularly important in drug delivery systems, where surface chemistry controls drug binding to targeted cells or tissues. For instance, nanoparticles can be functionalized to improve drug loading capability and release profiles and thus optimize therapeutic effects (Wahab et al., 2012) (Figure 10.7).

There is considerable interest today in the development of intelligent biomaterials that react to physiological stimulus in an active way by altering their surface properties to enhance performance. It is this integration of nanotechnology and surface chemistry that will be employed in developing advanced biomaterials that are capable of fulfilling the demands of modern medicine. The significance of surface chemistry for biomaterials extends to the technology of biosensing, where accurate measurement of biomolecules is critical to diagnosis and monitoring. Surface properties of biosensors dictate their sensitivity, specificity, and response time, making optimization of these properties in design necessary. Techniques such as surface functionalization with specific ligands or antibodies enable selective capturing of the target analytes and enhance the sensors sensitivity (Liu, 2021). Progress in nanomaterials such as graphene and metal nanoparticles has also improved the biosensors ability to detect biomolecules at single-cell or even single-molecule levels. This increased sensitivity is most critical in handling the various health problems, from instantaneous diagnosis during epidemics to chronic monitoring of chronic diseases. Nanotechnology and surface chemistry coming together on this platform hold tremendous promise for the future of precision medicine and real-time monitoring of health.

### 10.3.1. Interaction of Biomaterials with Biological Systems

The interplay between biomaterials and biological environments lies at the heart of development in new therapeutic technologies, particularly in nanomedicine. As biomaterials come into contact with biological environments, their surface chemistry plays a critical role in dictating cell and tissue response. Surface properties such as hydrophilicity, charge, and roughness may influence protein adsorption, cell adhesion, and ultimately the material's biocompatibility and function. Understanding such interactions is central to the generation of materials that not only integrate well within the body but also enhance therapeutic efficacy. The precise tuning of surface chemistry can optimize drug delivery systems, in which the goal is to achieve target actions with minimal side effects on the system. Therefore, the study of surface properties of biomaterials is a fundamental component that bridges the gap between nanotechnology and biological applications, with it serving as the foundation for subsequent innovations to emerge. Among the best applications of such principles is the development of targeted drug delivery systems through the use of nanocarriers. The capability of nanotechnology to deliver drugs directly to the site of interest can potentially increase treatment outcomes significantly, especially for conditions such as cancer. While numerous pre-clinical studies have shown the feasibility of employing such systems, there are very few that have successfully progressed to clinical use, particularly for passively targeted nanocarriers. The difference in outcomes between in vitro experiments and clinical applications attests to the complexity of biological systems and complexity of problems in biomaterial interactions in vivo. Researchers are more interested today in understanding the mechanics of the interaction, establishing a basis for designing more effective actively targeted nanocarriers. With the delineation of these principles, the field aims to enhance the therapeutic methods towards better localization and efficiency of treatment, hence promoting advancements in cancer drug therapeutics and more (Patra et al., 2018). Despite potential advancements in the field, there are challenges that hinder the clinic translation of actively targeted nanocarriers. As highlighted by recent appraisals, guidelines behind targeted drug delivery are not universal due to variations in biological barriers, immune responses, and the inherent nature of tumor microenvironments. Failure of clinically effective nanocarriers could be triggered by poor preclinical models that fail to account for such biological nuances. Thus, it is critical to study outstanding successful and failed targets along with their interaction with biological systems in an attempt to develop stronger targeting strategies. The future agenda should entail the development of standardized criteria for evaluating nanocarriers towards the aim of designing them in an optimum manner and predicting clinical success. Through research on the cutting edge of surface chemistry and biological interactions, advances in nanotechnology can fulfill their promise in altering therapeutic approaches (Rosenblum et al., 2018).

## 10.3.2. Surface Modification Techniques

Surface modification techniques are the secrets to enhancing properties and functionality of materials at the nanoscale, in line with advancements in nanotechnology. They are vital for designing surface properties to achieve desired phenomena in a wide range of applications, particularly in electronics, biotechnology, and catalysis. Among the widely applied techniques are chemical vapor deposition, atomic layer deposition, and self-assembled monolayers, each of which has certain advantages in respect to controlling thickness, uniformity, and composition. The methods allow not only tuning of physical properties, e.g., conductivity and corrosion resistance but also introduction of specific chemical functions that can interact with biological systems or facilitate catalytic processes. With emerging science, exploration of novel surface modification techniques is continuously finding new paths in nanocomposite and nanostructured material design and behavior (Chahal, 2024). Atomic layer deposition (ALD) is emerging as a groundbreaking technique in the field of surface modification, particularly renowned for its precision in the deposition of ultrathin films. This method enables conformal and uniform coverage of complex geometries, thus prefabricated electrodes as well as ultrafine powders can be used. ALD is demonstrated to be versatile in its application to further enhance the performance of nickel-rich lithium-ion battery cathodes. Research demonstrates that ALD possesses the ability to improve ionic conductivity effectively and create a protective coating to quench undesirable interfacial reactions and thereby improve mechanical stability and integrity (Meng et al., 2023) (Figure 10.8).

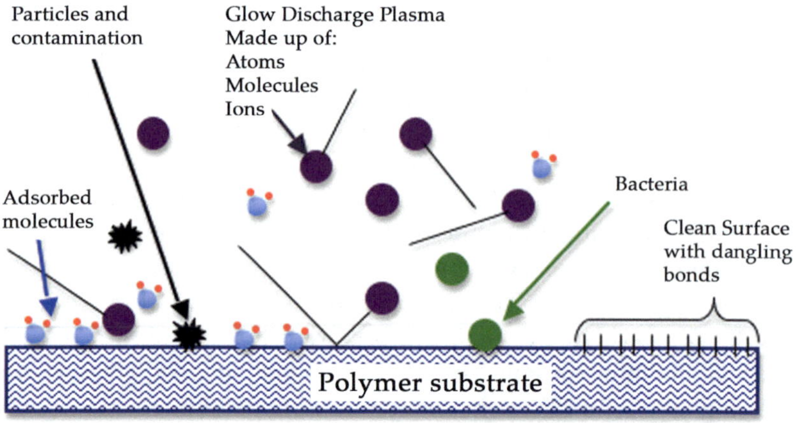


**Figure 10.8.** Schematic of cleaning of a polymer surface using glow plasma discharge.

*Source:* https://upload.wikimedia.org/wikipedia/commons/3/37/Glow_Plasma_Discharge_Schematic_Polymer_Chemistry.png.

As demonstrated in thorough studies, Li-containing oxide coatings were found to outperform their Li-free counterparts, opening up new avenues for next-generation energy storage technology while innovatively overcoming the issues of high-nickel content materials. This means that ALD not only addresses surface issues but also has an impact on the entire lifecycle and efficiency of energy storage devices. The economic and environmental impacts of surface modification techniques cannot be overstressed, given the rising imperative for sustainable practices in the field of materials science. By optimizing the surface characteristics, scientists can significantly enhance the performance metrics of materials and thus introduce more efficiency and longevity with a smaller amount of resource consumption and waste generation. Techniques such as self-assembly and biomimetic pathways enable the construction of functional surfaces with minimal energy expenditure and material usage, in accordance with the principles of green chemistry. In addition, these modifications render nanomaterials more relevant in fields such as environmental remediation and energy conversion, where their unique properties can be leveraged to address challenges confronting the world. Thus, as surface modification techniques evolve, their integration into larger economic and environmental systems becomes vital to the shaping of the future of materials science and nanotechnology (Chahal, 2024).

## 10.3.3. Applications in Drug Delivery Systems

The pursuit of advanced drug delivery has pushed significant advances in surface chemistry in nanotechnology and materials science. Through the regulation of nanoparticle surface properties, researchers can enhance therapeutic drug targeting and release profiles. Surface modification methods, such as functionalization using biomolecules or polymers, allow for the formation of engineered nanocarriers that selectively engage target cells or tissues. This ability to deliver therapeutics in a targeted manner minimizes systemic side effects and increases the therapeutic index of biomolecules. As demonstrated by recent studies, biomimetic integration of responsive nanocarriers enables controlled release mechanisms that respond to biological stimuli, which increases treatment efficacy and personalizes therapy. Such ideas are critical in the design of effective therapies for complex disorders, including cancer and chronic disease. On a biopharmaceutical level, the advances in drug delivery systems have led to the advent of colloidal self-assembly systems that exploit biocompatible nanoparticles. Such platforms are required to deliver diverse bioactive molecules like proteins, nucleic acids, and small-molecule therapeutics. By exploiting principles of surface chemistry, such systems can be formulated to achieve improved solubility and stability while facilitating the transit of therapeutics through biological barriers. This targeting capability not only improves patient results but also guarantees lower healthcare costs through minimized dosages of drug and side effects. The creation of such colloidal systems communicates their worth in the swiftly growing field of biotechnology as they take advantage of the unique physicochemical characters of nanomaterials to revolutionize drug delivery methods in medical procedures (Antunes et al., 2017). Recent research is emphasizing the central role of plasma technology for enhancing the surface properties of drug delivery systems. With non-equilibrium plasmas, scientists can now treat nanoparticles' surface to increase their compatibility, stability, and drug loading. The new process is consistent with pharmaceuticals' increased demand for sustainable and cost-efficient production processes. Surface modifications achieved by plasma treatment provide the ability to control hydrophilicity, charge, and roughness with high precision, all of which are essential in order to maximize nanoparticle interactions with biological environments. As discussed in current literature, the integration of plasma technologies into design within drug delivery systems presents a captivating future research area with the union of cutting-edge material science and therapeutic applications (Arias et al., 2019). These innovations will likely result in the availability of the next-generation drug delivery systems which are effective and environmentally friendly.

## 10.4. NANOSTRUCTURED COATINGS AND THIN FILMS

The field of nanostructured thin films and coatings has emerged as a key component in advancing the frontiers of surface chemistry in material science and nanotechnology. Nanostructured materials consist of physicochemical properties distinct from their bulk equivalents due to their reduced sizes, making their surface area-to-volume ratios greater. This feature is particularly advantageous for usage that entails controlled control over reactivity at the surface and interaction with near surroundings. For instance, by adjusting the deposition conditions and methods, scientists can achieve controlled thin film development with tailored surface features, which are particularly crucial for sensor devices and catalysis. Consequently, the understanding and use of nanostructured coatings can lead to significantly improved material performance and efficiency in a wide variety of materials utilized across various industries. Also relevant is the transition metal oxide (TMO) function of, say, nickel oxide (NiO) in nanostructured coatings, especially in electrochemical applications. As placed into evidence by ongoing research, the inherent electroactivity of TMOs combined with capacitance behavior renders them ideal candidates to utilize on photoelectrochemical cell (PECs) electrodes. Tailoring their photoelectrocatalytic properties through doping or dye-sensitization adds value to their use in energy conversion technology. This versatility is vital to the effective conversion of light energy to chemical energy, hence addressing energy needs of the world. Furthermore, studies of nanostructured TMO electrodes in thin film configurations have reported encouraging results that indicate they may be applied to efficient water photoelectrolysis and carbon dioxide conversion, hence developing towards sustainability in energy production (Bonomo et al., 2018) (Figure 10.9).

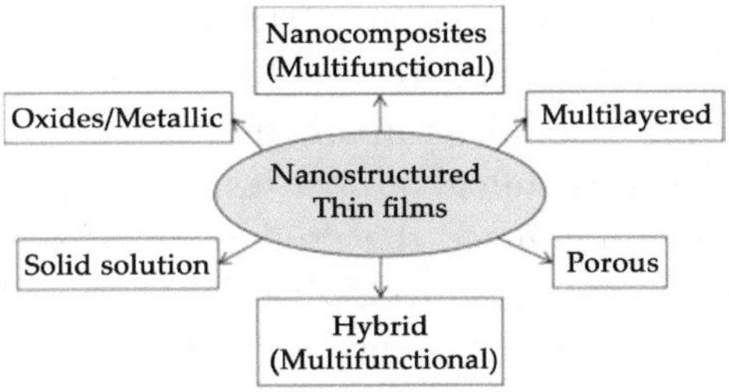


**Figure 10.9.** Nanostructured thin films.

*Source:* https://ars.els-cdn.com/content/image/1-s2.0-B9780081025727000015-f01-03-9780081025727.jpg.

In addition, innovation in plasma processing technologies is a milestone in the production and alteration of nanostructured thin films and coatings. As described in nascent literature, these processes enable the design of surfaces at the nanoscale, improving material performance and longevity in various applications. The non-equilibrium nature of plasma provides for efficient interaction at the micro and nanoscale, enabling the formation of coatings not only eco-friendly but low-cost and rapid as well. This makes plasma technology a prime facilitator in the era of nanotechnology, particularly in the application of plastic and textiles. Through the use of such advanced methodologies, researchers are able to develop novel coatings with specially tailored functionalities that respond actively to environmental signals, thus expanding the boundaries of materials science (Arias et al., 2019).

### 10.4.1. Techniques for Coating Deposition

The field of coating deposition is replete with varied methods that, each in their own manner, contribute to promoting surface chemistry in materials science and nanotechnology. Such methods include atomic layer deposition (ALD) due to its unmatched ability for providing precise and uniform coatings on complex surfaces. ALD operates based on sequential layer growth through chemical precursors such that one can experience controlled thin film development. This technique is especially applied to applications in lithium-ion batteries, as evidenced by recent studies that employ ALD to enhance the structural stability and performance of nickel-rich cathode materials. With ALD, researchers have proven the capability of coatings to act as protective films, inhibiting undesirable interfacial reactions without compromising ionic conductivity and overall electrochemical stability, thus proving the versatility of the method in the development of sophisticated energy storage materials (Meng et al., 2023). Apart from ALD, chemical vapor deposition (CVD) is another significant technique applied in coating deposition. CVD is characterized by the chemical reactions between gas precursors that lead to solid material deposition on substrate surfaces. CVD works best in creating coatings with better adhesive properties and specialist chemical properties. The application of CVD in nanotechnology has revolutionized the production of semiconductor devices to enhance their performance via low defect densities and improved material properties. The most recent advancements in CVD processes led to the formation of multifunctional coatings that can impart unique mechanical, thermal, and electrical properties to various substrates. Scientists can engineer coatings that fit the specifications of the future generation of applications by adjusting the parameters in CVD, following the energetic development of surface chemistry in materials science (Prakash et al., 2021) (Figure 10.10).

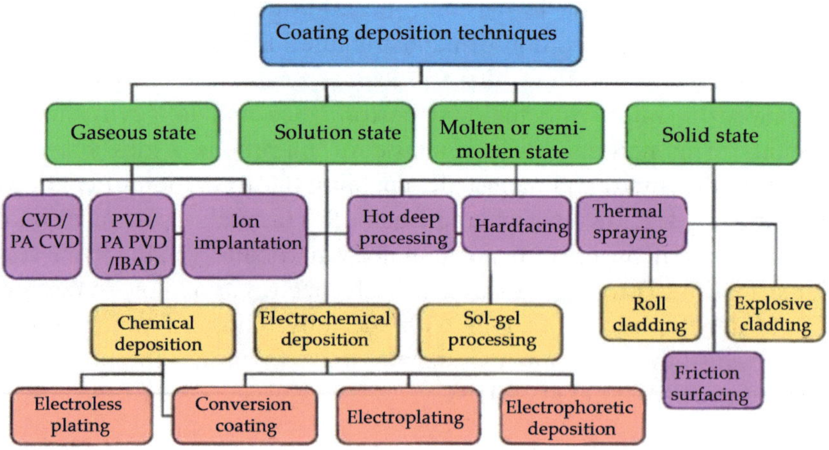


**Figure 10.10.** Classification of the main coating deposition techniques.

*Source:* https://www.researchgate.net/figure/Classification-of-the-main-coating-deposition-techniques_fig4_335163133.

Yet another important technique of coating deposition is sol-gel processing, which allows coatings to be synthesized by a solution route. Involves the conversion of a solution to a solid gel and then to a thin film by passing through a series of hydrolysis and polycondensation reactions. Sol-gel processing enjoys special renown for its versatility due to the allowance of the incorporation of a range of metal oxides and composite materials into the coatings. The process has benefits such as low processing temperatures and the ability to coat complex geometries in an economical manner. Recent developments have emphasized significant attention on the use of sol-gel coatings in protective uses due to its ability to enhance corrosion resistance as well as mechanical stability. Also, ongoing research still explores the potential of sol-gel derived materials for catalytic condition applications, which still remains to show its significance in the area of surface chemistry for material science and nanotechnology (Meng et al., 2023).

### 10.4.2. Properties of Nanostructured Coatings

Nanostructured coatings have gained significant attention over the past few years due to their new properties stemming from their significantly increased surface area and altered chemical composition at the nanoscale. Nanostructured coatings are found to enhance the mechanical strength, thermal stability, and corrosion resistance of materials, making them useful for a wide range of applications, including aerospace, automotive, and electronics.

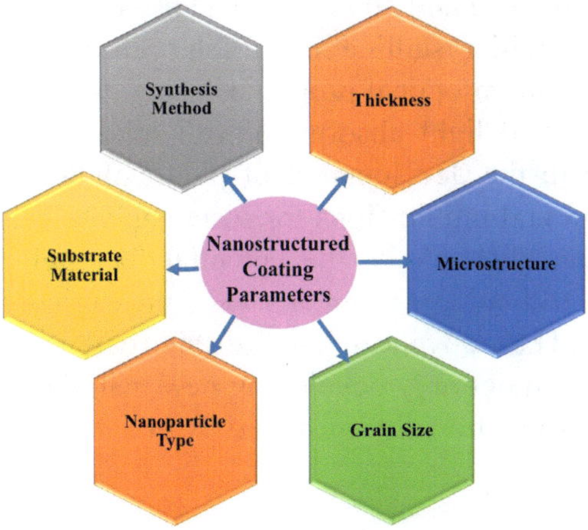


**Figure 10.11.** Various parameters influencing nanostructured coating formation.

*Source:* https://www.mdpi.com/2079-4991/12/8/1323.

Through regulation of the surface chemistry of such films, scientists can engineer their behavior upon interaction with other materials, and this in turn can lead to improved adhesion and compatibility. Functionalization of nanostructured surfaces can also impart additional properties like hydrophobicity or hydrophilicity, which are highly relevant for protective coatings and biomedical device use. The new advances in nanotechnology are further pushing the boundaries of materials science, demonstrating the importance of these coatings toward future technological progress (N/A, 2022).

In biomedicine, nanostructured coatings have opened a new frontier for better drug delivery systems as well as medical devices. Because of their inherent characteristics, they enable increased drug loading capacity with targeted delivery, which reduces the systemic side effects and enhances the therapeutic effectiveness. Iron (II, III) oxide nanoparticles, for instance, have been a prime example of nanostructured materials with differentiated functions in drug transport and magnetically guided delivery systems. The biocompatibility and high surface area to volume ratio of the nanoparticles additionally augment their applicability in medical devices. Additionally, their unique magnetic properties, such as superparamagnetism, enable ease in separation and targeted delivery, which is essential in preventing off-target effects. The study of such innovative materials highlights their potential to revolutionize therapeutic practices in medicine (Shawky et al., 2018) (Figure 10.11).

Further, the utilization of nanostructured coatings in renewable energy technologies, particularly solar cells, has been found to have immense potential for enhancing performance and efficiency. The use of low-temperature surface passivation layers in multicrystalline silicon solar cells is a good example of how one can achieve maximum optical properties using nanostructured approaches and thus gain improved light absorption with minimized energy loss. This is particularly true in the development of photovoltaics, where advances in efficiency have profound implications for sustainable energy technology. With advancements in nanotechnology, the potential for applying nanostructured coatings to energy technologies remains vast, presenting new ways to meet global energy needs. These advancements are a sign of the expanding importance of surface chemistry in not only developing new materials but also enhancing the performance and functionality of materials in a wide variety of industries (N/A, 2022).

### 10.4.3. Applications in Electronics and Optics

The integration of surface chemistry in nanotechnology has become the driving force behind the creation of applications in electronics and optics. Surface modification processes on the nanoscale have the power of greatly improving the functional properties of materials, providing ways to create improved electronic devices and optical systems. For instance, the modification of the surface chemistry of semiconducting materials can optimize electron mobility and device efficiency, which is a prime necessity in the design of transistors and photovoltaic cells. Further, the modification of the surface features such as hydrophobicity or conductivity enables the development of advanced coatings that trigger superior performance of electronic devices, which in turn results in smaller, faster, and more efficient devices. These developments underscore the necessity of knowing surface interactions on nanoscale dimensions to create the material properties typical of such applications. Furthermore, photonics innovation is increasingly reliant on the subtle use of surface chemistry in nanotechnology, particularly in the development of optical sensors and imaging systems. The interaction of light with engineered surfaces has the capability of producing exceptional sensor sensitivity and resolution. For example, plasmonic nanoparticles, in which the localized surface plasmon resonance phenomenon is being exploited, are discovering increasing uses to improve the efficacy of biosensors. These developments have severe consequences for diagnostic medicine and environmental sensors. Further, surface chemistry-inspired concepts and strategies enable the design of the photonic properties of materials, enabling revolutionary developments in next-generation optical components, such as lasers and waveguides. Because new surface techniques are being found with research consistently, the future of optical technology promises increased precision and functionality. Furthermore, interdisciplinary research conducted at facilities like the Boston University Photonics Center demonstrates the complexity of surface chemistry applications in electronics and optics. By combining research across various scientific disciplines, these facilities produce innovations that can be applied in everyday life. Existing efforts, reported in annual reports, mirror the successful search for financing of project exploitations of surface chemistry towards practical applications, like noninvasive optical monitoring technology for the detection of early-stage cancer (The Boston University Photonics Center, 2013). The strategic focus on cross-disciplinary research not only enhances education opportunities but also drives significant advances in the evolution of biophotonics and medical devices, thereby guaranteeing that surface chemistry developments remain aligned with priority societal needs (The Boston University Photonics Center, 2016). Therefore, synergy between surface chemistry and new applications in electronics and optics will continue to be a focal area of research.

## 10.5. CHALLENGES AND FUTURE DIRECTIONS

The field of materials science and nanotechnology is extremely challenged, particularly within surface chemistry, which is critical for the operation and usage of nanomaterials. The main challenge is understanding complex interactions at the nanolevel, where alteration in surface properties can have a deeply influential effect on material conduct. For instance, conventional drug delivery systems are generally plagued by low bioavailability and irregular plasma drug levels, compromising therapeutic performance. By focusing on controlled drug delivery, researchers aim to develop systems that will reach the targeted site specifically, enhancing safety and efficacy. Still, the intricate relationship of the surface characteristics of nanocarriers with their interactions in biological environments remains to be examined and engineered (Adepu et al., 2021). As challenges are being met by researchers, the future directions of nanotechnology will be focused on designing more sophisticated materials with enhanced surface functions. The advent of nanocellulose, for instance, represents the potential of biobased materials as omnipresent building blocks for novel applications. The hydrophilic character and high density of hydroxyl groups in nanocellulose render it susceptible to a variety of functionalizations, which liberate the potential of novel composites and biomedical applications.

Besides, optimization of these properties through chemical modification places into perspective the necessity for ongoing research to optimize surface interactions in a wide range of applications, from environmental cleanup to advanced drug delivery systems, so as to meet performance as well as sustainability requirements (Trache et al., 2020). Overcoming the new challenges will demand interdisciplinary endeavor. By encouraging cooperation among chemists, material scientists, and biomedical engineers, it is possible to innovate multifunctional nanomaterials more efficiently. Collaboration can encourage the creation of new fabrication techniques so that materials are not only efficient but also safe for application on human subjects and environmentally friendly. Future studies should aim to scale up the nanomaterials while ensuring stability and degradability over the long term, particularly when used in biomedical applications. With advances in surface chemistry in nanotechnology, an inter-disciplinary strategy integrating scientific inquiry and engineering practices will be essential to unlock the maximum potential of such new materials (Adepu et al., 2021; Trache et al., 2020).

## 10.5.1. Limitations in Current Surface Chemistry Research

The study of surface chemistry in nanotechnology is faced with extraordinary challenges that hinder progresses and applications. One of the major difficulties is describing surface characteristics of nanomaterials. Traditional approaches might not be readily applicable given the nanoscale feature and characteristic special chemical behavior of such materials. For instance, existing methods cannot provide accurate measurements of surface chemistry, which in turn can have a final impact on the efficiency of applications such as pollutant removal and material synthesis. The dynamic relation among surface chemistry, reactivity, and conditions of the environment renders analysis reliable, requiring the creation of new methodologies appropriate to nanoscale analysis. Also, current practices in surface chemistry rarely adequately take into account the environmental context of engineered nanomaterials. Nanoparticles, while they offer numerous advantages in materials science, also raise questions about their life and path in ecosystems. One of the serious limitations is the difficulties of analysis and detection of such particles in multi-component environmental samples. For example, surface-enhanced Raman spectroscopy (SERS) is an emerging technique for nanoparticle analysis but still lags behind in applicability factors such as material type, size, and surface chemistry. These hindrances limit not just the understanding of the potential hazard that nanoparticles can pose but also their effective integration into existing technologies for the uses in the environment, and it thus indicates a significant gap in existing research activities (Guo et al., 2017) (Figure 10.12).

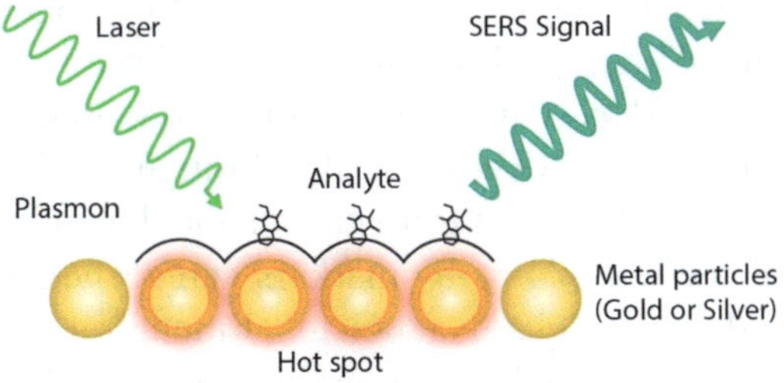


**Figure 10.12.** Surface-enhanced Raman scattering (SERS).

*Source:* https://static.horiba.com/fileadmin/Horiba/_processed_/7/5/csm_Raman_FAQ_SERS_principle_a4d2533210.jpg.

In addition, the lack of in-depth studies of adsorption properties of nanomaterials presents difficulties in their application to pollution control and other purposes. For instance, despite the fact that techniques such as adsorption and ion exchange are efficient means to remove heavy metals from water, specific conditions in which the process operates optimally—such as pH, ionic strength, and competition among various metal ions—are typically inadequately studied. Such lack is indicative of the need to standardize and optimize adsorption parameters specific to nanomaterials, as proper comprehension of the factors is vital for the facilitation of the practical utilization of nanotechnology in environmental contamination mitigation. These present limitations equally highlight the imperative of developing towards a more scientific research approach that harmoniously balances theoretical reflection and experimental practice such that developments in surface chemistry can fulfill their potential in solving the world's most pressing issues (Khaligh and Johan, 2020).

## 10.5.2. Emerging Trends in Nanotechnology

Recent advances in nanotechnology have paved the way for a novel generation of material science, particularly through the study of two-dimensional (2D) materials. Graphene, the first 2D material, has been at the center of intense scrutiny owing to its extraordinary electrical, mechanical, and thermal properties. Yet the research boom in other 2D materials, such as transition metal dichalcogenides (TMDs) and hexagonal boron nitride, indicates there is more widespread interest in the tuning of properties for desired applications. Surface chemistry is also critical for the enhancement of functionalities of such materials, since modifications at the surface have a considerable influence on their reactivity, stability, and compatibility in media. Such increased focus on surface chemistry enables the fabrication of heterostructures, where different 2D materials are stacked upon one another, leading to new electronic and optoelectronic applications, pushing the boundaries of what is achievable in nanotechnology (Haider et al., 2025). In biosensing, nanotechnology is revolutionizing the detection of biomolecules through the development of next-generation electrochemical sensors from nanochannel membranes. The need for precise and rapid detection of key biomarkers cannot be emphasized enough, especially in clinical diagnostics. Nanochannel membrane confinement of the target molecules leads to signal amplification and overall sensitivity (Figure 10.13).

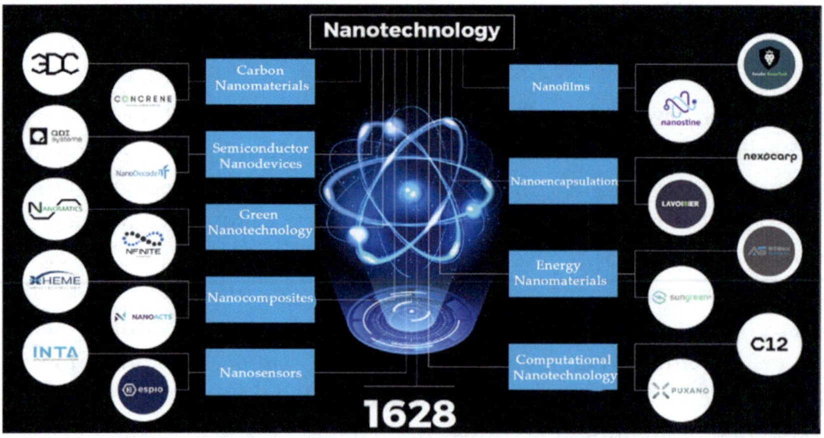


**Figure 10.13.** Emerging trends in nanotechnology.

*Source:* https://www.startus-insights.com/wp-content/uploads/2022/11/Nanotechnology-trends-InnovationMap-StartUs-Insights-noresize-2.webp.

Besides, such membranes can effectively exclude interferents, thus facilitating accurate measurements even in complicating biological media. The relationship between sensory performance and surface chemistry is crucial because the structure and stability of such sensors largely determine their efficiency. Innovations have also been made with regard to designing mesoporous nanochannel membranes, which highlight the importance of controlling surface geometries and properties to optimize utilization in biomedical monitoring as well as disease diagnosis (Zeng et al., 2025). With developments in nanotechnology, addressing environmental and scalability challenges involved is important in the generation of novel nanomaterials. Green chemistry approaches and sustainable synthesis pathways are becoming increasingly important as researchers attempt to minimize the environmental footprint of nanomaterials production. Reproducibility and stability complications must also be solved in order to facilitate commercial applications for emerging nanoscale materials. Studies on the study of surface interactions at the nanoscale are driving the trend, enabling the development of more stable and multi-functional nanostructures. These emerging trends illustrate a significant crossover of surface chemistry and materials science that will lead to novel solutions to the pressing issues of our time, such as clean energy technologies and environmental remediation, thus opening new avenues for future research and applications at the nanotechnology level (Haider et al., 2025).

### 10.5.3. Potential Impact on Materials Science

The possible impact of nanotechnology surface chemistry on materials science is profound, particularly in terms of the development of new materials with emergent nanoscale properties. On this scale, the volume to surface area ratio shifts dramatically so that materials can be more reactive and interaction-rich. This phenomenon has important advantages in applications ranging from catalysis to biomedical engineering where material properties are engineered to produce improved performance. Interestingly, scientists have been studying the use of nanoparticles for various dental applications, where surface chemistry facilitates enhanced bonding and antimicrobial activity compared to traditional materials. The development of nanotechnology in dentistry offers a fascinating case of how surface modification can revolutionize material properties and performance, enhancing patient care and procedural effectiveness in clinical settings (Umapathy et al., 2022). Also, the field of nanomedicine shows the pivotal role played by surface chemistry in materials science, its ability to revolutionize drug delivery being highlighted. Nanoparticles developed for a specific therapeutic application can cross biological barriers and deliver drugs in a targeted manner, accessing disease sites with utmost specificity. The approach not only maximizes the efficacy of existing therapy but also reduces side effects, a tremendous boost for personalized medicine. Surface engineering in a strategic manner enhances the stability and blood circulation time of nanoparticles, thus improving therapeutic efficacy. As nanomedicine research continues to progress, cross-talk between materials science and surface chemistry becomes increasingly evident, compelling interdisciplinary approaches towards solving complex medical disorders (Patra et al., 2018). Finally, as we expand our understanding of surface chemistry in nanotechnology, applications to the field of materials science ripple into the areas of environment and sustainability. Improvements in nanomaterials have the potential to develop effective solutions to pollution treatment, energy conversion, and storage technologies. For instance, nanostructured catalysts designed via rigorous surface chemistry optimization have the capability of enhancing the efficiency of chemical reactions significantly, facilitating cleaner industrial processes. In addition, nanotechnology offers opportunities to enhance the efficiency of renewable energy technologies such as solar cells and batteries by improving energy absorption and storage capabilities. As the world faces mounting environmental issues, proper utilization of nanotechnology in material science will be essential. Therefore, capitalizing on such advancements will have the potential to establish a sustainable future and satisfy various global needs (Umapathy et al., 2022).

## SUMMARY

The study of surface chemistry in nanotechnology and materials science reveals considerable impacts on environmental as well as technological advancement. The unique characteristics of nanomaterials, which to a large degree are dictated by their surface characteristics, support enhanced functionality in the form of agriculture and pollution removal. The convergence of nanotechnology and environmental sustainability reveals potential for alternative methods of agriculture that utilize nanoscale material to enhance crop yields with no implication in environmental degradation. As noted, research shows that the use of nanomaterials can improve the quality of the environment and enable the identification and cleanup of contaminated sites (Bacchini et al., 2013). These advancements are a significant intersection of surface chemistry and practical solutions to pressing world challenges. In addition, the intricate phenomena of interactions of engineered nanomaterials with biological systems establish the crucial role of surface chemistry. Findings of recent nanotoxicity studies show that the size and surface engineering of nanomaterials exert a significant influence on their bioavailability and toxicological profiles. A demonstration of how nanoparticles could be altered in cellular uptake and biocompatibility by coating is evidenced by comparing cerium and iron oxide nanoparticles' effect on NIH/3T3 mouse fibroblasts (Bohren et al., 2010). This underlines the importance of adequate surface engineering in nanomaterial design for specific applications, which is of utmost significance in order to safeguard human health and ecological environments. Last but not least, the destiny of nanotechnology hangs on further understanding and control of surface chemistry. By utilizing sophisticated synthesis and characterization techniques, scientists can tailor-designed nanomaterials with defined properties that maximize beneficial effects and reduce potential risks. The double-faced aspect of surface chemistry as it is a facilitator of technological advancements as well as an intervenor in environmental interactions provides key questions for further research. As we move deeper into the world of applications and impacts of nanotechnology, it becomes imperative to use a responsible research model that ensures safety and efficacy in all material science facets, hence ensuring a sustainable future.

The interfacing of nanotechnology, materials science, and surface chemistry is essential in optimizing the performance and efficiency of various materials at the nanoscale level. By controlling the surface properties of nanomaterials, researchers can transform substantially their reactivity, stability, and intermolecular interaction with other materials. This capability is essential in applications like catalysis where chemical reactions are enhanced by surface area enhancement and manipulated surface chemistry. Strategic surface engineering not only provides access to control of material properties but also facilitates the development of applications like drug delivery, sensors, and energy storage. Continuing research into nanotechnology surface chemistry reveals the transformative potential of such materials, highlighting the need for understanding into surface phenomena in order to maximize their potential, as indicated by ongoing Congressional discussion over the importance of nanotechnology research and development (Sargent et al., 2009). In addition, environmental, health, and safety (EHS) implications associated with nanotechnology have become increasingly scrutinized, requiring strict regulation and oversight. With increased use of nanomaterials in industry, ascription of their unique surface properties is requested to contemplate potential risks to human health and ecosystems. Methodological improvements aimed at characterization of surface properties are crucial not only for determining the functional worth of nanomaterials but also for evaluating safety impacts. Mitigation of these EHS concerns is of paramount importance; therefore, regulatory systems must evolve in tandem with nanotechnology innovation to bolster public faith and sustainability (Sargent et al., 2009). Surface chemistry knowledge in depth ensures that informed choices are made by the regulation bodies for nanotechnological innovations, thereby ensuring an interdisciplinarity in interactions among chemists, engineers, and policymakers. Finally, the greater socioeconomic advantages expected to emerge as a result of nanotechnology developments further illustrate the importance of continued surface chemistry research. As nanotechnology continues to evolve at a rapid pace, the role of surface interactions increasingly becomes more important when material structure and functionality are concerned. Major sectors, such as electronics, pharmaceuticals, and renewable energy, can greatly be revolutionized on the foundation of developments obtained through intensified surface chemistry studies. Besides, raising public consciousness about the advantages and risks of nanotechnologies can facilitate acceptance and engagement with these emerging technologies. A consolidated strategy to nanomanufacturing and material science, emphasizing both new applications and safety concerns, can enable the construction of the desired social benefits of nanotechnology and minimize possible risk, as highlighted in a variety of congressional hearings on this emerging technology (Sargent et al., 2009).

# REFERENCES


1. Adepu, S., & Ramakrishna, S. (2021). Controlled drug delivery systems: Current status and future directions. Molecules, 26(19), 5905. https://doi.org/10.3390/molecules26195905 (accessed on 3rd October 2025).
2. Aljabali, A. A. A., & Evans, D. (2015). Bionanomaterials from plant viruses. https://core.ac.uk/download/151161135.pdf (accessed on 3rd October 2025).
3. Antunes, F. E., Coelho, J. F. J., Dapas, B., Farra, R., et al. (2017). Recent advances in smart biotechnology: Hydrogels and nanocarriers for tailored bioactive molecules depot. https://core.ac.uk/download/322918616.pdf (accessed on 3rd October 2025).
4. Bacchini, F., Greppi, G., Mura, S., Roggero, G., et al. (2013). Advances of nanotechnology in agro-environmental studies. https://core.ac.uk/download/16749559.pdf (accessed on 3rd October 2025).
5. Berret, J. F., Bostelmann, T., Bée, A., Courtois, J., et al. (2012). Poly (acrylic acid)-coated iron oxide nanoparticles: Quantitative evaluation of the coating properties and applications for the removal of a pollutant dye. http://arxiv.org/abs/1212.1580 (accessed on 3rd October 2025).
6. Bohren, C. F., Sarrouj, H., Hafelli, U. O., Berret, J. F., Jordan, A., Safi, M., Mignet, N., et al. (2010). Interactions between sub-10 nm iron and cerium oxide nanoparticles and 3T3 fibroblasts: The role of the coating and aggregation state. http://arxiv.org/abs/1011.1670 (accessed on 3rd October 2025).
7. Bonomo, M., Decker, F., & Dini, D. (2018). Electrochemical and photoelectrochemical properties of nickel oxide (NiO) with nanostructured morphology for photoconversion applications. https://core.ac.uk/download/188829827.pdf (accessed on 3rd October 2025).
8. Campagnolo, L., Lacconi, V., Filippi, J., & Martinelli, E. (2024). Twenty years of *in vitro* nanotoxicology: How AI could make the difference. Frontiers in Toxicology, 6. https://www.semanticscholar.org/paper/de4f75917c250e01474f657c92f21d8c9dfe068a (accessed on 3rd October 2025).
9. Canal Arias, J. M., Canal Barnils, C., Cernák, M., Cvelbar, U., et al. (2019). White paper on the future of plasma science and technology in plastics and textiles. https://core.ac.uk/download/187265494.pdf (accessed on 3rd October 2025).
10. Chahal, P. (2024). Molecular precision at the nanoscale: Chemical synthesis, functionalization, and multidisciplinary applications of

nanomaterials in modern chemistry. International Journal for Research Publication and Seminar. https://www.semanticscholar.org/paper/dc8ebd2a0549cc0cecc307aa4e88295133a39822 (accessed on 3rd October 2025).

11. Chen, C., Berglund, L. A., Burgert, I., & Hu, L. (2021). Wood nanomaterials and nanotechnologies. Advanced Materials, 33. https://doi.org/10.1002/adma.202006207 (accessed on 3rd October 2025).
12. Das, S. K., & Marsili, E. (2009). A green chemical approach for the synthesis of gold nanoparticles: Characterization and mechanistic aspect. https://core.ac.uk/download/11308671.pdf (accessed on 3rd October 2025).
13. Debackere, K., Monard, E., Raspoet, D., Ratinckx, J., et al. (2025). Technologie En Innovatie in Vlaanderen: Prioriteiten. https://core.ac.uk/download/pdf/6468460.pdf (accessed on 3rd October 2025).
14. El Shafey, A. M. (2020). Green synthesis of metal and metal oxide nanoparticles from plant leaf extracts and their applications: A review. Green Processing and Synthesis, 9, 304–339. https://doi.org/10.1515/gps-2020-0031 (accessed on 3rd October 2025).
15. Ganapathe, L. S., Mohamed, M. A., Yunus, R. M., & Berhanuddin, D. (2020). Magnetite ($Fe_3O_4$) nanoparticles in biomedical application: From synthesis to surface functionalization. Magnetochemistry. https://www.semanticscholar.org/paper/d363b475cf434ce127caf5edee2e54d145e165ab (accessed on 3rd October 2025).
16. Guo, H., He, L., & Xing, B. (2017). Applications of surface-enhanced Raman spectroscopy in the analysis of nanoparticles in the environment. Environmental Science: Nano, 4, 2093–2107. https://www.semanticscholar.org/paper/c176edb299015c934d7f0d3a2fce9bf55c090a0c (accessed on 3rd October 2025).
17. Haider, A., Rasheed, M. F., Abrar, A., Mumtaz, A., Zeb, A., Rehman, M. U., Noor, G. M., et al. (2025). The chemistry of 2D materials from graphene to beyond. Kashf Journal of Multidisciplinary Research. https://www.semanticscholar.org/paper/8327cf5672c871ca7169dc72f79702c57ab32586 (accessed on 3rd October 2025).
18. Huang, J., Ma, W., Sun, H., Wang, H., He, X., Cheng, H., Huang, M., et al. (2020). Self-assembled DNA nanostructures-based nanocarriers enabled functional nucleic acids delivery. ACS Applied Bio Materials, 3, 2779–2795. https://doi.org/10.1021/acsabm.9b01197 (accessed on 3rd October 2025).

19. Jeevanandam, J., Barhoum, A., Chan, Y. S., Dufresne, A., & Danquah, M. K. (2018). Review on nanoparticles and nanostructured materials: History, sources, toxicity and regulations. Beilstein Journal of Nanotechnology, 9, 1050–1074. https://doi.org/10.3762/bjnano.9.98 (accessed on 3rd October 2025).
20. Joudeh, N., & Linke, D. (2022). Nanoparticle classification, physicochemical properties, characterization, and applications: A comprehensive review for biologists. Journal of Nanobiotechnology, 20. https://doi.org/10.1186/s12951-022-01477-8 (accessed on 3rd October 2025).
21. Khaligh, N. G., & Johan, M. R. (2020). Recent advances in water treatment using graphene-based materials. Mini-Reviews in Organic Chemistry, 17, 74–90. https://www.semanticscholar.org/paper/fed94283c8052d75998f3fc1930cd0005ca7e19e (accessed on 3rd October 2025).
22. Liu, G. (2021). Grand challenges in biosensors and biomolecular electronics. Frontiers in Bioengineering and Biotechnology, 9. https://www.semanticscholar.org/paper/ac999a3db5205cfe9b80f91f679e64d1ab4b4f95 (accessed on 3rd October 2025).
23. Meng, X., Wang, X., Liu, Y., Xiao, X., & Chen, Z. (2023). (Invited) Nanoscale surface coatings for high-performance nickel-rich $LiNixMnyCozO_2$ cathodes via atomic layer deposition. ECS Meeting Abstracts. https://www.semanticscholar.org/paper/9b5034ec8fabe6fe5cf2ec260ec81a8d05c91afe (accessed on 3rd October 2025).
24. Mujahid, M. H., Upadhyay, T. K., Khan, F., Pandey, P., Park, M. N., Sharangi, A. B., Saeed, M., et al. (2022). Metallic and metal oxide-derived nanohybrid as a tool for biomedical applications. Biomedicine & Pharmacotherapy, 155, 113791. https://doi.org/10.1016/j.biopha.2022.113791 (accessed on 3rd October 2025).
25. N/A. (2013). The Boston University Photonics Center Annual Report 2013–2014. https://open.bu.edu/bitstream/2144/29752/1/Photonics-Center-Final-Annual-Report-2013-2014-v2.pdf (accessed on 3rd October 2025).
26. N/A. (2016). The Boston University Photonics Center Annual Report 2015–2016. https://open.bu.edu/bitstream/2144/29754/1/Photonics-Center-Annual-Report-2016-final.pdf (accessed on 3rd October 2025).
27. Naresh, V., & Lee, J. Y. (2021). A review on biosensors and recent development of nanostructured materials-enabled biosensors. Sensors, 21(4), 1109. https://doi.org/10.3390/s21041109 (accessed on 3rd October 2025).

28. Patra, J. K., Das, G., Fraceto, L. F., Campos, E. V. R., Rodriguez-Torres, M. del P., Acosta-Torres, L. S., Díaz-Torres, L. A., et al. (2018). Nano based drug delivery systems: Recent developments and future prospects. Journal of Nanobiotechnology, 16. https://doi.org/10.1186/s12951-018-0392-8 (accessed on 3rd October 2025).
29. Prakash, A., Singh, S., & Davim, J. (2021). Functional Materials and Advanced Manufacturing. https://www.semanticscholar.org/paper/17fbe5f260ba87ed3f1e53a9e710ed4b87a5feb0 (accessed on 3rd October 2025).
30. Rosenblum, D., Joshi, N., Tao, W., Karp, J. M., & Peer, D. (2018). Progress and challenges towards targeted delivery of cancer therapeutics. Nature Communications, 9. https://doi.org/10.1038/s41467-018-03705-y (accessed on 3rd October 2025).
31. Sánchez-López, E., Gomes, D., Esteruelas, G., Bonilla, L., López-Machado, A., Galindo, R., Cano, A., et al. (2020). Metal-based nanoparticles as antimicrobial agents: An overview. Nanomaterials, 10, 292. https://doi.org/10.3390/nano10020292 (accessed on 3rd October 2025).
32. Sargent, J. F. (2009). Nanotechnology: A Policy Primer. https://core.ac.uk/download/5133471.pdf (accessed on 3rd October 2025).
33. Shawky, S., Awad, A. A., Abugable, A. A., & EL-Khamisy, S. F. (2018). Ultrasensitive optical biosensor for absolute RNA quantification using gold to probe genomic stability for cancer and neurological disorder diagnosis. Journal of Nanomedicine & Nanotechnology, 2018. https://www.semanticscholar.org/paper/535fa347199802d66e4c40737d8ec7c67d16d350 (accessed on 3rd October 2025).
34. Trache, D., Tarchoun, A. F., Derradji, M., Hamidon, T. S., Masruchin, N., Brosse, N., & Hussin, M. H. (2020). Nanocellulose: From fundamentals to advanced applications. Frontiers in Chemistry, 8. https://doi.org/10.3389/fchem.2020.00392 (accessed on 3rd October 2025).
35. Umapathy, V. R., Natarajan, P. M., Jones, S., Swamikannu, B., Johnson, W. M. S., Alagarsamy, V., & Milon, A. R. (2022). Current trends and future perspectives on dental nanomaterials–An overview of nanotechnology strategies in dentistry. Journal of King Saud University–Science, 34, 102231. https://doi.org/10.1016/j.jksus.2022.102231 (accessed on 3rd October 2025).
36. Wahab, R., Hwang, I., Shin, H. S., Kim, Y. S., Musarrat, J., Al-Khedhairy, A., & Siddiqui, M. (2012). Zinc oxide nanostructures and their applications. https://www.semanticscholar.org/paper/55fc596f837a22d41ef1320891e116fb5265ce07 (accessed on 3rd October 2025).

37. Zeng, H., Liang, K., Jiang, L., Zhao, D., & Kong, B. (2025). Electrochemical sensing mechanisms and interfacial design strategies of mesoporous nanochannel membranes in biosensing applications. Accounts of Chemical Research. https://www.semanticscholar.org/paper/401195e7906b357768fdd3de6a8ed010a2aa68c0 (accessed on 3rd October 2025).
38. Zhu, J., Hu, L., Zhao, P., Lee, L. Y. S., & Wong, K. Y. (2019). Recent advances in electrocatalytic hydrogen evolution using nanoparticles. Chemical Reviews, 120, 851–918. https://doi.org/10.1021/acs.chemrev.9b00248 (accessed on 3rd October 2025).

# INDEX

## C

**F**



**G**



**H**



**I**